The MPEG-21 Book

The MPEG-21 Book

Edited by

Ian S Burnett, *University of Wollongong, Australia*
Fernando Pereira, *Instituto Superior Técnico, Portugal*
Rik Van de Walle, *Ghent University, Belgium*
Rob Koenen, *MPEG Industry Forum, USA*

John Wiley & Sons, Ltd

Other Wiley Editorial Offices

John Wiley & Sons Inc., 111 River Street, Hoboken, NJ 07030, USA

Jossey-Bass, 989 Market Street, San Francisco, CA 94103-1741, USA

Wiley-VCH Verlag GmbH, Boschstr. 12, D-69469 Weinheim, Germany

John Wiley & Sons Australia Ltd, 42 McDougall Street, Milton, Queensland 4064, Australia

John Wiley & Sons (Asia) Pte Ltd, 2 Clementi Loop #02-01, Jin Xing Distripark, Singapore 129809

John Wiley & Sons Canada Ltd, 22 Worcester Road, Etobicoke, Ontario, Canada M9W 1L1

Wiley also publishes its books in a variety of electronic formats. Some content that appears
in print may not be available in electronic books.

Library of Congress Cataloging-in-Publication Data:

The MPEG-21 book / edited by Ian S Burnett . . . [et al.].
    p. cm.
 Includes bibliographical references and index.
 ISBN-13: 978-0-470-01011-2 (cloth : alk.paper)
 ISBN-10: 0-470-01011-8 (cloth : alk.paper)
 1. MPEG (Video coding standard) 2. Computer animation – Standard.
 3. Digital video. I. Burnett, Ian S
 TK6680.5.M686 2006
 006.7 – dc22
                                                    2005032766

British Library Cataloguing in Publication Data

A catalogue record for this book is available from the British Library

ISBN-13: 978-0-470-01011-2
ISBN-10: 0-470-01011-8

Typeset in 10/12pt Times by Laserwords Private Limited, Chennai, India
Printed and bound in Great Britain by Antony Rowe Ltd, Chippenham, Wiltshire
This book is printed on acid-free paper responsibly manufactured from sustainable forestry
in which at least two trees are planted for each one used for paper production.

Contents

Introduction to MPEG-21

Leonardo Chiariglione, CEDEO.net

MPEG was born in response to the industry needs for a digital television standard. Such a big project was accomplished in two steps, for 'low-resolution' pictures with stereo sound first, and for 'standard-definition' and even 'high-definition' pictures with multi-channel sound next.

It is interesting to observe that MPEG-1 and MPEG-2, the first two standards spawned by the initial MPEG efforts have given rise to products that are counted by billions of devices, pieces of content that are counted by tens of billions and services that are counted by millions of content hours. There should be no doubt that this result is due to the technical excellence of the standards, the timely availability of a standard that responded to industry needs, and to the provision – outside of MPEG – of licensing terms for the patented technologies that responded to industry practices.

The next standard, MPEG-4, had a more difficult story. Born as a project to provide solutions for 'very low bitrate channels' it eventually became a full multimedia standard comprising a video coding scheme covering 5 orders of magnitude in bitrate, audio coding schemes for all types of natural audio material covering 2 orders of magnitude in bitrate, coding schemes for synthetic audio and video information, techniques to compose audio and video media objects in a 3D space, file format, and so on.

Use of the MPEG-4 standard has been slow in coming. In spite of version 1 of the standard being approved in 1998, only now are we witnessing a mass use of some parts of the standard – video, audio and file format being probably the most important. Very much in line with the spirit of its development, MPEG-4 is being used in a variety of application domains some of which are totally unexpected.

It is legitimate to ask why there has been such a different response from the industry, but the answer is not straightforward. On the one hand MPEG-4 did not (and still does not) have a homogeneous constituency and the application domains are still hard to identify. But more importantly, the licensing terms offered by the licensors charged pay services usage fees; something that the affected industries were not prepared to accept. There should be no surprise if some proprietary technologies, unencumbered by licensing terms considered hostile by users, have made headway in MPEG-4's stead.

With its fourth standard – MPEG-7 – MPEG left the traditional video and audio compression field to address a technology deemed to be an important business enabler of digital media, that is, media description. In spite of being a huge body of technology, MPEG-7 version 1 was completed in less than five years but mass adoption is still to come four years after version 1 was completed. There is no reason to panic, though, as three years ago the use of MPEG-4 was very much in the early phases and the TV Anytime specifications, developed by an industry body targeting a specific service concept and making use of a broad range of MPEG-7 technologies, are being taken up by the industry.

With its fifth standard – MPEG-21 – that gives the title to this book, MPEG has undertaken a project even bolder than MPEG-7: the integration of multimedia technologies to support a variety of businesses engaged in the trading of digital objects that MPEG-21 calls Digital Items. The project has been running for six years and an impressive list of standard parts has already been released. The readers of this book will have a privileged access to a comprehensive description of the many multimedia technology facets covered by the standard provided by leading figures in the development of MPEG-21.

It is certainly not too early to ask about the prospects of use of MPEG-21, and the answers are many. On the one hand, MPEG-21 is a collection of technologies more 'independent' from one another than past MPEG standards. This means that it is known that some parts are already being taken up by some companies, if not industries. On the other hand, there are efforts – like those of the Digital Media Project – that have borrowed a range of MPEG-21 technologies and are integrating them with other technologies to provide complete solutions for future-looking digital media businesses.

A look back at the 17 years of the history of MPEG teaches us a few lessons. The first is the gradual decoupling of digital media standards and applications, while business remains as anchored as ever to the coverage of narrow application domains. The second is the increasingly difficulty of defining requirements for digital media standards. The third is the complexity of developing licensing terms for the digital media standards. The fourth and last is the demystification of the old adage that the development of standards is slow while business is fast. The history of MPEG proves the opposite – at least for MPEG standards.

Leonardo Chiariglione
Convenor, MPEG

Preface

The past ten years have seen computer technology make significant inroads into the daily lives of consumers; no longer is the technology the domain of the computer nerd. Today, digital music can be carried anywhere, with players becoming increasingly small and carrying increasing numbers of songs. As for videos, digital television has taken hold and portable DVD is now within easy consumer reach. The Internet has also become part of everyday experience with advertising invariably carrying the ubiquitous www.moniker. Behind all these successful multimedia products are standards that provide interoperability and thus generate a marketplace in which consumer equipment manufacturers can produce competitive, yet conformant, products.

The Moving Picture Experts Group (MPEG) has played a key role in developing the standards behind the explosion of multimedia-enabled consumer devices. The MPEG-1 and MPEG-2 standards are the core technologies behind digital TV, MP3s, AAC, and DVDs. The MPEG-4 standard has also seen success in its use in Internet video content, and the recent MPEG-4 part 10 (AVC) standard is making inroads into the mobile content and broadcasting arenas. MPEG has also produced an extensive standard for the description of multimedia content using XML metadata: MPEG-7. The latter promises interoperability in search and retrieval as users move beyond consumption of media to more interactive exploration. Thus, MPEG has produced a powerful and also rather popular set of media coding/compression standards and a complementary metadata framework for description. However, the multimedia marketplace continues to be hampered by a set of missing standards; this results in lack of content, inability to secure rights, multiple, unrelated content formats and terminal types, and overall, a confusing array of technologies. It was this scene that motivated MPEG to create MPEG-21 as a solution that would offer users transparent and interoperable consumption and delivery of rich multimedia content. The aim was to create a standard that would link together the media coding and metadata standards with access technologies, rights and protection mechanisms, adaptation technology, and standardised reporting so as to produce a complete 'multimedia framework'.

In standardising the set of technologies that comprise MPEG-21, MPEG has built on its previous coding and metadata standards such that MPEG-21 links these together to produce a 'package' (and, if required, a 'protected package') of multimedia content for users. To facilitate this vision, MPEG has introduced a new 'carriage' – the Digital Item that acts as a universal package for collecting, relating, referencing and structuring content. This has the potential to move the marketplace forward significantly as users and devices shift from the delivery and consumption of media files to the exchange of rich Digital

Item multimedia experiences. The latter combine multiple media formats, adaptation to devices and transparent handling of, for example, Digital Rights Management (DRM) to ensure that users can experience digital media even more seamlessly than they can with the familiar analogue and physical media formats.

MPEG-21 is thus a major step forward in multimedia standards. It collects together, for the first time, the technologies needed to create an interoperable infrastructure for protected digital media consumption and delivery. The standard is, unsurprisingly, large and there have been contributions from many different sectors of industry ranging from engineers through to those concerned with artistic and legal implications. However, the standard, while apparently complex, has a clear structure that can be readily understood. The purpose of this book is to detail MPEG-21 at a level that makes the standard available to users. A central concept of MPEG-21 is that the user is 'central', and in this book, we have tried to ensure that the text is accessible to as many of those users as possible who are interested in understanding the underlying technologies. The book covers all the parts of MPEG-21 that were complete or nearing completion at the time of writing. The chapters have all been written by experts who were involved in the writing of the MPEG-21 Parts themselves and thus they all contain insights that are only born out of long hours of collaboration involved in completing an international technical standard.

Standards are technical documents that define precisely the conformance required of users if interoperability is to be achieved. The chapters of this book are much more accessible than these normative documents and explain in detail the standard and the underlying concepts. However, as always, there is much, much more that cannot be squeezed into the book. Thus, each chapter contains a significant set of references that will allow readers to explore topics in further detail. While we have tried to avoid references to MPEG documents that are not always available to the public, some, inevitably, were necessary; they are at least useful for all those who have access to these documents. There are a number of ways to access such documents: first, the MPEG public pages at www.chiariglione.org/mpeg provide links to many public versions of the documents. Alternatively, it may be possible to obtain 'publicly available' documents direct from the MPEG 'Head of Delegation' for your country of residence. A list of the national standard bodies is maintained at http://www.iso.org/iso/en/aboutiso/isomembers/index.html and these should be able to put interested parties in contact with their MPEG 'Head of Delegation'.

ORGANISATION OF THE BOOK

The first two chapters of this book offer an introduction to MPEG and then to MPEG-21 itself. The discussion of MPEG gives a background of other MPEG standards (MPEG-1, MPEG-2, MPEG-4 and MPEG-7), which builds a context for the discussion of MPEG-21. It also considers the processes of MPEG and explains how technical contributions are built through a process of consensus and voting into honed international standards. The MPEG-21 introduction explains how the standard parts evolved from initial discussions on the nature of the framework. It provides a summary of important terminology and an individual discussion of the motivation, objectives and content of each part of the MPEG-21 collection of standards. The chapter also includes a use case that introduces many of the MPEG-21 concepts.

The main chapters of the book describe the normative technology of MPEG-21, as specified in the parts of the standard. The first three of these chapters consider the Digital Item (DI) in both 'clear' and 'protected' format as well as detail the use of identification and description mechanisms. Chapter 3 details the Digital Item model and the declaration of a DI using the XML schema-based Digital Item Declaration Language (DIDL). It explains the elements that are used within the model and the syntax and semantics of their expression in XML. The chapter also extends to a discussion of how DIs can be identified through the Digital Item Identification (DII) mechanisms. Chapter 4 builds on the 'clear' expression of Digital Items by detailing the 'protected' DI that can be created using the Intellectual Property Management and Protection (IPMP) Components tools. These include a 'protected' Digital Item XML schema known as IPMP DIDL, which allows the protection of hierarchical sections of a DI. The chapter then details the IPMP Info metadata that is used to describe the protection mechanism employed within a given Digital Item.

Having considered the fundamental unit of transaction, the DI, the next two chapters look at the standard parts related to rights. Chapter 5 provides a detailed treatment of the MPEG Rights Expression Language (REL), which is a distinct and new functionality offered by MPEG-21. The treatment explains the creation of licenses using the XML-based REL and uses simple examples to lead the reader through the language. The REL is a versatile language intended to offer a 'complete' approach to rights expression and is not restricted to limited functionality platforms. The treatment of expression of Rights using the REL is complemented by Chapter 6 on the Rights Data Dictionary (RDD), which details the structure and approach to the ontology of the standardised dictionary of terms. It explains the background to the RDD and details the context model that underlies the approach; this provides the reader with a clear description of how a coherent semantic structure of terms and relationships can be used to provide semantic interoperability for metadata-based rights expressions.

Up to this point the chapters have not considered the carriage and delivery of the DI, but Chapters 7 and 8 provide the opportunity to describe how the normative specifications allow DIs to be adapted and delivered. Chapter 7 details the Universal Multimedia Access (UMA) aspects of the framework. In particular, it shows how Digital Items can be adapted to user preferences, device characteristics and natural environments while effective delivery is maintained over a wide variety of networks. Chapter 8 describes a set of tools that allow MPEG-21 to handle content formats in an independent manner. Through the use of a set of XML-based Bitstream Description (BSD) tools that can describe binary streams, the MPEG-21 framework offers users a versatile approach to handling media. The basic bitstream tools are complemented by a system that allows the combination of constraints, adaptation and Quality of Service (QoS) measurements to create a normative format agnostic approach to adaptation for UMA.

The early chapters treat the Digital Item as a 'static' container, but Chapter 9 introduces Digital Item Processing, which seeks to make the DI a 'smart' item. The chapter considers the normative specification of an API of base operations (DIBO), which are then used in ECMAScript-based methods (DIMs) that can be incorporated into Digital Items. A second normative approach for more complex functionality is available through the use of Java-based Digital Item Extension Operations (DIXOs) and these are also explained.

Chapter 10 brings the core technical chapters to a close with a treatment of the mechanisms offered by MPEG-21 for the normative expression of requests for event reports and the format for transmission of those reports. This is a unique part of the standard and brings new mechanisms for the management of multimedia content.

The book ends with a discussion on the future of MPEG-21 and, more generally, MPEG standards in Chapter 12. It considers the new work that allows DIs to be bound to streams and looks at the new Multimedia Application Format and MPEG Multimedia Middleware standardisation activities; the relationships of these latter activities to MPEG-21 is highlighted.

ACKNOWLEDGEMENTS

MPEG-21 can be traced back to mid 1999 as an activity within the MPEG community. Since then, many people from different parts of the world have worked to create a coherent and valuable framework for multimedia delivery and consumption. At times, this has required trust that the original vision was correct, and often the novelty and different directions of the MPEG-21 work have led to new and valuable tests for long-standing MPEG processes. However, throughout this period there have been fruitful contributions, discussions and debates, and the editors would like to acknowledge those involved, as without that technical activity, this book would not have been started or reached publication.

The editors would like to thank the many people who reviewed the chapters on this book; their careful reading and suggestions for improvements have been invaluable in creating the best possible book for you, the reader. Of course, we also need to thank the authors of the chapters who squeezed another set of hours from their already overcommitted lives. Often this was a commitment of personal time and we need to also thank their families for supporting such dedication.

Finally, we hope that this book will provide readers with a valuable tool and resource when working in multimedia generally and, in particular, when using the MPEG-21 multimedia framework.

Ian S Burnett, Fernando Pereira, Rob Koenen and Rik Van de Walle

Acronyms and Abbreviations

3D	Three-Dimensional
3GPP	3rd Generation Partnership Project
AAC	Advanced Audio Coding
AC-3	Adaptive Transform Coder 3
ADTE	Adaptation Decision-Taking Engine
AFX	Animation Framework eXtension
AHG	Ad Hoc Group
AMD	Amendment
API	Application Programming Interface
AQoS	Adaptation Quality of Service
ARIB	Association of Radio Industries & Businesses
ASN.1	Abstract Syntax Notation number One
ASP	Application Service Provider
ATM	Asynchronous Transfer Mode
ATSC	Advanced Television Systems Committee
AU	Access Unit
AV	Audio-Visual
AVC	Advanced Video Coding
BC	Backward Compatible
BER	Bit Error Rate
BIEM	Bureau International des Sociétés Gérant les Droits d'Enregistrement et de Reproduction Mécanique
BIFS	BInary Format for Scenes
BiM	Binary format for Metadata
BintoBSD	Binary to Bitstream Syntax Description
BPP	Bit Per Pixel
BS Schema	Bitstream Syntax Schema
BSD	Bitstream Syntax Description
BSDL	Bitstream Syntax Description Language
BSDLink	Bitstream Syntax Description Link
BSDtoBin	Bitstream Syntax Description to Binary
CD	Committee Draft or Compact Disc
CE	Core Experiment or Consumer Electronics
CGD	Computational Graceful Degradation
CIBER	Cellular Intercarrier Billing Exchange Roamer

cIDf	content ID forum
CIF	Common Intermediate Format
CISAC	Confédération Internationale des Sociétés d'Auteurs et Compositeurs
CLL	Central Licensing Label Information
COA	Contextual Ontology Architecture
COR	Technical Corrigendum
CPU	Central Processing Unit
CRC	Cyclic Redundancy Code
CRF	Content Reference Forum
CSS	Cascading Style Sheets
CSV	Comma Separated Values
D	Descriptor
DAM	Digital Asset Management
DCOR	Draft Technical Corrigendum
DCT	Discrete Cosine Transform
DDL	Description Definition Language
DI	Digital Item
DIA	Digital Item Adaptation
DIAC	Digital Item Adaptation Configuration
DIBO	Digital Item Base Operation
DID	Digital Item Declaration
DIDL	Digital Item Declaration Language
DII	Digital Item Identification
DIM	Digital Item Method
DIML	Digital Item Method Language
DIP	Digital Item Processing
DIS	Digital Item Streaming
DIXO	Digital Item eXtension Operation
DMIF	Delivery Multimedia Integration Framework
DMP	Digital Media Project
DNS	Domain Name System
DoC	Disposition of Comments
DOI	Digital Object Identifier
DOM	Document Object Model
DPCM	Differential Pulse Code Modulation
DR	Defect Report
DRM	Digital Rights Management; also Digital Radio Mondiale
DS	Description Scheme
DSM-CC	Digital Storage Media – Command and Control
DTR	Draft Technical Report
DVB	Digital Video Broadcasting
DVD	Digital Versatile Disc
EAI	Enterprise Application Integration
EBNF	Extended Backus–Naur Form
EPG	Electronic Program Guide

ER	Event Report or Event Reporting
ER-R	Event Report Request
ES	Elementary Stream
FCD	Final Committee Draft
FDAM	Final Draft Amendment
FDIS	Final Draft International Standard
FF	File Format
FIPA	Foundation of Intelligent and Physical Agents
FPDAM	Final Proposed Draft Amendment
gBS Schema	generic Bitstream Syntax Schema
gBSD	generic Bitstream Syntax Description
gBSDtoBin	generic Bitstream Syntax Description to Binary
GFX	Graphics Framework eXtensions
GOP	Group of Pictures
HDTV	High Definition Television
HTML	HyperText Markup Language
HTTP	HyperText Transfer Protocol
IBC	International Broadcasting Convention
IDF	International DOI Foundation
IEC	International Electrotechnical Commission
IETF	Internet Engineering Task Force
IFPI	International Federation of the Phonographic Industry
IOPin	Input Output Pin
IP	Internet Protocol; also Intellectual Property
IPMP	Intellectual Property Management and Protection
IS	International Standard
ISAN	International Standard Audio-visual Number
ISBN	International Standard Book Number
ISMN	International Standard Music Number
ISO	International Organization for Standardization
ISRC	International Standard Recording Code
ISTC	International Standard Text Code
ISWC	International Standard Musical Work Code
IT	Information Technology
ITTF	Information Technology Task Force
ITU	International Telecommunication Union
ITU-R	International Telecommunication Union – Radio Standardization Sector
ITU-T	International Telecommunication Union – Telecommunication Standardization Sector
J-DIXO	Java DIXO
JPEG	Joint Photographic Experts Group
JSR	Java Specification Request
JTC	Joint Technical Committee
JTPC	ISO/IEC Joint Technical Programming Committee
JVM	Java Virtual Machine

JVT	Joint Video Team
JWG	Joint Working Group
LUT	Look-Up Table
M3W	MPEG Multimedia Middleware
MAC	Media Access Control Address
MAF	Multimedia Applications Format
MDS	Multimedia Description Schemes
MHP	Multimedia Home Platform
MI3P	Music Industry Integrated identifier Project
MIME	Multipurpose Internet Mail Extensions
MP3	MPEG-1 Audio Layer 3
MP4	MPEG-4 File Format
MPEG	Moving Picture Experts Group
MPEGIF	MPEG Industry Forum
NAB	National Association of Broadcasters
NB	National Body
NBC	(MPEG-2) Non-Backward Compatible coding; now Advanced Audio Coding (AAC)
NP	New Work Item Proposal
NSP	Network Service Provider
NTSC	National Television Standards Committee
ODRL	Open Digital Rights Language
OeBF	Open eBook Forum
OMA	Open Mobile Alliance
OMS	Ontology Management System
OS	Operating System
OWL	Web Ontology Language
PC	Personal Computer
PDA	Personal Digital Assistant
PDAM	Proposed Draft Amendment
PDTR	Proposed Draft Technical Report
QoS	Quality of Service
RA	Registration Authority
RAND	Reasonable and Non-Discriminatory
RDD	Rights Data Dictionary
RDF	Resource Description Framework
REL	Rights Expression language
RIAA	Recording Industry Association of America
ROI	Region-Of-Interest
RPN	Reverse Polish Notation
RSA	Rivest, Shamir and Adleman (RSA) Public-key Cryptosystem
RTP	Real-Time Transport Protocol
RTSP	Real-Time Streaming Protocol
SC	Sub-Committee
SF	Stack Function
SGML	Standard Generalized Markup Language

SMIL	Synchronized Multimedia Integration Language
SMPTE	Society of Motion Picture and Television Engineers
SNHC	Synthetic Natural Hybrid Coding
SNR	Signal-to-Noise Ratio
SOAP	Simple Object Access Protocol
STX	Streaming Transformations for XML
SVC	Scalable Video Coding
TAP	Transferred Account Procedure
TCP	Transmission Control Protocol
TR	Technical Report
TS	Transport Stream
TV	Television
UCD	Universal Constraints Description
UDDI	Universal Description, Discovery, and Integration of Web Services
UED	Usage Environment Description
UEP	Unequal Error Protection
UF	Utility Function
UMA	Universal Multimedia Access
UML	Universal Modeling Language
UMTS	Universal Mobile Telecommunications System
URI	Uniform Resource Identifier
URL	Universal Resource Locator
UTC	Coordinated Universal Time
VCD	Video CD
VHDL	Very High speed integrated circuit hardware Description Language
VM	Virtual Machine
VOP	Visual Object Plane
VRML	Virtual Reality Modeling Language
W3C	World Wide Web Consortium
WD	Working Draft
WG	Working Group
WMA	Windows Media Audio
WS-I	Web Services Interoperability Organization
XHTML	Extensible HyperText Markup Language
XInclude	XML Inclusions
XML	eXtensible Markup Language
XMT	eXtensible MPEG-4 Textual
XPath	XML Path Language
XPointer	XML Pointer Language
XSLT	Extensible Stylesheet Language Transformation
YM	sYstems Model

List of Contributors

Chris Barlas
Rightscom Ltd., UK

Ian S Burnett
School of Electrical, Computer and Telecommunication Engineering,
University of Wollongong, NSW, Australia

Leonardo Chiariglione
CEDEO.net

Sylvain Devillers
France Télécom, Issy les Moulineaux, France

Frederik De Keukelaere
Ghent University – IBBT, Ghent, Belgium,

Rik Van de Walle
Ghent University – IBBT, Ghent, Belgium

Thomas DeMartini
ContentGuard Inc., Los Angeles, USA

Martin Dow
Rightscom Ltd., UK

Gerrard Drury
University of Wollongong, Wollongong, Australia

Jill Kalter
ContentGuard Inc., Los Angeles, USA

Rob Koenen

Kyunghee Ji
Seoul University of Venture and Information, Seoul, Korea

Shane Lauf
School of Electrical, Computer and Telecommunication Engineering,
University of Wollongong, NSW, Australia

Fernando Pereira
Instituto Superior Técnico – Instituto de Telecomunicações, Lisbon, Portugal

Mai Nguyen
ContentGuard Inc., Los Angeles, USA

FX Nuttall
IT consultant, CISAC (International Confederation of Societies of Authors
and Composers)

Eva Rodriguez
Universitat Pompeu Fabra, Barcelona, Spain

Godfrey Rust
Rightscom Ltd., UK

Christian Timmerer
Klagenfurt University, Klagenfurt, Austria

Andrew Tokmakoff
Telematica Instituut, Enschede, The Netherlands

Edgar Valenzuela
ContentGuard Inc., Los Angeles, USA

Anthony Vetro
Mitsubishi Electric Research Labs, Cambridge, USA

Xin Wang
ContentGuard Inc., Los Angeles, USA

1

MPEG: Context, Goals and Working Methodologies

Fernando Pereira and Rob Koenen

1.1 INTRODUCTION

At the end of the eighties, a number of key technologies were coming to a point of maturity, which brought revolutionary ways to deliver content to end-users within technical reach. Advances in sound and image compression, in very-large-scale integration (VLSI) technology, in optical storage, combined with the work on high-speed delivery of digital information over phone lines together would enable the CD-interactive, the videophone, video on demand, the DVD and a multimedia-enabled Internet. As usual, things go slower in the real world than technology optimists would hope for, and bringing new services to market is a tough nut to crack. The CD-i was never a runaway success, but by all accounts the DVD is, albeit a decade later. Early in the 21st century, 10 years after the first trials, video on demand is starting to become a viable service. Videotelephony has some success as a business tool in the form of videoconferencing, but never had become the consumer success it was destined to be even halfway through the nineties. Perhaps the service gets a second life on 3G mobile phones, and as an intuitive add-on to the immensely popular Instant Messaging. While some people may have dreamt of a ubiquitous multimedia network, spanning the world and delivering many services, the Internet in its current form was at that time indeed just that – a distant dream.

At the basis of all these dreams and recent successes lies the compressed digital representation of audio and video signals. The consumer electronics (CE) and telecommunications companies that recognized the possibilities of these new technologies also recognized the benefits of standardizing them. Both these industries were well acquainted with the ins and outs of standardization. Voice and video compression standards were available or being developed in the International Telecommunications Union (ITU). Those standards were targeted at the needs of the telecommunications industry, while the opportunities

The MPEG-21 Book   Ian S Burnett, Fernando Pereira, Rik Van de Walle, Rob Koenen

were increasingly in the entertainment realm, and there were other requirements not supported by telecommunications standards, such as the support for random access in bit-streams and for trick modes or the need to support audio signals rather than just voice. It was understood that to launch the required new formats, it was necessary to have the right standards underpinning them, so that many players could create compatible products.

This understanding, shared by a handful of CE and Telecommunications companies, was the basis for the creation of the Moving Picture Experts Group (MPEG) [1] in 1988. They had the vision that by setting standards for audio and video representation, they would create a market from which they could all benefit. They understood that creating compat-ibility was crucial, and that it would not harm their chances to create successful products, but rather enhance them. And they would create this compatibility without eliminating the option for competition and excellence, as this chapter will explain further down.

Since its establishment, MPEG has set many widely used standards. The group has expanded its scope from basic coding technologies to technologies that support audio and video compression formats. In a sense, this already started with MPEG-1, when the group realized that just setting an audio and a video compression standard would not suffice. Something more was needed to support synchronization, storage and delivery. This was called *MPEG-1 Systems*. An account of how MPEG standards developed, starting from MPEG-1, is given below. This chapter will only briefly touch on MPEG-21, because it is the topic of discussion in the other chapters of this book.

This chapter will also explain how the ISO/IEC MPEG working group operates. It will look at the mission of MPEG, give a brief overview of the existing MPEG standards and, even more briefly, touch upon the goals of the MPEG-21 standard and how they relate to the other MPEG standards.

1.2 MPEG MISSION

MPEG [1] was established in 1988 with the mandate to develop standards for the coded representation of moving pictures, audio and their combination. MPEG operates in the framework of ISO/IEC's[1] Joint Technical Committee on Information Technology (JTC 1) [2] and is formally Working Group 11 (WG11) of Subcommittee 29 (SC29) of JTC1, which results in MPEG's official label being *ISO/IEC JTC1/SC29/WG11*. JTC1's mission is standardization in the field of Information Technology (IT) including the development, maintenance, promotion and facilitation of IT standards required by global markets meet-ing business and user requirements concerning design and development of IT systems and tools, performance and quality of IT products and systems, security of IT systems and information, portability of application programs, interoperability of IT products and systems, unified tools and environments, harmonized IT vocabulary, and user-friendly and ergonomically designed user interfaces. According to JTC1, '*IT includes the specification, design and development of systems and tools dealing with the capture, representation, pro-cessing, security, transfer, interchange, presentation, management, organization, storage and retrieval of information*' [2].

MPEG exists to develop standards that can be widely deployed. Deployment is, of course, the measure of success for any standard. According to its Terms of Reference, MPEG's area of work is the '*development of international standards for compression,*

---

[1] International Standards Organization or International Organization for Standardization and International Electrotechnical Commission

decompression, processing and coded representation of moving pictures, audio and their combination, in order to satisfy a wide variety of applications' [3]. MPEG's programme of work targets:

- serving as a responsible body within ISO/IEC for recommending a set of standards consistent with the area of work;
- cooperation with other standardization bodies dealing with similar applications;
- consideration of requirements for inter working with other applications such as telecommunications and broadcasting, with other image coding algorithms defined by other SC29 working groups and with other picture and audio coding algorithms defined by other standardization bodies;
- definition of methods for the subjective assessment of quality of audio, moving pictures and their combination for the purpose of the area of work;
- assessment of characteristics of implementation technologies realizing coding algorithms of audio, moving pictures and their combination;
- assessment of characteristics of digital storage and other delivery media target with the standards developed by WG11;
- development of standards for coding of moving pictures, audio and their combination taking into account quality of coded media, effective implementation and constraints from delivery media;
- proposal of standards for the coded representation of moving picture information;
- proposal of standards for the coded representation of audio information;
- proposal of standards for the coded representation of information consisting of moving pictures and audio in combination;
- proposal of standards for protocols associated with coded representation of moving pictures, audio and their combination.

To fulfil its programme of work, MPEG has developed a large number of standards that will be briefly presented in the following. Moreover, Section 1.5 will present the methodologies used by MPEG to timely develop high-quality technical specifications.

1.3 MPEG STANDARDS PRECEDING MPEG-21

To date, MPEG has developed five major sets of technical standards as listed below. Work on most of these is still going on, including maintenance on the earlier MPEG standards. To better understand the context in which the MPEG-21 project was launched, it is not only necessary to look to the general multimedia landscape but also to the past work of the MPEG itself.

Since its start in May 1988, MPEG has developed the following sets of technical standards[2]:

- ISO/IEC 11172 (MPEG-1), entitled 'Coding of Moving Pictures and Associated Audio at up to about 1.5 Mbps'

---

[2] Although MPEG-3 was once envisioned to address high bitrates and high definition TV, the standard was never developed because its requirements could be met by MPEG-2. MPEG-5 and MPEG-6 have never been defined, leaving it up to the reader to guess why the MPEG-7 number was chosen. Finally, MPEG-21 is the first MPEG standard to be set in the 21st century.

- ISO/IEC 13818 (MPEG-2), entitled 'Generic Coding of Moving Pictures and Associated Audio'
- ISO/IEC 14496 (MPEG-4), entitled 'Coding of Audio–Visual Objects'
- ISO/IEC 15938 (MPEG-7), entitled 'Multimedia Content Description Interface'
- ISO/IEC 21000 (MPEG-21), entitled 'Multimedia Framework'

MPEG follows the principle that a standard should specify as little as possible while guaranteeing interoperability. This increases the lifetime of the standard by allowing new technology to be introduced and encouraging competition. For example, MPEG coding standards only specify bitstream syntax and semantics as well as the decoding process; nothing is said about how encoders are supposed to create compliant bitstreams. This leaves developers much freedom to optimize their choice of encoding tools, for example, motion estimation, rate control, psychoacoustic model, to maximize the performance for their target applications and content. It also allows the standard's performance to increase over time, following the developers' increasing knowledge on how to best exploit the coding tools. Lastly, it enables the various players to compete in quality and price, while providing mutual interoperability.

1.3.1 THE MPEG-1 STANDARD

The MPEG-1 standard [4] represents the first generation of the MPEG family; it was set in the period from 1988 to 1991. At the end of the eighties, with ITU-T recommendation H.261 [5] on video coding for communication purposes almost finalized, it became clear that the same coding technology could provide a digital alternative to the widely spread analogue video cassette player. MPEG-1's goal was to provide a complete audio-visual digital coding solution for digital storage media such as CD, DAT, optical drives and Winchester discs. Since CD was the major target, the standard was optimized for 1.5 Mbps bitrate range, but the standard works at lower and higher bitrates as well.

To compete with analogue videotape recorders, the MPEG-1 solution had to provide the special access modes typical of these devices such as fast forward, fast reverse and random access, and the video quality had to be at least comparable to VHS quality. This highlighted that coding efficiency is not the only requirement of a coding scheme; there are other important functionalities, depending on the targeted area of use (e.g., random access, low delay and the object-based interactive functionalities built into the MPEG-4 standard).

MPEG usually develops audio and video coding standards in parallel, together with the multiplexing and synchronization specifications. Although designed to be used together, the individual specifications can also be used independently and with other tools, for example, a different video format can be used together with the MPEG-1 Systems and Audio solutions. For this reason, the MPEG standards are organized in Parts, each one defining a major piece of technology. MPEG-1 has five Parts:

- *Part 1 – Systems*: Addresses the combination of one or more data streams (MPEG-1 Video and Audio) with timing information to form a single stream, optimized for digital storage or transmission.

- *Part 2 – Video*: Specifies a coding format (video stream and the corresponding decoding process) for video sequences at bitrates around 1.5 Mbps. The target operational environment was storage media at a continuous transfer rate of about 1.5 Mbps, but the coding format is generic and can be used more widely. It supports interactivity functionalities or special access modes such as fast forward, fast reverse and random access into the coded bitstream. The coding solution is a typical hybrid coding scheme based on block-based Discrete Cosine Transform (DCT) applied to a single picture or to a prediction error obtained after temporal prediction (on the basis of one or two pictures) with motion compensation. DCT is followed by quantization, zigzag scan and variable length coding (see Figure 1.1). MPEG-1 Video only supports progressive formats and the number of lines is flexible.

- *Part 3 – Audio*: Specifies a coding format (audio stream and the corresponding decoding process) for monophonic (32–192 kbps) and stereophonic (128–384 kbps) sound. This standard specifies three hierarchical coding layers – I, II and III – which are associated to increasing complexity, delay and efficiency. Layer III is more commonly known as *MP3*. These coding solutions are designed for generic audio, and exploit the perceptual limitations of the human auditory system, targeting the removal of perceptually irrelevant parts of the signal. They do not rely on a model of the signal source, like voice coders do.

- *Part 4 – Conformance Testing*: Specifies tests to check if bitstreams (content) and decoders are correct according to the specifications in Parts 1, 2 and 3.

- *Part 5 – Software Simulation*: Consists of software implementing the tools specified in Parts 1, 2 and 3; this is a Technical Report (see Section 1.5.3) that has only informative value. In later MPEG standards, the so-called *Reference Software* became a normative Part of the standard.

MPEG-1 is still a very popular format for Internet video streaming and downloading. It did not become widely used in the area of digital storage media, partly because the DVD (using MPEG-2) soon followed it; still, the MPEG-1–based Video CD (VCD) is selling to the millions in China. It is also well known that the success of MP3, needless to explain, set the stage for the ongoing revolution in digital music distribution.

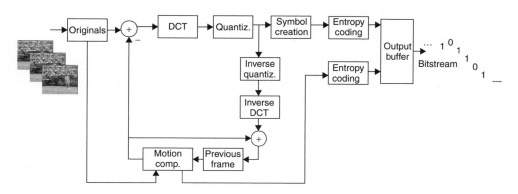

Figure 1.1 Simplified MPEG-1 Video encoder architecture

1.3.2 THE MPEG-2 STANDARD

The success of solutions on the basis of H.261 and MPEG-1 whetted the appetite for more widely usable digital coding schemes, wider range of bitrates, more choice in video resolution and support for interlaced signals. The requirements for a common video representation solution in the area of audiovisual (AV) entertainment, both broadcasting and storage, brought together the coding experts of ITU-T and ISO/IEC (through MPEG), who jointly developed a generic video coding standard targeting medium and high-quality applications, including high definition TV (HDTV). This joint specification was published as ISO/IEC 13818 Part 2, ('MPEG-2 Video'), and as recommendation ITU-T H.262 by ITU-T. The MPEG-2 Systems, Video and Audio specifications are largely based on the corresponding MPEG-1 specifications [6, 7]. Later on, another Audio Part was added, providing a significant performance increase over the backward compatible Part 3. This new audio codec, initially known as *non-backward compatible audio* was dubbed Advanced Audio Coding (AAC) just before the standard was published.

The different Parts of MPEG-2 are as follows:

- *Part 1 – Systems*: Addresses the same requirements as MPEG-1 Systems but adds support for the following:
  (i) error-prone environments such as broadcasting;
  (ii) hardware-oriented processing and not only software oriented processing;[3]
  (iii) carrying multiple programs simultaneously without a common time base;
  (iv) transmission in ATM environments.
  These requirements resulted in the specification of two types of Systems streams: the Program Stream similar to, and compatible with, MPEG-1 Systems streams, and the new Transport Stream to carry multiple independent programs.
- *Part 2 – Video*: Specifies a generic coding format (video stream and the corresponding decoding process) for video sequences up to HDTV resolutions. The basic coding architecture is the same as for MPEG-1 Video, adding support for interlaced video formats and scalable (hierarchical) coding formats (e.g., temporal, spatial, Signal to Noise Ratio (SNR)). To be sufficiently generic, MPEG-2 Video comprises a large set of tools. In some applications, some of these tools are too complex for their utility, and MPEG-2 Video defines so-called *profiles* to be able to provide coding solutions with appropriate complexity. Profiles are tool subsets that address the needs of a specific class of applications. Also, a certain number of *levels* are defined for each profile, to limit the memory and computational requirements of a decoder implementation. Levels provide an upper bound on complexity of the bitstream, and minimum capabilities for the decoder, so that compliant bitstreams will always play on compliant decoders. An MPEG-2 Video decoder is able to decode MPEG-1 Video streams, which is called *forward compatibility*. Backward compatibility, implying that (subsets of) MPEG-2 Video are decodable by MPEG-1 Video decoders, is only provided for specific profiles through scalability.
- *Part 3 – Audio*: Specifies a coding format (audio stream and the corresponding decoding process) for multichannel audio. Part 3 provides backward and forward compatibility with MPEG-1 Audio streams, which is why it is also known as *backward*

---

[3] To accommodate hardware implementations, the large variable length packets that MPEG-1 Systems use to minimize the software overhead may no longer be used in MPEG-2.

compatible (BC) audio. Because of the requirement for backward compatibility (an MPEG-1 stereo decoder should be able to reproduce a meaningful version of the original multichannel MPEG-2 Audio stream), MPEG-2 Audio is technically similar to MPEG-1 Audio, with the same three layers. The most important difference is the support for multichannel audio, typically in 5 + 1 combinations.

- *Part 4 – Conformance Testing*: Specifies tests allowing to check if bitstreams (content) and decoders are correct according to the specifications in the technical Parts of MPEG-2. For video streams, conformance is specified to a profile@level combination; this means conformance is defined for streams complying with a certain level of a certain profile.
- *Part 5 – Software Simulation*: Consists of software implementing the tools specified in Parts 1, 2 and 3; this is an informative Technical Report (see Section 1.5.3).
- *Part 6 – Digital Storage Media – Command and Control (DSM-CC)*: Specifies generic control commands independent of the type of the DSM. This allows MPEG applications to access local or remote DSMs to perform functions specific to MPEG streams without the need to know details about the DSMs. These commands apply to MPEG-1 Systems, and MPEG-2 Program and Transport streams.
- *Part 7 – Advanced Audio Coding (AAC)*: Because MPEG-2 Audio (Part 3) was specified with a backward compatibility constraint, coding efficiency had to be compromised quite significantly. MPEG-2 AAC Audio (also known as *Non-backwards* Compatible, NBC) specifies a multichannel coding format without an MPEG-1 backward compatibility requirement, yielding similar qualities at much lower bitrates than MPEG-2 BC Audio.
- *Part 8*: Withdrawn (initially targeted 10-bit video coding)
- *Part 9 – Real-time Interface*: Specifies additional tools for use of MPEG-2 Systems in real-time interchange of data, for example, as required in many telecommunications applications.
- *Part 10 – DSM-CC Conformance Extension*: Specifies tests to check if bitstreams (content) and decoders are correct according to the specification in Part 6.
- *Part 11 – Intellectual Property Management and Protection (IPMP) on MPEG-2 Systems*: A recent addition specifying further Systems tools that allow the IPMP capabilities developed in the context of MPEG-4 to be used with MPEG-2 Systems streams.

The MPEG-2 standard is undoubtedly the most successful AV coding standard to date. The number of MPEG-2–based devices is hard to estimate, but must be counted by the hundreds of millions. MPEG-2 Audio, Video and Systems are present in Digital Versatile Discs (DVDs [8], the CE device with the fastest sales growth ever), in Digital Video Broadcasting (DVB) [9] set-top boxes, and also in Advanced Television Systems Committee (ATSC) [10] set-top boxes. In general, digital TV and digital AV storage *speaks* MPEG-2.

1.3.3 THE MPEG-4 STANDARD

While the MPEG-1 and MPEG-2 standards follow a representation model, which has its roots in analogue television using the so-called *frame-based model*, MPEG-4, launched by MPEG in 1993, embodies a significant conceptual jump in audio-visual content representation – the object-based model. The object-based model avoids the blindness of the frame-based model by recognizing that audio-visual content aims at reproducing a world that is made of elements, called the *objects*. By adopting the object-based model, MPEG-4 [11, 12] starts a new generation of content representation standards, where the audio-visual scene can be built as a composition of independent objects with their own coding, features and behaviours. Multimedia developments on the Internet showed that users wanted the same interaction capabilities in the audio-visual world that they were used to in terms of text and graphics. Figure 1.2 illustrates a simplified object-based audio-visual coding architecture showing that the reproduced scene is a composition according to a script, the so-called *composition information*, of a number of audio and visual objects that have been independently coded and are thus independently accessible. This architecture allows a full range of interactions, automatic or user driven, from the simple, local composition interaction, where an object is removed or its position changed in the scene to the remote interaction with the sending side asking for an object not to be sent, to avoid wasting bandwidth and computational resources with low priority objects.

Among the major advantages of the object-based approach, it is worthwhile to highlight the following:

- *Hybrid natural and synthetic coding*: In the MPEG-4 representation playground, audio and visual objects can be of any origin, natural or synthetic, for example, text, speech, frame-based video, arbitrarily shaped video objects, 3D models, synthetic audio, 2D meshes. In MPEG, this goes by the acronym 'SNHC' for Synthetic-Natural Hybrid Coding.

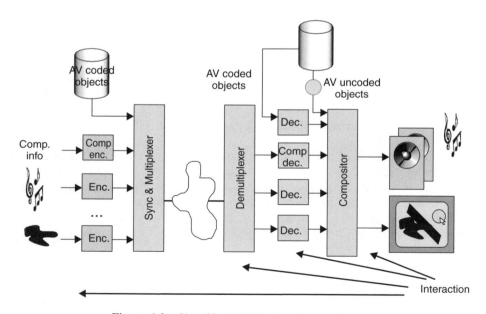

Figure 1.2   Simplified MPEG-4 coding architecture

- *Content-based interaction and reusing*: Because, by definition, objects are independently represented from each other and thus independently accessible, it is possible to directly interact with the various objects in the scene changing their properties or behaviour. Moreover, objects may be reused from a scene to another, which is also an important feature to decrease costs in content production.

- *Content-based coding*: As objects are independently coded, and many types of objects are allowed in the MPEG-4 representation playground, it is a natural option to code each type of object by taking benefit of the intrinsic characteristics of the object; this means that a text object will be coded with a text coding tool while a 3D object will be coded with a 3D model coding tool. This close adjustment of the coding tools to the type of data brings significant coding efficiency gains automatically, which may be rather substantial for the types of content involving several data types, as it typically happens nowadays, for example, with news programs.

- *Universal access*: With the growing usage of mobile and wireless terminals, access to audio-visual content, anywhere and at anytime, has become the most expected feature. The intrinsic representation flexibility and granularity of the object-based approach in terms of selective error robustness and scalability creates the ideal conditions to provide, for each consumption conditions, a content variation that is the most adequate to provide the best experience for the relevant usage environment.

The power and advantages of the object-based representation paradigm make MPEG-4 a standard that does not target a specific application domain, but may find application from low bitrate personal mobile communications to high-quality studio production. This broad range of applications has associated with it a large set of requirements which has, with passage of time, made the MPEG-4 standard grow to become rather large and currently organized in 21 Parts. When some of the initial Parts – notably Part 1, Systems – grew too large through quite a few additions, MPEG adopted a more fine-grained approach by developing smaller Parts to facilitate referencing by the industry. This resulted in the following Parts:

- *Part 1 – Systems*: Specifies the systems architecture and the tools associated to scene description – both the BInary Format for Scenes (BIFS) and eXtensible MPEG-4 Textual (XMT) formats, as well as multiplexing, synchronization, buffer management and management and protection of intellectual property (IPMP). It also specifies the MP4 file format that is designed to be independent of any particular delivery protocol while enabling efficient support for delivery in general, and to contain the media information of an MPEG-4 presentation in a flexible, extensible format that facilitates interchange, management, editing and presentation of the media. It also specifies MPEG-J, which is a Java application engine defining how applications may be contained in a bitstream and executed at the client terminal.

- *Part 2 – Visual*: Specifies all the coding tools associated with visual objects, both of natural and synthetic origin; for example, this implies specifying video coding solutions from very low bitrates to very high-quality conditions as well as generic dynamic 3D objects and specific objects for human faces and bodies.

- *Part 3 – Audio*: Specifies all the coding tools associated with aural objects, both of natural and synthetic origin; for example, this implies specifying coding solutions for music and speech data for a very wide bitrate range, from transparent music to very low bitrate speech, as well as synthetic audio including support for 3D audio spaces.

- *Part 4 – Conformance Testing*: Defines tests that allow to check if bitstreams (content) and decoders are correct according to the specifications in the technical Parts of MPEG-4. For visual and audio streams, conformance is specified to a profile@level combination where a profile is defined as a set of object types [12].

- *Part 5 – Reference Software*: Includes software corresponding to most Parts of MPEG-4, notably video and audio encoders and decoders; this software is copyright free (not patents free) for conformant products. Unlike in MPEG-1 and MPEG-2, MPEG-4 reference software for decoders is considered normative, which means that it helps implementers' understanding of the textual parts.

- *Part 6 – Delivery Multimedia Integration Framework (DMIF)*: Specifies a delivery–media independent representation format to transparently cross the borders of different delivery environments.

- *Part 7 – Optimized Software for MPEG-4 Tools*: Includes optimized encoding software for visual coding tools such as fast motion estimation, fast global motion estimation and fast and robust sprite generation (this Part is a technical report).

- *Part 8 – Carriage of MPEG-4 Content over IP Networks (4onIP)*: Specifies the mapping of MPEG-4 content into several IP-based protocols.

- *Part 9 – Reference Hardware Description*: Includes very high speed integrated circuit hardware description language (VHDL) descriptions of MPEG-4 tools that are portable, synthesizable and simulatable.

- *Part 10 – Advanced Video Coding (AVC)*: Specifies more advanced frame-based video coding tools that provide up to 50 % higher coding efficiency than the best video coding profile in MPEG-4 Part 2 for a wide range of bitrates and video resolutions; however, compared to previous standards, its decoder complexity is about four times that of MPEG-2 Video and two times that of the MPEG-4 Visual Simple profile. This Part has been developed by the Joint Video Team (JVT) created to formalize the collaboration between the ISO/IEC Moving Picture Experts Group (MPEG) and the ITU-T Video Coding Experts Group (VCEG) for the development of this codec. MPEG-4 AVC is known within ITU-T as *Recommendation H.264*.

- *Part 11 – Scene Description and Application Engine*: Includes tools already specified in Part 1 related to scene composition – both the binary (BIFS) and textual (XMT) formats, MPEG-J and multi-user worlds.

- *Part 12 – ISO Base Media File Format*: Specifies the ISO base media file format, which is a general format forming the basis for a number of other more specific file formats; it contains the timing, structure and media information for timed sequences of media data, such as audio/visual presentations. This Part is applicable to MPEG-4 and also to JPEG2000 (joint specification).

- *Part 13 – IPMP Extensions*: Specifies tools to manage and protect intellectual property on audio-visual content and algorithms, so that only authorized users have access to it.

- *Part 14 – MP4 File Format*: Defines the MP4 file format as an instance of the ISO base media file format (Part 12); this was previously included in Part 1, but is now published separately.

- *Part 15 – AVC File Format*: Defines a storage format for video streams compressed using AVC; this format is on the basis of, and compatible with, the ISO base media file format (Part 12), which is used by the MP4 file format, the Motion JPEG2000 file format and the 3GPP file format, among others.

- *Part 16 – Animation Framework eXtension (AFX)*: Specifies tools for interactive 3D content operating at the geometry, modelling and biomechanical levels and includes tools previously defined in MPEG-4 in the context of 'SNHC'. AFX offers a unified standardized 3D framework providing advanced features such as compression, streaming and seamless integration with other audio-visual media, and allowing the building of high-quality creative cross media applications.

- *Part 17 – Streaming Text Format*: Specifies text streams, notably the concatenation of text access units, format of text streams and text access units and signalling and decoding of text streams; in terms of text formats, this Part complements the tools already specified in Part 1.

- *Part 18 – Font Compression and Streaming*: Specifies tools allowing the communication of font data as part of an MPEG-4 encoded audio-visual presentation.

- *Part 19 – Synthesized Texture Stream*: Specifies synthesized textures, which result from the animation of photo-realistic textures by describing colour information with vectors; these vectors are animated over time, producing very low bitrate movies, called *synthesized textures*.

- *Part 20 – Lightweight Application Scene Representation (Laser)*: Provides a scene representation targeting a trade-off between expressivity, compression efficiency, decoding and rendering efficiency, and memory footprint.

- *Part 21 – Graphics Framework eXtensions (GFX)*: Provides a fully programmatic framework with flexible compositing and rendering operations of natural and synthetic media assets. GFX reuses industry-approved (and deployed) APIs as well as MPEG-4 specific APIs used for audio-visual and gaming applications, thereby allowing a wide range of entertainment applications, from mobile to desktop devices.

It is clear from this list that MPEG-4 is not only very innovative but also a large standard, and still growing. As applicable for most standards, some tools – Parts – will be more successful than others. The technical specification is not the last stage on the road to widespread adoption; licensing needs are to be addressed as well. The delay in setting up licensing conditions for MPEG-4 technology and the negative response to the initial licensing conditions significantly impacted the adoption of the MPEG-4 standard. Section 1.6 will discuss some of the post-standardization activities required to get an uptake of open standards like the MPEG.

1.3.4 THE MPEG-7 STANDARD

With the availability of the MPEG-1, MPEG-2 and MPEG-4 and other digital coding standards, it became easier to acquire, produce and distribute audio-visual content. However, the abundance of digital content poses a formidable challenge for content management. The more content there is, the harder it becomes to manage, retrieve and filter it in order to find what you really need; and content has value only if it can be found quickly and efficiently. After making important contributions to the explosion of digital audio-visual content, MPEG recognized that it needed to address the problem of audio-visual content management.

In 1996, MPEG launched the MPEG-7 project, formally called *Multimedia Content Description Interface* [13, 14] with the goal to specify a standard way of describing various types of audio-visual information such as elementary pieces, complete works and repositories, irrespective of their representation format or storage medium. Like previous MPEG standards, MPEG-7 is designed to meet a set of requirements gleaned from relevant applications. Unlike previous MPEG standards, the audio-visual representation format to be developed would no longer target the compression and reproduction of the data itself, but rather of the data about the data, so-called *metadata*.

MPEG-7 'descriptions' provide metadata solutions for a large set of application domains. They are media and format independent, object-based, and extensible. They can operate at different levels of abstraction, from describing low-level, automatic and often statistical features, to representing high-level features conveying semantic meaning. The provision of a description framework that supports the combination of low-level and high-level features in a single description is a unique feature of MPEG-7, which, together with the highly structured nature of MPEG-7 descriptions, constitutes an essential difference between MPEG-7 and other available or emerging multimedia description solutions.

Following the principle that MPEG standards must specify only the minimum that is necessary, MPEG-7 only specifies the description format and its decoding but not the description creation and consumption engines, thus leaving much freedom to application developers. MPEG-7 specifies two major types of tools:

- *Descriptors (D)*: A descriptor is a representation of a feature that defines the syntax and the semantics of the feature representation; a feature is a 'distinctive characteristic of the data that signifies something to somebody'. Examples are a time-code for representing duration, colour moments and histograms for representing colour, and a character string for representing a title.
- *Description Schemes (DS)*: A description scheme specifies the structure and semantics of the relationships between its components, which may be both descriptors and description schemes. A simple example is a description of a movie, temporally structured as scenes and shots, including some textual descriptors at the scene level and colour, motion and audio amplitude descriptors at the shot level.

MPEG-7 descriptions may be expressed in two ways: textual streams, using the so-called *Description Definition Language (DDL)*, and binary streams, using the binary format for MPEG-7 data (BiM), which is basically a DDL compression tool. To address the MPEG-7 requirements [14], MPEG-7 specifies technology in 11 Parts:

- *Part 1 – Systems*: Specifies the tools for the following:
  (i) transporting and storing MPEG-7 descriptions in an efficient way using the BiM binary representation format (besides compressing the DDL descriptions, the BiM can also be seen as a general XML compression tool);
  (ii) synchronizing MPEG-7 descriptions with the content they describe (MPEG-7 descriptions may be delivered independently or together with the content they describe);
  (iii) managing and protecting the intellectual property associated with the descriptions.

- *Part 2 – Description Definition Language (DDL)*: Specifies a language for creating new description schemes as well as extending and modifying existing ones. The DDL is based on the W3C's XML (eXtensible Markup Language) Schema Language; some extensions to XML Schema were developed in order to address all the DDL requirements.

- *Part 3 – Visual*: Specifies visual description tools, notably basic structures and descriptors or description schemes, for the description of visual features and for the localization of the described objects in the image or video sequence. The MPEG-7 visual descriptors cover five basic visual features that include colour, texture, shape, motion and localization (a face recognition descriptor is also defined).

- *Part 4 – Audio*: Specifies audio description tools which are organized in areas such as timbre, melody, silence, spoken content and sound effects.

- *Part 5 – Multimedia Description Schemes (MDS)*: Specifies the description tools dealing with generic as well as multimedia entities. Generic entities are those that can be used in audio, visual and textual descriptions, and therefore can be considered 'generic' to all media; multimedia entities are those that deal with more than one medium, for example, audio and video. MDS description tools can be grouped into six different classes according to their functionality (see Figure 1.3): (i) content description: structural and semantic aspects; (ii) content management: media, usage, creation and production; (iii) content organization: collections and models; (iv) navigation and access: summaries, variations and views; (v) user: user preferences and usage history; and (vi) basic elements: datatype and structures, schema tools, link and media localization and basic DSs.

- *Part 6 – Reference Software*: Consists of software for implementing the tools specified in Parts 1–5; as in MPEG-4, this software is considered normative and can be used free of copyright for implementing applications that conform to the standard.

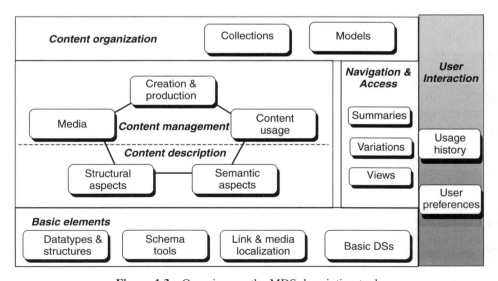

Figure 1.3 Overview on the MDS description tools

- *Part 7 – Conformance Testing*: Specifies procedures allowing to check if description streams are according to the specifications in Parts 1–5; conformance for descriptions is specified to a profile@level combination.
- *Part 8 – Extraction and Use of MPEG-7 Descriptions*: Provides useful information on the extraction and use of descriptions, notably referencing the Reference Software (this is a technical report).
- *Part 9 – Profiles and Levels*: Specifies description profiles and levels. A description profile generally describes a subset of all the description tools defined in MPEG-7. The description tools in a description profile support a set of functionalities for a certain class of applications; a level of a description profile defines further constraints on conforming descriptions, constraining their maximum complexity.
- *Part 10 – Schema Definition*: Specifies the schema definition across the Parts of MPEG-7; it collects together the description tools specified in MPEG-7, assigns a namespace designator, and specifies the resulting syntax description in a single schema using the DDL.
- *Part 11 – Profile Schemas*: Includes the schemas for the various profiles.

Although MPEG-7 is a rich, powerful and substantially complete standard, it has yet to find the major application uptake it deserves. One of the reasons could be that it addresses a different target than the previous MPEG standards, where the need for interoperability is less obvious.

1.4 THE MPEG-21 STANDARD

MPEG-21 is the subject of this book, and all the other chapters are dedicated to this standard. Notably, Chapter 2, 'An Introduction to MPEG-21', explains the how and why of MPEG-21 in detail. This introductory chapter will limit itself to explaining how MPEG-21 fits in with the other MPEG standards.

MPEG started the discussions that ultimately led to the start of MPEG-21 in 1999. A workshop was held at the 51st meeting in Noordwijkerhout (NL), which brought together many speakers from different fields who were directly or tangentially related to MPEG standardization.

The start of MPEG-21 was marked by the realization that in spite of the existence of MPEG-1, MPEG-2, MPEG-4 and even MPEG-7, full interoperability in multimedia distribution and consumption, however laudable, was still a largely unattained goal. While many elements were in place to build an infrastructure for the delivery and consumption of multimedia content, there was no 'big picture' to describe how these elements – both ready and under development – were related and would fit together to create an interoperable multimedia infrastructure. These discussions were seeded by the paper 'Technologies for e-content' by Leonardo Chiariglione [15]. The paper listed a number of fields in which new standardization or 'glue-standardization' was required to further approach the goal of full and seamless interoperability in the multimedia arena.

From face-to-face and online discussions that followed in MPEG, the aim of early MPEG-21 efforts was further refined as in the quotes from [16]:

(i) to understand if and how the various available components fit together, and

(ii) to discuss which new standards may be required, if gaps in the infrastructure exist and, once the above two points have been reached,

(iii) to actually accomplish the integration of different standards.

Although not stated explicitly in the quotes, this was understood to include the development of new standards where needed – either inside or outside of MPEG.

The fundamental underlying premise of MPEG-21 was that the universal availability of digital networks (notably the Internet) is changing traditional business models, adding the new aspect of the electronic trade of digital content to the trade of physical goods – and in the long run even replacing it. This makes the trade of intellectual property a fundamentally different proposition, and it also does away with traditionally existing borders between different media types (text, video, music, etc.)

This assessment led to the definition of the requirements for a multimedia framework, to support this new type of commerce. From the very start, it was recognized that MPEG would need to collaborate with other standards bodies in order to make the dream of an interoperable multimedia framework into reality. The start of that collaboration was the workshop in Noordwijkerhout, which had very broad representation across industry.

The first results were a list of seven areas of attention and a document describing relevant use cases, as well as the first edition of a more formal document called a *Technical Report*, which lists the vision and these areas of attention [17]. These areas of attention form a central concept to MPEG-21. The next chapter will discuss all about them in considerable detail.

For a complete introduction to MPEG-21, this list is included below. For a detailed explanation, please refer to Chapter 2. Before listing these seven areas, two concepts that are fundamental to MPEG-21 need to be introduced. The first is a *User*, who interacts with *Digital Items*. The concept of a User transcends that of the traditional consumer – it encompasses every participant in the value chain, or, quoting from the Technical Report [17]: *'A User is any entity that interacts in the MPEG-21 environment or makes use of a Digital Item'*. A Digital Item is, roughly speaking, any piece of content or information that is structured according to the Schema for a Digital Item (see Chapter 3). The Technical Report defines a Digital Item as follows [17]: *'A Digital Item is a structured digital object with a standard representation, identification and metadata within the MPEG-21 framework. This entity is also the fundamental unit of distribution and transaction within this framework'*.

The seven key areas of attention for MPEG-21 are as follows:

(1) *Digital Item Declaration*: provides a uniform and flexible abstraction and interoperable schema for declaring Digital Items;

(2) *Digital Item Identification and Description*: sets a framework for identification and description of any entity regardless of its nature, type or granularity;

(3) *Content Handling and Usage*: interfaces and protocols that enable creation, manipulation, search, access, storage, delivery and (re)use of content across the content distribution and consumption value chain;

(4) *Intellectual Property Management and Protection*: provides the means to enable Digital Items and their rights to be persistently and reliably managed and protected across a wide range of networks and devices;

(5) *Terminals and Networks*: provides the ability to provide interoperable and transparent access to content across networks and terminals;

(6) *Content Representation*: deals with representation of the media resources;

(7) *Event Reporting*: deals with the metrics and interfaces that enable Users to manage events within the framework.

Just as MPEG-4 was not created to replace MPEG-2, and MPEG-7 was not intended (nor suited) to replace any of the foregoing MPEG standards, MPEG-21 does not seek to replace any of the existing MPEG standards. Its nature is very different, and to make a multimedia framework actually 'work', conventions for data representation (similar to what MPEG-1, MPEG-2 and MPEG-4 provide) and metadata representation (similar to what MPEG-7 gives) are required, in addition to the issues that MPEG-21 addresses.

MPEG-21's grand vision is one of seamless exchange of content between any two or more Users, in a world where the interests of all Users (consumers, producers, distributors and others alike) are balanced, a world where anyone can be a content author and have their content published, and a world where physical boundaries do not impede the flow of digital content.

1.5 MPEG'S STANDARDIZATION PROCESS

After introducing the MPEG standards, this chapter will now describe the process that guides their creation. Over the last decade, MPEG has been an efficient and successful standardization body, setting off a revolution in the area of digital audio-visual representation. A significant part of this success is due to the MPEG standardization process, which will be briefly presented here.

1.5.1 MEMBERSHIP AND LEADERSHIP

Attendance of MPEG meetings, virtually a requirement for efficient participation in MPEG's work, requires accreditation by a National Standards Body or by a standards committee in liaison with MPEG (e.g. SMPTE, EBU).[4] Information about National Standards Body can be found in [18]. Requirements for accreditation vary; each National Standards Body sets its own rules. Each national delegation has a Head of Delegation, appointed by the corresponding National Body; heads of delegation meet with the Convenor during MPEG meetings to address matters of general interest to the group.

While 25 experts participated in the first meeting, the MPEG community has grown to be fairly large; typically, about 300 experts from some 200 companies, universities, research institutes and other organizations, representing about 25 countries, take part in MPEG meetings. A large part of the MPEG membership is made up of individuals operating in company research, product development, product marketing and academia. Expert

---

[4] Typically, the MPEG Convenor may also make a (small) number of personal invitations per meeting.

representatives from industry interest groups also participate. These experts represent many different domains including telecommunications, broadcasting, content creation, cinema, equipment manufacturing, publishing, intellectual property management.

MPEG has a fairly informal environment, where technical contributions are discussed in minute detail, and technology choices are driven by technical and functional criteria. MPEG operates under the regulation of the 'Directives' issued by ISO/IEC and the 'Procedures for the Technical Work' issued by JTC1 [2].

Since its first meeting, the same 'Convenor', Dr Leonardo Chiariglione, a well-known polyglot born in Italy, chairs the MPEG workgroup [19]. Activities are structured in several subgroups. Some of the subgroups have a clear mandate to produce standards, the Systems, Description, Video, Audio and Synthetic and Natural Hybrid Coding (SNHC) subgroups. The remaining subgroups support the specification process. Table 1.1 shows the list of MPEG subgroups with their goals.

1.5.2 MEETINGS

MPEG started to meet in May 1988 (Ottawa, Canada) and held its 73rd meeting in July 2005 (Poznan, Poland). MPEG used to meet three times a year, but since 1995 the rule is more four meetings a year to be able to deal with the workload.

With very rare exceptions, MPEG meetings last 5 days, Monday through Friday. During MPEG meetings, participants bring input contributions, which generally seek to influence MPEG's specifications, whether it is their directions, the choice of technology, or the exact text. While MPEG carries out a lot of work by email, the most difficult or contentious

Table 1.1   MPEG structure as of July 2005 [3]

Subgroup	Goal
Convenor	General coordination and vision
Systems	Combination of visual, audio and associated data and their delivery
Multimedia Description Schemes (MDS)	Generic description tools and languages
Video	Representation of video data
Audio	Representation of audio data
Synthetic and Natural Hybrid Coding (SNHC)	Representation of synthetic visual data
Requirements	Identification of applications, functionalities, and requirements for MPEG standards
Integration	Conformance testing and reference software
Test	Coordination and performance of evaluation tests
Implementation	Feasibility studies, implementation issues
Liaison	Relations with external bodies
Joint Video Team (Joint with ITU-T)	Development of MPEG-4 Part 10 (Advanced Video Coding) and future joint ITU/MPEG coding standards

issues are best addressed in face-to-face discussions. Ultimately, it is at these meetings that all decisions are taken, and all (output) documents are approved. In order to be able to run these meetings efficiently, MPEG delegates are requested to submit (electronic) input contributions prior to the MPEG meeting, to a repository that is accessible to all MPEG members. Delegates are then able to study the contributions in advance and come to the meetings well prepared. Some 300–500 such *input contributions* are uploaded to MPEG's FTP site at each meeting.

The result of every MPEG meeting is a collection of some 150 documents called *output documents*. These documents record the agreements and conclusions of the meeting. Particularly important output documents are the (draft) standards and the meeting's resolutions. The latter document contains the main conclusions and records the approval of the output documents.

In principle, all MPEG documents, input and output, are for MPEG participants only. However, quite a few of the output documents are designated as public documents and can be freely distributed and accessed. The MPEG FTP site stores all MPEG documents (input and output) and access is restricted to MPEG members. Public MPEG documents can be found on MPEG's website [1].

An MPEG meeting is structured in plenary meetings (typically, 4 hours on Monday morning, 2 hours on Wednesday morning and all afternoon on Friday until late in the evening) and in subgroup meetings. Subgroups may have plenary meetings and may in turn have several 'breakout' meetings on specific issues, which may run in parallel. Also, subgroups may meet together, for example, the Systems and the Video groups discussing how to packetize video streams. All these meetings have a specific topic and agenda. The fact that so much is going on in parallel may make an MPEG meeting confusing and a bit of an intimidating experience for a newcomer. Table 1.2 shows the typical evolution of an MPEG meeting during the week.

Although much work gets done and approved during MPEG meetings, most of the technical development, experimentation and evaluation takes place during the period between MPEG meetings. To facilitate this work between meetings, MPEG sets up *Ad hoc groups* (AHGs) at the end of each MPEG meeting. AHGs are established with mandate, membership, one or more chairmen, a duration (usually until the next meeting) and occasionally a face-to-face meeting schedule.

AHGs are open to any MPEG member and they typically gather delegates working on some specified area of work[5]; AHGs last from one MPEG meeting to the next, when they may or may not be reinstated. AHGs work through emails, and under certain rules they are authorized to hold physical meetings. Many AHGs meet on the Saturday and/or Sunday prior to the MPEG meeting. AHGs issue a report on completion of their task; these reports are presented to the general MPEG membership at the Monday plenary meeting. An AHG makes recommendations for approval at an MPEG meeting; it never takes any decisions by itself.

1.5.3 TYPES OF STANDARDIZATION DOCUMENTS

MPEG exists to produce standards and the ones produced by ISO are numbered with five digits. For example, the ISO number for MPEG-1 is 11172 and for MPEG-21, it is 21000.

---

[5] AHGs are listed in a public area of MPEG's website, which means that everybody can subscribe to the emailing lists. This allows MPEG to benefit from as broad technical inputs as possible.

Table 1.2 A typical MPEG meeting

Time	Event	Attendance	Purpose
Monday morning	MPEG plenary	All	• Presentation of National Body positions • Presentation of work done by the ad hoc groups since the last meeting • Agree on the goals for the week
Monday pm through Tuesday night	Subgroups meetings and breakouts	Interested participants	• Process input contributions • Edit technical specifications
Monday, 6 pm	Heads of Delegation meeting	Heads of National Delegations	• Address issues of general interest to MPEG (no specific technical discussions) • Consider non-technical National Body documents • Consider strategic issues for MPEG
Tuesday, 6 pm	Liaison subgroup meeting	Interested participants	• Discuss and approve communication with bodies
Tuesday, 7 pm	Chairmen meeting	Subgroups chairs and Convenor	• Present work done so far • Coordinate work between subgroups • Agree on joint meetings for the rest of the week
Wednesday morning	MPEG plenary	All	• Hear progress reports from subgroups • Discuss major technical issues • Set goals for the remainder of the week
Wednesday after plenary until Friday plenary	Subgroups meetings	Interested participants	• Process input contributions • Edit technical specifications • Draft resolutions • Approve output documents
Thursday, 6 pm	Chairmen meeting	Subgroups chairs and Convenor	• Present work done during the week and the associated resolutions • Resolve sticky issues
Friday afternoon	MPEG plenary	All	• Hear full report from each of the subgroups • Approve resolutions by the various subgroups • Approve general resolutions • Charter ad hoc groups to work until next meeting

MPEG, through JTC1, can produce the types of documents presented in Table 1.3 [2]; this table lists, for each type of document, its purpose, the various successive stages, and the minimum voting period that National Bodies must be allowed for each stage (this corresponds to the very minimum time the promotion of a certain document may take from one stage to the next).

As the various Parts of the MPEG-21 standard evolve, successive versions of the relevant Parts are issued when new tools need to be added to that Part of the standard. Versions serve to specify new tools, either offering new functionalities or bringing a significant improvement to functionalities already supported. Formally speaking, later versions correspond to the amendments to that Part of the standard. New versions of a Part of the standard do not substitute or redefine tools; they can only add more tools, so that a new version never renders old implementations non-compliant. At any given point in time, a Part of the standard consists of the original text as altered by all approved amendments and corrigenda.

After any ballot involving comments by the National Bodies, MPEG produces a document called *Disposition of Comments* (*DoC*), which lists MPEG's responses to every single comment and accounts for how they were accommodated in the updated draft or why they were rejected. All types of standardization documents listed in Table 1.3 have been used in MPEG-21 standardization as shown in Table 1.4.

1.5.4 WORKING PRINCIPLES

A few working principles lay the foundation of MPEG's way of working. These principles all seek to make MPEG standards as widely useful as possible, while minimizing the burden of implementing them and allowing maximal freedom to the market parties.

The first is *one functionality, one tool*, which expresses that only the best technology is chosen to provide a certain functionality, and no alternatives are accepted. This promotes interoperability and should keep the standards as lean as possible.

The second is to *only specify the minimum*. This means that standards must specify the minimum to ensure interoperability, and no more. This leaves maximum room for innovation and competition.

The third is to specify *not systems but tools*. As MPEG must satisfy the needs of various industries, it is best to provide building blocks (such as video coders, audio coders) instead of giving a complete end-to-end, rigid specification. The standardized tools may be used and integrated by different industry segments in different ways.

The fourth and last principle is *verification of performance*, which means that the performance of all developed tools is checked against the initial performance requirements. For audio and video coding schemes, this happens in the form of rigid, subjective tests, the so-called *Verification Tests*.

1.5.5 STANDARDS DEVELOPMENT PROCESS

The publication of a standard is the last stage of a long process that starts well before a new item of work is formally approved. After an assessment stage, a proposal for new work is submitted and approved at the subcommittee and then at the technical committee level (SC29 and JTC1 respectively, in the case of MPEG). The assessment stage involves a proactive exploration by MPEG, by researching application domains with a need for

Table 1.3 Types of standardization documents

Type of document	Purpose	Evolution process
International Standard The FDIS and IS copyrights are owned by ISO.	Most Parts of MPEG standards are developed as International Standards (IS); International standards contain technical specifications. New editions may be published after the first edition, and they wrap all amendments and corrigenda till that date into a single document. If a new edition also contains changes of substance, an approval process similar to amendments applies.	• New Work Item Proposal (NP) – 3 months ballot (with comments) • Working Draft (WD) – no ballot; MPEG decides when to produce • Committee Draft (CD) – 3 months ballot (with comments) • Final Committee Draft (FCD) – 4 months ballot (with comments) • Final Draft International Standard (FDIS) – 2 months binary (only yes/no) ballot – failing this ballot (this means a 'no' vote) implies going back to WD stage • International Standard (IS)
Amendment The FDAM and DAM copyrights are owned by ISO.	Amendments include technical additions or technical changes (but not corrections) to an International Standard; amendments are edited as a delta document to the edition of the International Standard they amend. Each Amendment has to list the status of all Amendments and technical Corrigenda to the current edition of the standard it is amending. Amendments are published as separate documents.[5] MPEG Amendments very rarely – if ever – render pre-amendment editions of the standard incompatible because this would render existing and deployed systems non-compliant.	• New Work Item Proposal (NP) – 3 months ballot (with comments) • Working Draft (WD) – no ballot • Proposed Draft Amendment (PDAM) – 3 months ballot (with comments) • Final Proposed Draft Amendment (FPDAM) – 4 months ballot (with comments) • Final Draft Amendment (FDAM) – 2 months binary (only yes/no) ballot – failing this ballot (no vote) implies going back to the WD stage • Amendment (AMD)
Corrigendum The COR copyright is owned by ISO.	Corrigenda serve to correct technical defects in International Standards (or Amendments).[6] Usually, the correction of editorial problems does not justify issuing a technical Corrigenda; editorial corrections may wait until a technical corrigenda has to be issued (if any). Technical Corrigenda never serve to specify new technology.	• Defect Report (DR) – no ballot • Draft Technical Corrigendum (DCOR) – 3 months ballot (with comments) • Technical Corrigendum (COR)

(continued overleaf)

Table 1.3   (*continued*)

Type of document	Purpose	Evolution process
Technical Report The DTR and TR copyright is owned by ISO.	Technical Reports provide standards-supporting information, like a model/framework, technical requirements and planning information, a testing criteria methodology, factual information obtained from a survey carried out among the national bodies, information on work in other international bodies or information on the state-of-the-art regarding national body standards on a particular subject [2].	• New Work Item Proposal (NP) – 3 months ballot (with comments) • Working Draft (WD) – no ballot • Proposed Draft Technical Report (PDTR) – 3 months ballot (with comments) • Draft Technical Report (DTR) – 3 months ballot (with comments) • Technical Report (TR)

---

[5] When additions to a standard are produced, they may be published as an Amendment or as a new edition. A new edition incorporates the additions and all the changes in prior Amendments and Corrigenda.

[6] An Amendment specifies new technology and is often similar to an International Standard; an amendment should never serve to make corrections to previously specified technology; this is the purpose of Corrigenda.

an MPEG standard, combined with a technology exploration activity to investigate if there are one or more technologies available to fulfil these needs. When there is sufficient understanding of the landscape in terms of applications, functionalities and technologies, a new work item is proposed. If it is approved, a formal standards development process starts. This exploration phase is essential to guarantee that MPEG standards are published timely, not too early (and technologically immature), and not too late (after the market has made its proprietary technological choices). Ideally, the standard should arrive between the 'camel humps', each of which correspond to research and product development. MPEG frequently creates AHGs, targeting the study of specific application areas or technologies to assess their relevance in terms of standardization. The group also organizes open seminars to invite new perspectives, ideas and visions from experts outside MPEG (frequently drawing them into the fold).

The MPEG standardization process is dynamic, and relatively fast, very much the opposite of the typical image of standardization. It includes these five major steps [32]:

1. *Identification of Applications*: Identification of relevant application domains through member inputs and AHG study.

2. *Identification of Functionalities*: Identification of the functionalities relevant to the applications identified in the previous step.

3. *Extraction of Requirements*: Definition of technical and functional requirements from the functionalities identified in the previous step. Emphasis is put on requirements that are common to different applications as MPEG does not seek to develop specific tools for specific applications, but rather more generic tools to be used across application domains.

Table 1.4 Temporal evolution of MPEG-21 milestones

Date, meeting	Event
July 2002 Beijing, CN	• Part 1 – Vision, Technologies and Strategy – PDTR
October 2000 La Baule, FR	• Part 2 – Call for Proposals for Digital Item Declaration
	• Part 3 – Call for Proposals for Digital Item Identification and Description[7]
January 2001 Pisa, IT	–
March 2001 Singapore, SG	–
July 2001 Sydney, AU	• Part 1 – Vision, Technologies and Strategy – DTR
	• Part 2 – Digital Item Declaration – CD
	• Parts 5/6 – Call for Proposals for a Rights Data Dictionary and Rights Description Language
December 2001 Pattaya, TH	• Part 2 – Digital Item Declaration – FCD
	• Part 3 – Digital Item Identification – CD
March 2002 Jeju, KR	• Part 3 – Digital Item Identification – FCD
	• Part 7 – Call for Proposals for Digital Item Adaptation
May 2002 Fairfax, VA, US	• Part 2 – Digital Item Declaration – FDIS [20]
July 2002 Klagenfurt, AT	• Part 3 – Digital Item Identification – FDIS [21]
	• Part 5 – Rights Expression Language – CD
	• Part 6 – Rights Data Dictionary – CD
October 2002 Shanghai, CN	–
December 2002 Awaji, JP	• Part 5 – Rights Expression Language – FCD
	• Part 6 – Rights Data Dictionary – FCD
	• Part 7 – Digital Item Adaptation (CD)
	• Part 10 – Call for Proposals on Digital Item Processing: Digital Item Base Operations and Digital Item Method Language
March 2003 Pattaya, TH	–
July 2003 Trondheim, NO	• Part 5 – Rights Expression Language – FDIS [22]
	• Part 6 – Rights Data Dictionary – FDIS [23]
	• Part 7 – Digital Item Adaptation – FCD
October 2003 Brisbane, AU	–
December 2003 Waikaloa, HI, US	• Part 1 – Vision, Technologies and Strategy – PDTR, second edition
	• Part 7 – Digital Item Adaptation – FDIS [24]
	• Part 10 – Digital Item Processing – CD
	• Part 11 – Evaluation Tools for Persistent Association – PDTR
	• Part 12 – Test Bed for MPEG-21 Resource Delivery – PDTR
	• Part 4 – Call for Proposals on MPEG-21 IPMP
	• Part 15 – Call for Proposals on MPEG-21 Event Reporting
	• Part 13 – Call for Proposals on Scalable Video Coding Technology[8]
March 2004 Munich, DE	• Part 1 – Vision, Technologies and Strategy – DTR, second edition [17]

(continued overleaf)

Table 1.4   *(continued)*

Date, meeting	Event
	• Part 2 – Digital Item Declaration – CD, second edition • Part 4 – Updated Call for Proposals for MPEG-21 IPMP • Part 8 – Reference Software – CD • Part 9 – File Format – CD • Part 11 – Evaluation Tools for Persistent Association – DTR [25]
July 2004 Redmond, WA, US	• Part 2 – Digital Item Declaration – FCD, second edition • Part 8 – Reference Software – FCD • Part 9 – File Format – FCD • Part 12 – Test Bed for MPEG-21 Resource Delivery – DTR [26] • Part 16 – Binary Format – CD
October 2004 Palma de Mallorca, ES	• Part 4 – IPMP Components – CD • Part 10 – Digital Item Processing – FCD • Part 15 – Event Reporting – CD • Part 16 – Binary Format – FCD
January 2005 Hong Kong, CN	• Part 2 – Digital Item Declaration – FDIS, second edition [27] • Part 9 – File Format – FDIS [28] • Part 17 – Fragment Identification for MPEG Media Types – CD
April 2005 Busan, KR	• Part 4 – IPMP Components – FCD • Part 8 – Reference Software – FDIS [29] • Part 10 – Digital Item Processing – FDIS [30] • Part 14 – Conformance – CD • Part 16 – Binary Format – FDIS [31]
July 2005 Poznan, PL	• Part 15 – Event Reporting – FCD • Part 17 – Fragment Identification for MPEG Media Types – FCD

[7] Later, only Digital Item Identification.
[8] Although Scalable Video Coding (SVC) was initially intended to become MPEG-21 Part 13, this work item was later moved to MPEG-4. After this move, no media coding tools are specified in the context of MPEG-21. MPEG-21 Part 13 is empty since then.

4. *Development of Technical Specifications*: Development of the technical specifications, in three distinct phases:
   (i) *Call for Proposals*: After specifying the requirements, MPEG issues a public call for (technology) proposals. Such a call describes the application domains and functionalities, lists the requirements, and also includes the rules and criteria to be used in the evaluation of the invited proposals.
   (ii) *Proposal Evaluation*: The proposals submitted for consideration by MPEG are evaluated through a well-defined and fair evaluation process in which MPEG members and proponents participate. The exact nature of the evaluation process very much depends on the type of requirements and technologies, and may comprise very different types of evaluation including formal subjective testing,

objective metrics and evaluation by (MPEG) experts. It is very important to understand that the target of the evaluation is not to pick absolute winners but rather to identify the best parts from the proposals with the goal to integrate them into a single specification. This represents the end of the competitive part of the process; from that moment on, MPEG experts will work in a collaborative process to improve a single specification, although that process may still entail some competition (see the definition of core experiments below).

(iii) *Specification*: Following the evaluation phase, the specification process starts with all MPEG experts working on the improvement of the solution built with the best evaluated tools. This is the start of a powerful collaborative process, involving tens and often hundreds of the best experts in the world. This combination of competition, 'cherry picking', and collaborative development is at the heart of the success of MPEG in setting the highest quality standards. Typically, this collaborative process involves the specification and continuous improvement of a so-called *working model*, which corresponds to early versions of the standard. The working model evolves over time; at any given moment it will include the best performing tools. It is represented in both text and software. To be able to carry out meaningful experiments, both normative tools (those to be specified in the standard) and non-normative tools (those that do not need normative specification to guarantee interoperability) are included in the working model. The working model constitutes a powerful and rather complete framework for performing experiments by multiple independent parties. The standard evolves on the basis of the outcome of these controlled experiments. The working model progresses through versions by including new tools for additional functionalities, by substituting some tools with better performing ones, and by simply improving the tools that are already in the working model. For historic reasons, such an experiment is called a *core experiment* (CE); a core experiment is a controlled process guided by precisely defined conditions and metrics so that the results are unambiguous and comparable [33]. When new tools are accepted for the working model, proponents have to provide a textual description and a working software implementation ready to be integrated in the working model. At the end of the specification process, the results are a textual specification, a software implementation fully in sync with the textual specification (the software version of the working model) and a set of content, usually bitstreams. The content (often *conformance bitstreams* representing e.g., audio, video) can be used by the implementers of the standard to check the compliance of their products.

5. *Verification of Performance*: Verification of performance is the last step in the standards development process and very much unique to MPEG. This step serves to verify, both to MPEG and to potential users of the new standard, that the specifications are able to meet the (market) requirements that were identified when the project started. This verification typically implies performing substantial tests, for example, subjective video or audio testing using formal methodologies with panels of subjects. (In fact, many new subjective testing methodologies have been developed within MPEG to test certain new functionalities, for example, error resilience and object-based video coding in MPEG-4). The results of these verification tests are a very important proof to the industry that the technology performs well.

This standards development process has been applied for all MPEG standards, with variations and iterations as required by the nature of the standard under development. It is not fully linear; the requirements identification is a continuous process and additional calls are often issued for missing pieces. While hundreds of experts are involved, the overall process keeps a fast pace in order to timely provide industry with the required standards. Keeping commitments, notably sticking to deadlines, is essential for industry to be able to rely on MPEG standards for their internal technology and products plans.

1.6 AFTER AN MPEG STANDARD IS READY

After an MPEG standard is ready, even after verification testing, industry is not necessarily ready to adopt it. A number of things may need to happen to facilitate such adoption, and a standards body like MPEG is not the natural focus of such activities. Some of them are not really in MPEG's mandate, and it is a fact of life that many people who worked hard to get a specific standard developed, stop attending standardization meetings to focus on implementation and adoption. A non-exhaustive list of such additional activities includes marketing and promotion, product interoperability tests, and encouraging licensing to happen.

The important issue of licensing will now be discussed in a bit more detail, and the next subsection will summarize the activities of the MPEG Industry Forum, which seeks to promote emerging MPEG standards (notably MPEG-4, -7 and -21).

1.6.1 LICENSING

As discussed, MPEG develops MPEG standards. MPEG does not (cannot, under ISO rules) deal with licensing.

There may be a (sometimes large) number of patent claims essential for an MPEG standard. It is virtually impossible to design a state-of-the-art coding algorithm – audio or video – that does not use some party's intellectual property. What matters most are so-called 'essential patents'. An essential patent (or claim) is one that is necessarily infringed by an implementation of the standard, or to put it in another way, a claim that no reasonable implementation would be able to work around. Oftentimes, a license is required in order to make and sell an implementation that infringes such essential patents.

JTC1 (see Section 1.2) rules that MPEG requires those companies that propose technologies which get adopted into the standard, to sign and submit a statement that they will license their essential patents on Reasonable and Non-discriminatory terms, also referred to as *RAND terms*. (There is no definition in MPEG or JTC1 of what constitutes RAND or what is 'Reasonable'.) An MPEG standard contains, as an annex, a list of companies that have submitted such statements. This list does not necessarily contain all parties that hold essential IP rights, as it is conceivable that a standard uses inventions from parties that were not involved in the standardization process, and also does not relate specific tools with specific parties.

After an MPEG standard is ready for deployment, the market generally hopes that as many holders of essential patents as possible offer a joint patent license, a one-stop shop, to the market. Such a one-stop shop is convenient to any party that needs to be licensed in order to be able to use the standard. It is very difficult for a potential user of the standard to talk to all patent owners individually, and some people have argued that even if this were feasible, the end result would be a myriad of different licensing terms and fees that would all add up.

In the case of MPEG-2 Video, the market was not ready to adopt the standard without some assurance that licenses could be obtained for reasonable rates, and discussions about this issue led to the establishment of MPEG LA [34]. In spite of its name, MPEG LA is an independent company, which has no official ties to, or mandate from, MPEG or any other organization involved in setting or adopting MPEG standards. (This includes ISO and MPEGIF; in the following text).

MPEG LA plays the role of an administrator who licenses patent claims that are essential to MPEG-2 Video and Systems. MPEG LA also licenses patents that are essential to MPEG-4 Visual and MPEG-4 Systems. Another licensing agent, Via Licensing Corporation [35] (an independent subsidiary of Dolby Laboratories), licenses patents essential to MPEG-2 AAC, and MPEG-4 Audio including AAC and High Efficiency AAC. Thomson licenses patents essential to MP3 (MPEG-1 Layer III Audio). Sisvel [36] and Audio MPEG [37][6] license MPEG Audio patents as well.

Such 'patent pools' generally get established after some party asks any interested license holder wishing to enter into a future patent pool to submit their patent for evaluation by independent experts. These experts then examine whether the patent is indeed essential to the standard. After an initial round of evaluations, the owners of patents, that have been evaluated as essential, meet to discuss the terms of the joint license and to choose an administrator. Patent holders (also called 'licensors') determine the licensing models and the fees; Licensing Agents such as MPEG LA collect royalties on behalf of the patent owners and distribute the proceeds. MPEG patent pools have continued to grow in size. A patent pool is never 'closed'; for example, in the case of MPEG-2 Video, an owner of a patent that may be relevant can still submit a patent for determination of essentiality today. If the patent is indeed deemed essential, this patent holder can join the pool. In a spring 2004 snapshot, the MPEG-2 Video Patent pool licensed 630 patents from 129 families (sets of related patents) for 23 licensors to over 650 licensees [34].

It is important to note that there is no guarantee that a pool includes all patents essential to a certain standard, as there is no obligation to join a pool, not even for those who have participated in the creation of the standard. Also, a pool provides a non-exclusive license; a licensor will usually still license the same patent directly as well.

Note that this chapter does not seek to give an exhaustive list of licensors of patents essential to MPEG standards; there may well be other licensing agents and/or companies, not mentioned here, that license patents essential to an MPEG standard.

These joint license schemes are not licensed on behalf of ISO, or MPEG (or MPEGIF, in the following text), nor are they (or do they need to be) officially 'blessed' by any standards-related organization. In other words, there is no 'authority' involved in licensing, it is a matter of private companies working together to offer convenience to the marketplace. In some instances, the licensing administrator seeks some reassurance from governments that no anti-trust liability exists, because a patent pool may involve dominant market players making a pricing agreement. For example, in the case of MPEG-2, MPEG LA obtained a 'comfort letter' from the US Department of Justice, and a similar letter was issued by the European authorities. In general, patent pools like the ones that are established for MPEG standards have been regarded as encouraging competition rather than hindering it.

---

[6] Not to be confused with the MPEG Audio subgroup

1.6.2 THE MPEG INDUSTRY FORUM

The MPEG Industry Forum [38] was established in the year 2000 as a nonprofit organization registered in Geneva, Switzerland. Originally, the Forum was known as the *MPEG-4 Industry Forum*, with a focus on MPEG-4. In 2003, at the request of many of its members with interests in plural MPEG standards, the Forum expanded its scope to include all emerging MPEG standards, adding the promotion of MPEG-7 and MPEG-21 to the activities of MPEGIF.

According to its Statutes, the goal of MPEGIF is *'To further the adoption of MPEG Standards, by establishing them as well accepted and widely used standards among creators of content, developers, manufacturers, providers of services and end users'*.

MPEGIF was established after the MPEG-4 standard was finalized. The first discussions were held in 1999, after many participants in MPEG realized that the standard was ready but that there was still more to be done to get it adopted in the marketplace, and also that there were required actions that went beyond the scope of MPEG. Such activities include the following:

- *Marketing and Promotion*: While MPEG undertakes some activities that can be labelled as 'marketing', such as the production and publication of promotional material, ISO standardization workgroups generally lack the means to create sophisticated marketing materials, attend trade shows, set up information clearing houses, organize large workshops and so on. The MPEG Industry Forum has a strong marketing activity, which started with MPEG-4 and is gradually extending into MPEG-7 and MPEG-21. MPEGIF exhibits, with members, at large events like the yearly National Association of Broadcasters (NAB) and International Broadcasting Convention (IBC) shows in Las Vegas and Amsterdam, respectively, and regularly organizes half- or full-day programmes dedicated to MPEG technology at such and other events. It also operates public mailing lists with thousands of subscribers interested in general news, in technical answers or in discussions about non-technical issues. MPEGIF members have the opportunity to brand themselves as participants in the MPEG community and to advertise their products on MPEGIF's home page [38].

- *Product Interoperability Tests*: MPEG carries out extensive bitstream exchanges and specifies how to test conformance of a standard, but this work is generally limited to singular Parts of a standard and not on their integration, and focuses on the pre-product stages. The goal is to make sure the standard is correctly documented. After this stage, more interoperability tests are required when initial products are being readied. MPEGIF's 'interop' tests focus on integrated products, often using more than one Part of the standard (e.g., Audio plus Video wrapped in a Systems layer of a file format). They may involve agreements on use of non-MPEG technology to be able to carry out end-to-end tests. MPEGIF is working on extending this programme into an interoperability logo programme, where successful participants can put a logo on their products, as a further reassurance to the marketplace that the products are indeed interoperable. Related to this programme is the effort to build a database of products with compatibility information.

- *Licensing*: As discussed above, after deployment of MPEG-2, the marketplace is keenly aware of the fact that licenses may be required to implement, deploy and use MPEG technologies. Because of the many essential patents that appear to be involved in

MPEG standards, patent pools are very important. The MPEG Industry Forum seeks to encourage the easy access to reasonable licensing. MPEGIF's statutes state that it shall not license patents or determine licensing fees (which, for anti-trust reasons, it would never be able to do anyway), nor require anything from its members in terms of licensing. MPEGIF has, however, discussed issues pertaining to licensing, and it hopes that some of its discussions and actions will lead to a timely and convenient access to licenses against reasonable and non-discriminatory terms and conditions. MPEGIF does not receive any part of the collected royalties, nor does it want to. MPEGIF has, among its members, licensors, licensees and entities that are neither of those. Collectively, the members have an interest in fair and reasonable licensing, because MPEG standards will fail without it.

MPEGIF is open to all parties that support the Forum's goals. Since about a year after its inception, MPEGIF has had about 100 Full members and some 5–10 Associate Members. Associate members pay a significantly lower fee, but cannot vote or be elected. This type of membership is only open to organizations that are not-for-profit; in practice, these would be other fora or educational institutions. MPEGIF deliberately keeps its fees low, in the order of a few thousand US Dollars annually for a Full member, and a few hundred US dollars for an Associate member. The philosophy is to have a low barrier for entry into the work of promoting MPEG standards.

1.7 FINAL REMARKS

MPEG has set a number of highly successful standards. Their technical quality is a result of a well-defined standardization process, which also served to bring together many companies and institutions around the globe. After the coding standards of MPEG-1, MPEG-2, and MPEG-4 and the metadata standard MPEG-7, the MPEG workgroup has been developing a standard of broader scope, aiming to further facilitate the seamless trade, involving the exchange and use of digital content across different services, devices networks and business models, this is MPEG-21. The challenge is formidable and complex, and it is the subject of discussion in the remainder of this book.

REFERENCES

[1] MPEG Home Page, http://www.chiariglione.org/mpeg/, accessed in 2005.
[2] JTC1 Home Page, http://www.jtc1.org/, accessed in 2005.
[3] MPEG Convenor, "Terms of Reference", Doc. ISO/MPEG N7300, MPEG Poznan Meeting, Poland, July 2005, http://www.chiariglione.org/MPEG/terms_of_reference.htm.
[4] ISO/IEC 11172:1991, "Coding of Moving Pictures and Associated Audio at up to About 1.5 Mbit/s", 1991.
[5] ITU-T Recommendation H.261, "Video Codec for Audiovisual Services at p×64 kbit/s", International Telecommunications Union – Telecommunications Standardization Sector, Geneva, 1990.
[6] ISO/IEC 13818:1994, "Generic Coding of Moving Pictures and Associated Audio", 1994.
[7] B. Haskell, A. Puri, A. Netravali, "Digital Video: An Introduction to MPEG-2", Chapman & Hall, 1997.
[8] DVD Forum Home Page, http://www.dvdforum.org/, accessed in 2005.
[9] Digital Video Broadcasting (DVB) Home Page, http://www.dvb.org/, accessed in 2005.
[10] Advanced Television Systems Committee (ATSC) Home Page, http://www.atsc.org/, accessed in 2005.
[11] ISO/IEC 14496:1999, "Coding of Audio-Visual Objects", 1999.
[12] F. Pereira, T. Ebrahimi (eds), "The MPEG-4 Book", Prentice Hall, 2002.
[13] ISO/IEC 15938:2002, "Multimedia Content Description Interface", 2002.

[14] B.S. Manjunath, P. Salembier, T. Sikora (eds), "Introduction to MPEG-7: Multimedia Content Description Language", John Wiley & Sons, 2002.

[15] L. Chiariglione, "Technologies for E-Content", talk for WIPO International Conference on Electronic Commerce and Intellectual Property, Switzerland, 1999, http://www.chiariglione.org/leonardo/publications/wipo99/.

[16] L. Chiariglione, K. Hill, R. Koenen, "Introduction to MPEG-21", in "Open workshop on MPEG-21 – Multimedia Framework", in Doc. ISO/MPEG M5707, MPEG Noordwijkerhout Meeting, The Netherlands, March 2000.

[17] ISO/IEC TR 21000-1:2004, "Information technology – Multimedia framework (MPEG-21) – Part 1: Vision, Technologies and Strategy", 2004.

[18] ISO International Organization for Standardization, ISO Members, http://www.iso.org/iso/en/aboutiso/isomembers/, accessed in 2005.

[19] W. Sweet, "Chiariglione and the Birth of MPEG", *IEEE Spectrum*, vol. **34**, no. 9, September 1997, pp. 70–77.

[20] MPEG MDS Subgroup, "MPEG-21 Multimedia Framework, Part 2: Digital Item Declaration", Final Draft International Standard, Doc. ISO/MPEG N4813, MPEG Fairfax Meeting, USA, May 2002.

[21] MPEG MDS Subgroup, "MPEG-21 Multimedia Framework, Part 3: Digital Item Identification", Final Draft International Standard, Doc. ISO/MPEG N4939, MPEG Klagenfurt Meeting, Austria, July 2002.

[22] MPEG MDS Subgroup, "MPEG-21 Multimedia Framework, Part 5: Rights Expression Language", Final Draft International Standard, Doc. ISO/MPEG N5839, MPEG Trondheim Meeting, Norway, July 2003.

[23] MPEG MDS Subgroup, "MPEG-21 Multimedia Framework, Part 6: Rights Data Dictionary", Final Draft International Standard, Doc. ISO/MPEG N5842, MPEG Trondheim Meeting, Norway, July 2003.

[24] MPEG MDS Subgroup, "MPEG-21 Multimedia Framework, Part 7: Digital Item Adaptation", Final Draft International Standard, Doc. ISO/MPEG N6168, MPEG Waikaloa Meeting, USA, December 2003.

[25] MPEG Requirements Subgroup, "MPEG-21 Multimedia Framework, Part 11: Evaluation Tools for Persistent Association", Draft Technical Report, Doc. ISO/MPEG N6392, MPEG Munich Meeting, Germany, March 2004.

[26] MPEG Implementation Subgroup, "MPEG-21 Multimedia Framework, Part 12: Test Bed for MPEG-21 Resource Delivery, Draft Technical Report, Doc. ISO/MPEG N6630, MPEG Redmond Meeting, USA, July 2004.

[27] MPEG MDS Subgroup, "MPEG-21 Multimedia Framework, Part 2: Digital Item Declaration", Final Draft International Standard, 2nd edition, Doc. ISO/MPEG N6927, MPEG Hong Kong Meeting, China, January 2005.

[28] MPEG Systems Subgroup, "MPEG-21 Multimedia Framework, Part 9: File Format", Final Draft International Standard, Doc. ISO/MPEG N6975, MPEG Hong Kong Meeting, China, January 2005.

[29] MPEG MDS Subgroup, "MPEG-21 Multimedia Framework, Part 8: Reference Software", Final Draft International Standard, Doc. ISO/MPEG N7206, MPEG Busan Meeting, South Korea, April 2005.

[30] MPEG MDS Subgroup, "MPEG-21 Multimedia Framework, Part 10: Digital Item Processing", Final Draft International Standard, Doc. ISO/MPEG N7208, MPEG Busan Meeting, South Korea, April 2005.

[31] MPEG Systems Subgroup, "MPEG-21 Multimedia Framework, Part 16: Binary Format", Final Draft International Standard, Doc. ISO/MPEG N7247, MPEG Busan Meeting, South Korea, April 2005.

[32] L. Chiariglione, "MPEG and Multimedia Communications", *IEEE Transactions on Circuits and Systems for Video Technology*, vol. **7**, no. 1, February 1997, pp. 5–18.

[33] MPEG MDS Subgroup, "MPEG-21 MDS CE Development Process", Doc. ISO/MPEG N4252, MPEG Sydney Meeting, Australia, July 2001.

[34] MPEG LA Home Page, http://www.mpegla.com/, accessed in 2005.

[35] Via Licensing Corporation Home Page, http://www.vialicensing.com/, accessed in 2005.

[36] Sisvel Home Page, http://www.sisvel.com/, accessed in 2005.

[37] Audio MPEG Home Page, http://www.audiompeg.com, accessed in 2005.

[38] MPEG Industry Forum Home Page, http://www.mpegif.org/mpegif/, accessed in 2005.

2

An Introduction to MPEG-21

Ian S Burnett and Fernando Pereira

2.1 INTRODUCTION

MPEG-21 is quite different from the MPEG standards, MPEG-1 [1], MPEG-2 [2], MPEG-4 [3] and MPEG-7 [4], that preceded it. The main change is that MPEG-21 is a framework standard with which MPEG sought to fill the gaps in the multimedia delivery chain (as perceived in 1999). Overall, the aim of the framework is to generate seamless and universal delivery of multimedia. The extent to which this goal has been achieved should be left to the reader of this book to decide, but this chapter is intended to overview the primary components of the vision and the standard. Initially, the chapter considers the objectives, motivation and background that led to the development of the MPEG-21 standard. As with all standards, MPEG-21 has developed a terminology and the chapter briefly considers the vital terms before considering a use case, which highlights the application of several parts of the standard. Finally, this chapter will summarize the current parts of MPEG-21, highlighting their motivation and objectives.

2.2 MOTIVATION AND OBJECTIVES

Content is increasingly available everywhere, whether consumers are walking, at home, in the office or on public transport. The content is digital and there is a growing desire from consumers to have rapid, selective access to that content (be it video, music or text) on whatever device they have available. The mobile phone, for instance, is becoming a universal portable portal, far exceeding its original purpose of mobile telephony. In homes, various manufacturers are bidding to create the central home portal that will store and distribute content, throughout the house, to a myriad of devices (even fridges). Meanwhile, user roles are also changing; consumers are rebelling against the very nomenclature that implies that they are just receiving content and becoming creators. Digital photography has led an explosion in personal content creation, collection and distribution. However, there

The MPEG-21 Book   Ian S Burnett, Fernando Pereira, Rik Van de Walle, Rob Koenen
© 2006 John Wiley & Sons, Ltd

is also a downside to the ease with which content creation and manipulation has become an everyday event for children. That downside is that the legal and rights mechanisms that we have used for physical content are increasingly being challenged and proving inadequate.

While some consumer groups would suggest that a content free-for-all is the way of the future, the attractions are soon dispelled once family and personal photos are 'accidentally' distributed and personal information suddenly becomes a public event. Hence, content owners and consumers share, to differing degrees, desires to manage rights and protection of content and information, repurpose their content to meet user preferences and device capabilities, track their content, and so on. MPEG-21 is an open standard framework that seeks to offer a solution to users ranging from consumers to professional content providers. As such, it is a complete framework for delivering and managing multimedia content throughout the chain between creation and consumption.

In 1999, MPEG had achieved dramatic results with its early video and audio codec standards (MPEG-1 [1] and MPEG-2 [2]); there had been significant adoptions in broadcasting and distributed media such as Video Compact Discs (VCDs) and Digital Versatile Discs (DVDs). The MP3 (MPEG-1/-2 Audio Layer III) revolution on the Internet was also in full swing, but with the impact of peer-to-peer yet to be seen. Another ambitious MPEG coding standard had just been completed while a metadata standard was still in progress; the former, MPEG-4, brought object-oriented approaches to multimedia coding with its scene composition, coding mechanisms and integration of natural and synthetic content, while MPEG-7 had grown into a significant body of metadata for the description of multimedia (more details on the previous MPEG standards are given in Chapter 1). However, while the previous MPEG standards offered a powerful set of tools for representing and coding multimedia, it was still proving difficult to create complete multimedia applications. In particular, a multimedia application based on a loose collection of standard formats lacks the 'glue' and 'interconnects' that will ensure the widespread adoption that the world has seen in, for example, telecommunications networking. Out of this scenario was born a vision of an MPEG activity that would create a 'big picture' of the existing multimedia content standards. This activity asked questions such as: Do the current multimedia representation–related standards fit together well? If so, how exactly do they interconnect and how should they be used together? When a user wishes to 'sell' or 'transact' multimedia content, are all the pieces in place to complete the transaction? If all the systems are not available, which ones do exist? Which of them are standardized? Who is going to be responsible for making the 'interconnects' between the pieces?

While there are a number of well-known standards in the multimedia space, nobody had created a solution that aimed at a 'big picture' view of the multimedia world. Such a standard would place users and content as the key components and ensure that the supporting technologies became transparent. MPEG believed that, if this could be achieved, the slow take-off of multimedia content as a profitable industry would be significantly boosted. In particular, a well-considered, complete framework would allow different and diverse communities (all with their own models, rules and formats) to interoperate efficiently, transacting content as a commodity rather than as a technical problem to be negotiated. To quote an example from the standpoint of 2005, as the authors write this chapter, the LIVE8 [5] concert had recently taken place and, amidst much publicity, live downloads of band performances were made available within the hour on the Internet. It would appear to the uninformed onlooker that everything in the world of live audio recordings is set for a booming, automated future; the reality though was somewhat different – even with today's networks and standards, manual delivery on physical media was still a vital part of that one-hour process. It is also worth remembering that solutions that do exist tend to be 'walled-garden', proprietary solutions that do not lend themselves to the interoperability required when users and organizations collaborate on a global event; that is the domain of open standards.

The music industry was the first multimedia space to really feel the 'pain' that widespread consumer ability to manipulate, manage and create content could bring. Owing to the combined popularity of one of the earlier MPEG standards (MP3), a rapidly improving Internet and increases in personal computer performance, users soon began to grab content from the physical world by 'ripping' tracks from CD and freely distribute that content. First, it was done via ftp servers and then through WWW sites, but soon peer-to-peer technologies took over and the trickle became a flood of piracy. The file sharers ignored and flouted the 'old world' licenses and copying restrictions that had been placed on the CD content. Today, the waters have calmed and products such as Apple's iTunes [6] have brought legal downloads and, vitally, Digital Rights Management (DRM) to the arena. While some perceive that the latter restricts users, if commerce is to be part of digital consumption of multimedia, it is clear that DRM will play a vital role. Thus, MPEG-21 has grown as a vision and standard over a period of turbulent growth in the DRM marketplace, and it has been vital to ensure that, in the framework, it is possible to protect content value and the rights of the rights holder. It is, however, equally important to ensure that DRM provisions do not prevent interoperability but contribute to the vision of 'transparent consumption' rather than hamper or prevent it.

Thus, with MPEG-21, MPEG has sought to create a complete framework that will be the 'big picture' of the existing multimedia and enabling standards. The vision of MPEG-21 is *to enable transparent and augmented use of multimedia resources across a wide range of networks and devices* [7]. The firm belief was, and is today, that by creating a full-featured overarching standard, this vision can be achieved and the result will be a substantial enhancement in the user's experience of multimedia. In turn, digital media will, as a market, grow faster and achieve the dominance that has long been expected.

2.2.1 OBJECTIVES

The 'big picture' primarily had to meet the needs of its users, but also, a vital constituency for MPEG is formed by the industry players who will produce the enabling technology and equipment for those users. To ensure that both sets of demands were met, MPEG followed a formal approach to the development of MPEG-21. This can be summarized as six, not necessarily consecutive, steps:

1. Creation of a basic structure of a multimedia framework that supports the vision of MPEG-21.
2. Identification of the critical set of framework elements.
3. Building of a clear understanding of the relationship between the framework elements; based on this work, identify gaps in the technology and missing elements.

4. Ensuring that other standardization bodies with relevant input are consulted and invited to participate in the activity.

5. Evaluation of missing framework elements and answer the question: "Does this technology fall within MPEG's expertise and remit?". If so, MPEG would proceed to call for technologies and standardize a solution. If not, other standardization bodies would be consulted and engaged.

6. Integration of the defined framework based on existing and new, MPEG and external standardized technologies.

Thus, in the first of these process tasks, MPEG set out to create a 'big picture' that would document the state of the art in multimedia and identify critical areas that required interoperability. These latter, critical areas are always the keys to standardization – it is only when interoperation or interworking is required that a standard really makes sense. Further, MPEG recognized that, while it had considerable breadth of expertise in the multimedia space, some areas would fall outside its domain and would thus not become MPEG standards. The hope was that other appropriate standards bodies would pick up these areas and create external standards that completed the canvas of the big picture.

So what were the pieces of the big picture? MPEG identified a set of initial areas that were parts of the picture: digitization of all carriers and all technologies; abstraction of applications from transports and user platforms; content description, identification and usage rules; secure distribution, metering and access to protected content; and content authoring [8]. Perhaps, the most notable facet of this collection of areas is the breadth of coverage from authoring of content through to user applications. Previous MPEG standardization has concentrated on the specification of bitstream formats and decoders, yet in MPEG-21 it is necessary to consider a broader usage definition that includes creation, modification as well as decoding. This distinction has remained in MPEG-21 to this day and influences the thinking and terminology usage throughout the standard.

Given that the parts of the picture had been identified, the task became one of creating the basis of a framework that would allow a world where users only need to be concerned with choosing content, determining how and when they wish to consume it and the cost that they are willing to pay. The underlying tenets of that framework were to be that it should remain agnostic to delivery mechanisms (broadcast, bidirectional), transport and the final delivery platform. It was recognized early in the work that most of the content representations themselves had already been standardized both internally, by MPEG, and externally, for example, by ISO/IEC SC29WG1, JPEG. This has led to a situation where there are no current parts of MPEG-21 dealing with compression of multimedia content, and instead the parts create the framework for delivery of, for example, MPEG-2 [2] or MPEG-4 [3] standardized content.

While content delivery was clearly central to the MPEG-21 standardization efforts, there were several other factors that were significantly influencing the direction of MPEG-21. The concept of users, their needs, content interactions and the usage rights pertaining to the content were all vital inputs to the process. Broad groups of users were identified, for example, authors and performers, producers, application developers, retailers, and end-users, and their particular needs and interactions with the framework analysed. Further requirements were identified for usage rights that would be attached to content and then modified as that content progressed along the delivery chain from originator to end-user.

The initial approach taken in the creation of MPEG-21 was that of generating an initial technical report (TR). The objectives of this TR were expressed as [9]:

'Today, there is no "big picture" to describe all of the standards activities which either exist or are currently under development that relate to multimedia content. By starting such an activity, we will begin to understand how the various components fit together, how they are related, and the possibility of discussing what new standard may be required where gaps in the framework exist.'

When work on the TR was started in 2000, it was not clear that there would be standard parts to follow and it was only as the report took shape that the holes in the landscape where MPEG standardization would make sense became apparent. The standard that exists today is very much in line with the early spirit and the 'big picture' that was recorded in that TR.

Having briefly discussed the motivation and thinking of the MPEG community that led to the commencement of the MPEG-21 project, the following section considers the basic elements of the multimedia framework that were identified. However, first it is necessary to introduce some unique MPEG-21 terms that will be used throughout this book.

2.3 TERMS – DIGITAL ITEMS, USERS AND RESOURCES

One of the important roles of the TR was to define some terms that could be consistently used throughout MPEG-21. We briefly review those that are important in the context of the standard prevailing today (Figure 2.1).

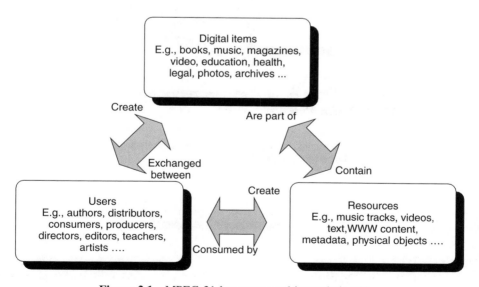

Figure 2.1   MPEG-21 key terms and interrelationships

Digital Item: For the MPEG-21 standard, it was necessary to introduce a term that could describe any 'identifiable asset' that might be delivered using the framework. This led to the introduction of a new term and object, the 'Digital Item' which is formally said to be 'a structured digital object with a standard representation, identification and metadata within the MPEG-21 framework' [7]. Digital Items can be considered as the 'what' of the Multimedia Framework (e.g., a video collection, a music album) and are the fundamental units of distribution and transaction within this framework. In practice, this means that any content that is to be used in MPEG-21 is placed within a 'virtual container': a Digital Item. This container places a structure around the content and allows the addition of other resources (see the next definition for the MPEG definition of an MPEG-21 resource) and metadata. An example resource is the ubiquitous MP3 track. In MPEG-21, an MP3 track is placed into a Digital Item along with any metadata for example, track and artist details, (Rights expressions) and other resources, for example, JPEG images of cover art (Figure 2.2).

Resource: This term is used to mean an individually identifiable asset such as a video or audio clip, an image or a textual asset. The key to this definition is that the authors of DIs can choose their own Resource granularity. Hence, for some, a Resource would

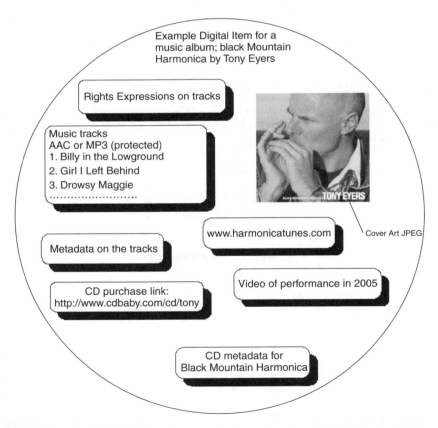

Figure 2.2 An example Digital Item for a music album [image and details used with permission from Tony Eyers]

be individual music and video tracks while for others, multiplexed, audio-visual content would be a single resource. A primary difference in such a choice would be that the encoding of audio-visual content (e.g., in MPEG-4) would usually include synchronization and timing information, whereas individual resources in MPEG-21 Digital Items are not temporally linked. Resources are broadly defined and may even be a physical object (by reference). The term Resource is used throughout MPEG-21 since the term 'content' is overloaded and it was vital to distinguish between Digital Items and the assets that they contain. While the term 'content' is used heavily in the TR to keep the document accessible, in later parts 'content' is almost always avoided.

User: Users can be considered the 'who' of the Multimedia Framework and throughout the MPEG-21 standards Users (with a capital U to signify their importance) are central. As a formal definition, the term defines any entity that interacts with or makes use of Digital Items and the MPEG-21 Multimedia Framework. Importantly, MPEG-21 does not differentiate between a 'content provider' and 'consumer' – both of them are Users and interact with Digital Items. Users assume certain rights and responsibilities depending on the type of interaction they have with other Users and particular Digital Items.

2.4 THE MPEG-21 VISION

The overall picture and vision of MPEG-21 presented in the TR divided the multimedia framework into seven loosely grouped architectural elements [7]. These were established on the basis of the evaluation process outlined in the previous sections. The boundaries between the elements are not strict and later standardization has demonstrated that technologies can span multiple elements. The elements identified are:

1. *Digital Item declaration* is regarding a uniform and flexible abstraction and interoperable schema for declaring Digital Items.
2. *Digital Item identification and description* is regarding the ability to identify and describe any entity regardless of its nature, type or granularity.
3. *Content handling and usage* is regarding interfaces and protocols that enable the creation, manipulation, search, access, storage, delivery and (re)use of content across the distribution and consumption value chain.
4. *Intellectual property management and protection* is regarding the enabling of intellectual property rights on content to be persistently and reliably managed and protected across a wide range of networks and devices – this also deals with a Rights Expression Language and a Rights Data Dictionary.
5. *Terminals and networks* is regarding the ability to provide interoperable and transparent access to content across networks and terminals.
6. *Content Representation* is regarding how the media resources are represented.
7. *Event Reporting* is regarding the metrics and interfaces that enable Users to understand precisely the performance of all reportable events within the framework.

The seven architectural elements and their role in facilitating Digital Item–related transactions between Users are depicted in Figure 2.3. The figure was created during the process of the deliberation and writing of the TR, which set the scene for MPEG-21's

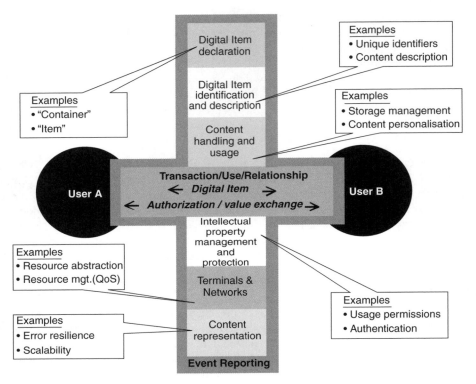

Figure 2.3 The seven elements of the MPEG-21 vision [7]. Copyright ISO. Reproduced by permission

standardization. It illustrates that Digital Items are the unit of transaction, in this case between two Users, and then shows how the various elements of MPEG-21 are 'loaded' on top of the Digital Item, providing the functionalities of the framework.

During the scoping discussions, the seven elements and their interrelationship were used as a way of identifying the areas of required standardization and to create the requirements for the new standards, not necessarily with a one-to-one relationship with the seven elements. They were explained in detail in the TR, which forms Part 1 of the standard. The following gives a short summary of each element.

1. Digital item declaration

While MPEG-21 was not intended to create new formats for content, it was found that it was vital that there should be an unambiguous way of declaring a 'Digital Item' [10]; this is a structured container for all the media resources and metadata that an author wishes to connect in a single 'object'. Observation of the real world shows that people buy structured collections and very rarely, if ever, purchase elementary streams. At a minimum, there is always some metadata that explains the contents of the stream. The media resources and descriptions were not to be limited to MPEG media types and the Digital Items were to be extensible so as to create sub-items that maintain the DI structure. The Descriptors held in a Digital Item were considered to be textual and/or media components and could be linked to the media resources or fragments thereof. The other key aspect of a DI was that it should be the foundation structure for transactions in MPEG-21 and thus the Digital Item Declaration became the core element of the framework. The Digital Item concept is a significant, differentiating factor of MPEG-21 and a complete set of technologies allowing, for example, standardized protection, adaptation and packaging of Digital Items that has been created. The Digital Item alters the handling of content from being that of transmitting elementary streams to that of the exchange of a package. This is a vital difference if commercial transactions are to be enabled. In the physical world, it would be difficult to sell any product without packaging or labelling, which relays some information about the product. The Digital Item brings that functionality to commerce in multimedia content.

2. Digital item identification and description

A key aspect of MPEG-21 is the need for all content to be identifiable and locatable. The aim was to ensure that all Digital Items and media resources in the MPEG-21 framework would carry unique and persistent identifiers. A large number of requirements for such identifiers and locators were laid down, and many existing media-specific identification schemes (such as ISBN (International Standard Book Number) [11], ISRC (International Standard Recording Code) [12], ISWC (International Standard Musical Work Code) [13]) were noted. It was also recognized that several initiatives were addressing the problem of cross-domain identification and were creating identifiers across broad industry sectors and not just for one field of endeavour (for example, ISAN (International Standard Audio-visual Number) [14], DOI (Digital Object Identifier) [15], and the cIDf (content ID forum)) [16]. Thus, MPEG-21 aimed to create an umbrella in which Users could embrace these identifiers in an interoperable framework.

Description of content was also recognized as being an important part of the story. However, it was found that the flexibility of Digital Item Identification, as it developed, meant that, in combination with existing metadata schemes (e.g., MPEG-7 and other industry-specific schemes), extra description mechanisms were unnecessary.

3. Content handling and usage

During the early stages of MPEG-21, some of the areas identified as lacking in standards were network, server and client handling of content. Further, there were concerns that no standards existed for the personalization of content and the use of agents to consume and filter content for users. This element is unique among the others in that in subsequent discussions MPEG has almost entirely considered this area to be outside of its remit and has relied on other organizations to standardize the mechanisms in this field. The area in which this element is most visible in the standard parts today is User personalization, which is incorporated into the adaptation mechanisms of Part 7 of the standard (Chapter 7).

4. Intellectual Property Management and Protection

Digital Rights Management (DRM) has been a growing area of interest since the mid-90s, and today the field is reaching maturity with several standardizations aiming at interoperability (Suns DReaM [17], Coral [18], DMP[1] [19]). MPEG identified that incorporating

---

[1] Digital Media Project

DRM mechanisms into the multimedia framework would be vital and that the approach would need to be one of interoperability and versatility. The field is known within MPEG as IPMP (Intellectual Property Management and Protection) owing to historic reasons dating from the formulation of MPEG-4, which encompasses two approaches to DRM (IPMP and IPMP-X or IPMP Extensions). In MPEG-21, this element includes the concepts of a rights language and dictionary of terms as well as protection mechanisms for content itself. For the latter, MPEG-21 adopts a new approach that allows multiple DRMs to be used as 'plug-in' protection mechanisms within the standard. Further, MPEG has developed a versatile 'protected' version of the Digital Item [20] that can utilize the 'plug-in' DRM approach.

5. Terminals and networks

Since the aim of the framework was to create transparent access to advanced multimedia content (such as described in the compound objects declared as Digital Items), creating an effective shield between Users and the terminal and network issues was vital. Connectivity, bandwidth and Quality of Service are all important issues and these areas became vital components of the adaptation framework, which was created as Part 7 of MPEG-21: Digital Item Adaptation [21].

6. Content representation

Content representation is a controversial area in MPEG-21 standardization. The MPEG-21 TR [22] highlighted that content could be (i) content represented by MPEG standards; (ii) content used by MPEG but not covered by MPEG standards, for example, plain text, HTML, XML, and SGML; (iii) content that can be represented by (i) and (ii) but is represented by different standards or proprietary specifications; and (iv) future standards for other sensory media. In short, content could be in any format and form as far as MPEG-21 was concerned. The TR also noted that the framework should support content representations to facilitate the mantra of 'content anytime, anywhere'. There is also mention that it is necessary that MPEG-21 should support synchronized and multiplexed media. In discussions at MPEG meetings, there have primarily been two views on content representation in MPEG-21; one is that scalable and versatile coding techniques is a vital component of the framework while the opposing view is that the latter are just other formats for which the framework should cater. In general, the latter view has prevailed and there are currently no explicit 'content representation' parts of the standard. This does not, in any way, reduce the importance of coding technologies in the overall MPEG-21 framework; rather, it means that, in general, the codecs have been and are being defined outside the MPEG-21 set of parts in, for example, MPEG-4. The reader should also refer to the final chapter of this book, which discusses a new organization of standardization within MPEG.

7. Event reporting

The need to report on interactions ('events') with Digital Items using a standard mechanism was recognized during the early stages of the project. The aim was to have an 'audit' trail created by Digital Items such that the trail is accessible using a standardized mechanism. This meant that there was a need for standardized metrics and interfaces for the interactions that might reasonably be expected for Digital Items. Further, the transport of the standard data attached to events as Event Reports would need to be standardized.

These seven elements were not intended to be exclusive and may have overlapped but rather were aimed at being distinct areas of the 'big picture' in which standardization goals and requirements could be set. Within each element, coherent areas of standardization were identified, requirements generated and technologies selected to meet those requirements. The results of the MPEG standardization processes that followed the initial phase associated with the creation of the TR have become the 18 Parts of MPEG-21 that constitute the final standard (as of July 2005).

2.5 DIGITAL ITEMS – WHAT IS NEW?

One of the difficulties of a 'framework' standard rather than, for example, a 'coding standard' is understanding exactly how one can use the framework to meet the needs of applications. The nature of a standardized framework is that users must choose how to use the framework in their application space, and, necessarily, different applications spaces will adopt different flavours of the framework. Thus, before diving into a particular application example, this chapter will consider the different approach of MPEG-21 and how the concept of a 'Digital Item' impacts multimedia applications.

Prior to MPEG-21, MPEG created standards for formats for one or multiple types of time-based media. The media coding formats (MPEG-1 [1], MPEG-2 [2], MPEG-4 [3]) all successfully couple time-based audio and video together to create a format that can be successfully transported to a user's terminal where the format is presented via a player. MPEG took an approach of standardizing the format of the bitstream and the behaviour of the decoder. Alongside the format standards, MPEG-7 was created as a standard for metadata that would describe the time-based media content held in the MPEG coding formats. While MPEG-7 [4] is broad in its applicability, the uniqueness of the standard is that it explicitly deals with the description using metadata of time-based media on a broad range of levels from low level (e.g., AudioSpectrumEnvelope) through to high level (e.g, Author) [4]. Thus, it provides, for example, metadata that describes in detail the low-level spectral behaviour of an audio waveform. MPEG-21, however, is a quite different standard compared to its predecessors; in terms of content, MPEG-21 is different in a number of key ways:

Content agnostic: MPEG-21 does not limit itself to one coding format, MPEG coding formats or even ISO coding formats. MPEG-21 is a framework that can handle multimedia content of any type or format.

Not a format: The fundamental concept behind MPEG-21 is not a format at all but a rather flexible framework. Rather, MPEG-21 refers to other formats and other metadata but does not mandate that the content is carried or transported in any particular format or bitstream.

A new core concept: MPEG-21 is built around a core concept, the Digital Item, which introduces a new container for content.

So, what is a Digital Item and what are the consequences of using them? While a definition of the Digital Item has been given, it is worthwhile exploring the concept further, as the consequences of using Digital Items rather than the current multimedia content formats are far reaching with regard to the MPEG-21 framework.

Most of today's personal computer hard disks are huge repositories for multimedia content. They contain a large number of files containing video, music and metadata that describe that content. These files are organized to varying degrees by the file and directory hierarchy. It is often all too easy to grab a file from that hierarchy and lose track of where the file belonged and the other files from the directory with which it was associated. In summary, most computers use a file-based approach to storing and cataloguing multimedia content.

If the computer is compared with physical media, the individual file-based approach of computer-stored multimedia seems rather odd. Consider a music CD. CDs come in cases that contain a booklet that provides information on the artists, the CD contents, the recording, and so on. The case liner notes also display information about the recording and the tracks on the CD. Thus, while the CD can be removed from its case when required, the overall package that the user purchases and places in his/her collection is a structured package of descriptive information about the CD ('metadata') and the CD that contains the content itself (the 'Resource').

At the core of MPEG-21 is the concept of moving away from the single media/single file paradigm to one where the physical packages of multimedia content with which consumers are familiar are emulated by electronic packages. These 'electronic packages' are the Digital Items at the core of the MPEG-21 standard. Rather than allowing Digital Items to be ill defined, MPEG chose to create a formal declaration mechanism for Digital Items. Hence, a Digital Item has a clear structure created by a Declaration file. The details of these files are given later in this book (see Chapter 3), but it suffices to say in this introduction that the files are written in a Digital Item Declaration Language (DIDL) that is based on XML Schema. The structure allows a user to add media content and metadata into the Digital Item and place them clearly within a hierarchical structure. The key is that, irrespective of how many media files (known as resources in MPEG-21) or sections of metadata are added to the Digital Item, the relationship of the Digital Item parts should be clearly understood from the structure of the Digital Item Declaration.

It would be possible to go into much more detail on the mechanisms behind Digital Items but that is dealt with in the later chapters of this book. Here, we will consider a simple example of Digital Items and develop that example to include some of the extended tools offered by MPEG-21. As an example, we will consider the creation of a television programme package by the broadcaster XYZ, who has recently decided to make television programme content available over the Internet and to mobile devices. The packages contain programme resources as well as a range of additional material.

2.6 WALKTHROUGH FOR A BROADCASTING USE CASE

In order to better understand the purpose and benefits of the MPEG-21 standard, this section offers a walkthrough of the most important components of the multimedia framework for a broadcasting use case. It is important to not that this is just one example since MPEG-21 does not target one specific application but rather targets as many applications as possible by shaping the framework to applications' individual needs.

The use case can be summarized as follows:

'A broadcaster intends to create a library of programmes as Digital Items. The intention is not only to use Digital items as a useful archival format but also to repurpose the extensive library to other transport and delivery mechanisms such as WWW pages as well as streaming to TV and mobile devices. The broadcaster also wishes to ensure expression of their rights to the content and ensure that it is protected in all formats'.

The TV broadcaster has been using digitized content for some time and thus has a recording of the TV documentary programme in a MPEG-2 format [2], so this is the basis of the Digital Item. The programme is the first 'sub-item' of the new Digital Item that contains the MPEG-2 version of the audio-visual content.

2.6.1 DIGITAL ITEM DECLARATION

Technically, in the MPEG-21 DIDL [8], there is a hierarchy of an *Item, Component* and *Resource* created for this sub-item; however, such details are held over to the discussion of Chapter 3, which considers DIDL in detail. In this section the concepts of DI construction are considered more important than the detailed element syntax and semantics. The 'credit' and 'title' metadata for the documentary is inserted as a description of the video and is attached to the audio-visual content item. Further, a script of the video content is attached to the sub-item and the timing of the script metadata is fixed to the video content by a series of anchors that allow metadata to be linked explicitly to sections of the video. A more compact set of subtitles are also held as anchored descriptive metadata in the sub-item. This basic 'Documentary' Digital Item is illustrated in Figure 2.4. The immediate advantage of the Digital Item at this point is purely one of organizing the individual metadata files and resources into a structure that could be kept common across the whole of the broadcaster's programme archive. However, it will be immediately apparent that, while this creates a structure for the content, it would be useful for the broadcaster to be able to identify both the complete Documentary Digital Item and the constituent metadata files and media resources.

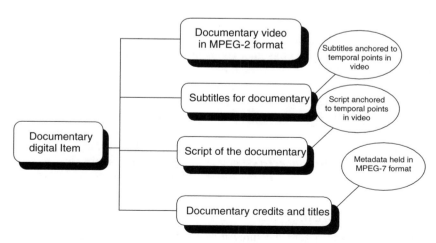

Figure 2.4 A simple documentary programme Digital Item

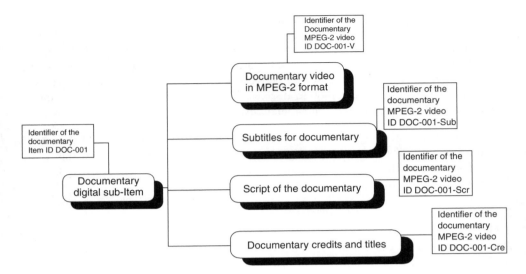

Figure 2.5   Identifiers placed as metadata on Resources and related metadata in a Digital Item

2.6.2 IDENTIFICATION

So as to allow identification of all components within the new Documentary Digital Item, a series of identifiers are attached to the resources (Figure 2.5). These identifiers are chosen by the broadcaster so as to be appropriate in their space and MPEG-21 provides a structure in which to place the chosen identification mechanism. In this case, the identifiers chosen would usefully be able to express derivation and inheritance such that a core identifier could be extended to describe all the resources in the Documentary item. Hence, a single identifier could be attached to the root of the Documentary item and then the derived versions of the identifier attached to the scripts and video or audio resources.

2.6.3 RIGHTS EXPRESSIONS

The broadcaster is also conscious that the content needs to have associated rights to express the broadcaster's intentions for consumer rights and redistribution rights. These rights are expressed using a Rights Expression Language (REL), which details the rights that are granted to particular users and such conditions as timing of playback and the permitted redistribution channels. The Rights expressions are considered as descriptive metadata in MPEG-21 and can be directly included in the Digital Item or included referentially. The latter is often more convenient as it allows complex Rights expressions to be held separate from the content itself; this makes for simpler updating and modification of the Rights expressions. In the example Documentary Digital Item, the broadcaster could place Rights expressions on all or some of the resources and metadata included in the item.

2.6.4 PROTECTED DIGITAL ITEMS

In a trusted environment where the integrity of playback devices can be guaranteed, it may be acceptable to only include clear expressions of access rights and to ignore direct

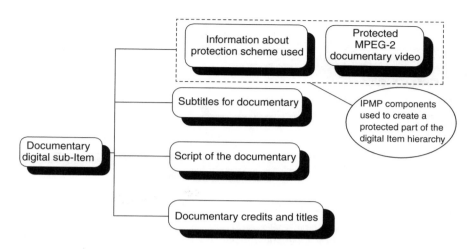

Figure 2.6 Insertion of a protected element into a Digital Item

protection of the Digital Item or its constituent parts. However, in broadcasting, content is widely distributed and thus protection of, at least, parts of the Digital Item is an important consideration. However, the broadcaster distributes content using more than one DRM and thus it is vital that any protected version of the Documentary Digital Item caters to this. This is possible using the IPMP Components part of the MPEG-21 standard. It allows Digital Items and the constituent resources to be protected from any point in the hierarchy of the item using a specific protection scheme (Figure 2.6 and [20]). The protection scheme is signalled using metadata and could be one of the many protection and, more broadly, DRM schemes in current usage. Hence, it is simple for the broadcaster to retain support for several DRM mechanisms while still gaining the advantage of the MPEG-21 multimedia framework.

2.6.5 ADAPTATION

It is possible to allow adaptation of a Digital Item and its resources in a number of ways. At the most basic level, multiple versions of content can be simply included into the Digital Item. This might be suitable if the broadcaster intends to make the documentary content available over the Internet and a second sub-item could then be added to the Documentary Digital Item. This would contain the content needed explicitly for the Internet site, for example, a series of WWW pages are held in the sub-item as resources.

However, it is clearly cumbersome to add to the Digital Item for every delivery platform and mechanism that the broadcaster wishes to use. Thus, a more flexible approach to adaptation could be used by the broadcaster. One such approach is to adapt resources during delivery, that is, on the fly, such that they match a user's preferences, terminal capabilities or the network conditions better (however, fully interactive adaptation is not feasible without the back channel provide by e.g. Internet television services). To achieve this, the broadcaster can include descriptive metadata in the Digital Item to indicate choices of content for certain terminal screen sizes or, perhaps, network bandwidths.

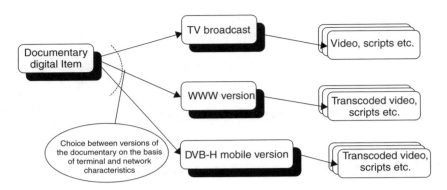

Figure 2.7   A Digital Item with a choice of sub-items based on characteristic descriptors

Thus, in this example, a choice can be introduced between versions of the video such that a specific, low-rate version is available for mobile devices. This is illustrated in Figure 2.7.

It is also possible for the Digital Item resources to be transparently adapted in a delivery network. In particular, on the basis of inserted descriptors, the format of a resource can be described and then used as a basis of adaptation at network nodes. These more complex facilities are detailed in Chapters 7 and 8 on Digital Item Adaptation.

2.6.6 EVENT REPORTS

When users receive Digital Items and the constituent resources, it is often useful if the owner of the licensed content can 'track' usage or 'activity'. Thus, a broadcaster may be interested in discovering which parts of a program are replayed or the most popular parts of a rich presentation of a TV programme. Thus, for the Documentary Digital Item, the broadcaster may create 'Event Report Requests', which will trigger data to be collected and transmitted on the basis of actions performed on the Digital Item or the user's terminal. MPEG-21 provides the format for such requests in Part 15 of the standard [23]. However, this technology would again only be appropriate when a back channel is available, for example, in IP TV systems.

This high-level scenario illustrates how the MPEG-21 multimedia framework can be utilized to meet a typical application. It will be clear that there are many ways to achieve a set of goals in a given application space. This is a facet of MPEG-21 – it is a set of standards and it will be vital in a given application space for users to agree on how they intend using that framework. Further, it is highly likely that users in any given application space would wish to build application infrastructure relevant to their space. As an example, it is unlikely that the broadcaster's customers would wish to view Digital Items being used to carry health records. Further, the latter example highlights the advantage of a protection system that allows application spaces to choose and signal appropriate DRM mechanisms. Overall, MPEG-21 is not targeting any specific application domain and, appropriately configured, can fit the needs of all applications in the multimedia landscape. Configuration may include selection of appropriate MPEG-21 parts and building appropriate applications on top of the framework.

2.7 MPEG-21 STANDARD ORGANIZATION

The MPEG-21 standard might only have been a single technical report! The work was intended to be one of collecting the state of the art in multimedia technology and creating a framework from those parts. Standardization by MPEG was to be undertaken only if holes in the framework within the MPEG remit were identified. However, it was soon apparent during the construction of that report that there were substantial holes in the framework that would require normative standardization and thus a report that only summarized the 'big picture' would be inadequate. The result is a standard that, at the time of writing, had 18 parts as shown in Table 2.1. The ISO numbers indicate the publication date following the colon. This allows differentiation of editions of the standard parts since different editions would have different publication dates (see Chapter 1). Currently, only the Digital Item Declaration (Part 2) has been significantly amended (this means new technology has been added to the first edition) and, in this case, a second edition has been published to integrate all the tools in a single document.

In the following section, brief summaries of each part of the standard are given. For the complete, published standard parts a more substantial treatment is given in later chapters of this book.

Table 2.1 Parts of MPEG-21 including dates of publication following the colon in the ISO standard number

Part	ISO number	Title
1 (TR)	ISO/IEC TR 21000−1:2002	Vision, Technologies and Strategy [7]
	ISO/IEC TR 21000−1:2004	Vision, Technologies and Strategy [24]
2	ISO/IEC 21000−2:2003	Digital Item Declaration [25]
	ISO/IEC 21000−2:2005	Digital Item Declaration Second Edition [8]
3	ISO/IEC 21000−3:2003	Digital Item Identification [26]
4	ISO/IEC 21000−4: TBP	IPMP Components [20]
5	ISO/IEC 21000−5:2004	Rights Expression Language [27]
6	ISO/IEC 21000−6:2004	Rights Data Dictionary [28]
7	ISO/IEC 21000−7:2004	Digital Item Adaptation [21]
8	ISO/IEC 21000−8:TBP	MPEG-21 Reference Software [29]
9	ISO/IEC 21000−9:2005	MPEG-21 File Format [30]
10	ISO/IEC 21000−10:TBP	Digital Item Processing [32]
11 (TR)	ISO/IEC TR 21000−11:2004	Persistent Association Technology [33]
12 (TR)	ISO/IEC 21000−12:2005	Test Bed for MPEG-21 Resource Delivery [34]
13	ISO/IEC 21000−13: −	Empty
14	ISO/IEC 21000−14: TBP	Conformance [35]
15	ISO/IEC 21000−15: TBP	Event Reporting [23]
16	ISO/IEC 21000−16: TBP	Binary Format [36]
17	ISO/IEC 21000−17: TBP	Fragment Identification of MPEG Resources [37]
18	ISO/IEC 21000−18: TBP	Digital Item Streaming [38]

Note: TR stands for 'Technical Report', and TBP stands for 'To Be Published by ISO/IEC', implying that it is at an earlier stage of development, prior to publication by ISO/IEC.

2.8 MPEG-21 STANDARD OVERVIEW

MPEG standards are generally not single documents but are created as a series of parts. This has a number of advantages but, most importantly, it allows industries and users to utilize the individual technologies as stand-alone solutions. One example of the usefulness of this approach can be seen in the US Digital Television standards where MPEG-2 video and systems are used but MPEG-2 audio is replaced by an alternate solution. For MPEG-21, the approach is particularly important as many users may find one or a few technologies appealing or relevant but would not be able to adopt all framework tools. MPEG-21 was, however, designed to be and grew as a 'complete' framework of tools. The interrelationships between parts are normatively expressed in many of the parts themselves (particularly those developed later in the standardization) and, thus, users who adopt a solution that utilizes a significant subset of parts will generally benefit most from the standard. Thus, while this treatment subdivides MPEG-21 into its constituent documentary parts, it should be remembered while reading this section that these are interlinked components of one framework for multimedia consumption and delivery.

2.8.1 VISION, TECHNOLOGIES AND STRATEGY (TECHNICAL REPORT)

2.8.1.1 Motivation

The TR was motivated by the need for MPEG to ensure an understanding of the space and identify the opportunities for standardization of the Multimedia Framework.

2.8.1.2 Objective

Part 1 should establish a defined vision for the MPEG-21 Multimedia Framework. This vision should allow the development of a coherent and structured set of standard parts to be integrated in the framework. The objective is also to create an architecture that can be used to identify missing technologies and hence formulate a strategy to standardize the framework.

2.8.1.3 Summary

The TR has been summarized earlier in this chapter. The second edition of the TR was published in 2004. It provides the user with an overview of the multimedia space, the need for the MPEG-21 framework, and the parts that were standardized at the time it was edited. The architecture of MPEG-21 is divided into the seven elements, which are discussed in Section 2.1 of this chapter [7, 24].

2.8.2 DIGITAL ITEM DECLARATION

The Digital Item Declaration (DID) was motivated by the need for users to have a clear model and way to express the structure of a Digital Item. With such a tool, users can exchange complex sets of mixed (audio, video, text etc.) media interoperably.

2.8.2.1 Objective

Part 2 of the standard should specify the model and instantiation of the core unit of transaction in MPEG-21, the Digital Item.

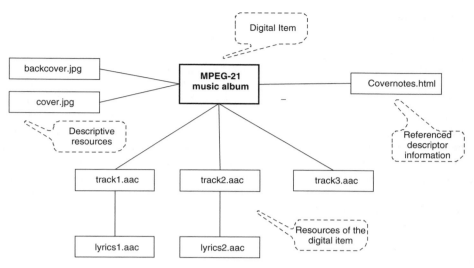

Figure 2.8 A figurative music album Digital Item

2.8.2.2 Summary

The DID was the first normative part of MPEG-21 to be completed and has recently been updated in a second edition. Digital Items are the fundamental unit of transaction in MPEG-21; the DID defines a clear model for the definition of those DIs. The basic model is not a language but provides a set of abstract concepts and terms that can be used to define a scheme. The model defines a Digital Item as a digital representation of an asset and this is the object of actions that occur within the framework (the User exchanged the DI). In practice, the representation of the model is as XML elements in the DIDL, which is normatively described in the standard part. The DIDL is normatively described by an included XML schema. Figure 2.8 gives a figurative view of a Digital Item and Chapter 3 of this book gives a detailed explanation and analysis of DID [8, 25].

2.8.3 DIGITAL ITEM IDENTIFICATION

2.8.3.1 Motivation

Digital Item Identification (DII) was motivated by the need for users to have a mechanism that allows the unique identification of Digital Items and their constituent resources. This allows users to easily store, access and reference Digital Items as well as utilize DRM mechanisms that require such identification.

2.8.3.2 Objective

DII should describe how to uniquely identify Digital Items (and parts thereof), IP related to the Digital Items (and parts thereof), and Description Schemes (sets of metadata descriptors defined by a schema) used in Digital Items. This standard (Part 3) should also explain the relationship between Digital Items (and parts thereof) and existing identification and description systems.

2.8.3.3 Summary

Part 3 is the smallest part of the standard as it was soon realized that DII only needed to offer a 'shell' whereby Users of Digital Items could choose identifiers relevant to their assets and resources. Thus, DII provides a schema that can be used to insert identifiers of many types (expressible as URIs) into Digital Item Declarations to identify the Digital Item and the constituent Resources. This Part also defines XML elements that allow for a Digital Item 'type' to be specified. This was envisaged as a mechanism for distinguishing between, for example, Digital Items carrying Resources and those carrying Rights expressions or Event Reports (ERs) alone. It is important to note that DII does not attempt to define new identification schemes or mechanisms. Rather, current identification schemes (e.g., International Standard Recording Code (ISRC)) may be registered for use in MPEG-21 and, hence, in the Digital item Identifier elements. Figure 2.9 shows DII in use in a figurative Digital Item, while Chapter 3 of this book gives full coverage of the DII syntax and approach [26].

2.8.4 IPMP COMPONENTS

2.8.4.1 Motivation

The motivation for the design of the IPMP components part was the need to create a 'protected' form of the Digital Item. It aimed to allow users to exchange content sets in Digital Items that were protected using a DRM mechanism or to just, for example, encrypt or watermark a particular media resource or section of the DID.

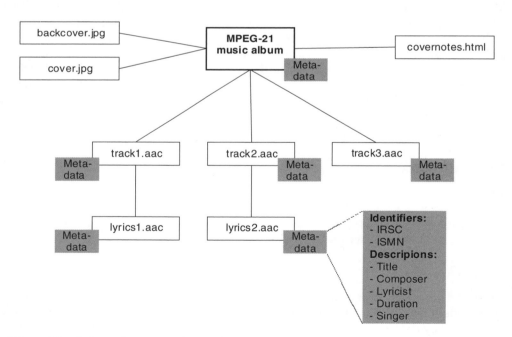

Figure 2.9 A figurative music album Digital Item with identifiers in use [26]. Copyright ISO. Reproduced by permission

2.8.4.2 Objective

Part 4 should specify how to include IPMP information and protected parts of Digital Items in a DIDL document.

2.8.4.3 Summary

The aim with IPMP Components was not to create a 'shiny, new' MPEG-21 DRM standard but, rather, to ensure that the MPEG-21 framework could interoperably exploit existing DRMs (including the IPMP work created by MPEG itself in MPEG-4). This was mainly due to the significant progress made by several DRMs (Windows DRM, Apple Fairplay etc.) in the marketplace and the lack of consensus, at that time, within MPEG to create 'yet another DRM'. IPMP Components consists of two parts:

– IPMP Digital Item Declaration Language, which provides for a protected representation of the DID model, allowing DID hierarchy that is encrypted, digitally signed or otherwise governed to be included in a DID document in a schematically valid manner.

–IPMP Information schemas, defining structures for expressing information relating to the protection of content, including tools, mechanisms and licenses

In combination, these two sections allow users to protect a Digital Item from any point in the hierarchy of the DI (as specified by the DID) and to indicate the mechanism of protection [20].

2.8.5 RIGHTS EXPRESSION LANGUAGE

2.8.5.1 Motivation

The MPEG REL was motivated by the need for a defined set of semantics to establish a user's Rights in the multimedia framework. Prior to this work, there was no standardized way to create Rights expressions for MPEG content; the closest previous work was the IPMP Extensions developed for MPEG-4.

2.8.5.2 Objective

Part 5 should define a REL and an authorization model that allows users to specify whether the semantics of a set of Rights Expressions permit a given Principal (i.e., the user who is receiving the expressions) to perform a given Right upon a given optional Resource during a given time interval based on a given authorization The REL was intended to offer a versatile but standard mechanism to exchange details about a user's rights related to certain content.

2.8.5.3 Summary

The MPEG-21 REL is explained in detail in Chapter 5 of this book; here we give a brief summary of the overall concept of the REL.

The REL is intended to provide a flexible, interoperable approach that enables systems to ensure that Resources and metadata are processed in accordance with their associated

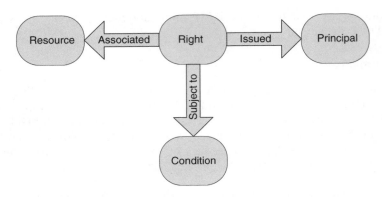

Figure 2.10   The REL data model [27]

Rights. The REL aims at guaranteeing end-to-end interoperability and consistency between different systems and services. It offers a rich and extensible framework for the declaration of rights, conditions and obligations that can all be easily associated with Resources.

The MPEG-21 REL data model (illustrated conceptually in Figure 2.10) for a Rights expression is constructed from four basic entities and the MPEG REL assertion 'grant' then defines the relationship between those entities. An MPEG REL grant is thus built out of the following elements:

- The Principal to whom the grant is issued
- The Right that the grant specifies
- The Resource to which the right in the grant applies
- The Condition that must be met before the right can be exercised [27].

2.8.6 *RIGHTS DATA DICTIONARY*

2.8.6.1 Motivation

Given the definition of the REL, it was necessary to create a set of terms with defined ontological relationships to be used in that language. One key factor with the Rights Data Dictionary (RDD) is that it is intended to be a dictionary of human readable terms and was not motivated by the definition of machine-to-machine terms exchange or negotiation.

2.8.6.2 Objective

Part 6 should specify a RDD that comprises a set of clear, consistent, structured, integrated and uniquely identified terms. These terms are used in support of the MPEG-21 REL (Part 5) [27].

2.8.6.3 Summary

The MPEG-21 RDD is explained in detail in Chapter 10 of this book and the interested reader is referred directly to that treatment. However, here we present a short summary of

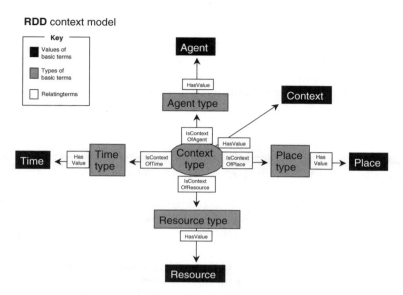

Figure 2.11   The context model on which the Rights Data Dictionary is based [28]. Copyright ISO. Reproduced by permission

the RDD. The Rights Data Dictionary was derived from the work of the Indecs consortium [39] and John Sowa [40] in the late 1990s. The RDD extends that work and is based around a Context model, which, like the Indecs work, is an event-based framework. The aim of the Context Model is to create a natural language ontology for the terms used in Rights processing. The Context Model (illustrated in Figure 2.11) is based around a Basic Term set of five terms: Agent, Context, Resource, Place and Time. The basic idea is that an Agent performs an action (the Context) on a Resource at a certain Place and Time. For example, a Listener performs the action 'to listen' on an MP3 in Trafalgar Square at midnight on December 31, 2006 [28].

This Context Model allows the introduction of terms into the ontology in a logical manner and the creation of terms associated to a Context that will fit within the Context Model. For example, an Agent 'Listener' listens to a 'Music Track' Resource at a given time (Today, 12 p.m.) in the United Kingdom.

The base set of terms in the RDD number some two thousand and include fourteen 'special' Act terms that are specifically designed to work with the REL as Rights verbs.

2.8.7 DIGITAL ITEM ADAPTATION

2.8.7.1 Motivation

Digital Item Adaptation (DIA) was motivated by the need to provide Universal Multimedia Access (UMA) and hence have a means of enabling interoperable adaptation of content. Typically, the vision was one of the universal Digital Item 'container' being transmitted through a network and adapted using environmental, user and terminal characteristic descriptors to maximize the potential user's experience on a given device.

2.8.7.2 Objective

Part 7 should specify the syntax and semantics of metadata tools that may be used to assist the adaptation of Digital Items, that is, the DID and Resources referenced by the declaration. The tools could be used to satisfy transmission, storage and consumption constraints, as well as Quality of Service management by Users.

2.8.7.3 Summary

DIA, which primarily contains metadata on which adaptation of content can be based, was a natural progression from the previous MPEG standardization of MPEG-7, which specified a significant body of multimedia description metadata. Within the latter, several parts of the work (e.g., VariationDS, TranscodingHints) scraped the surface of the field of UMA. They cover the presentation of information, with higher or reduced complexity, so as to suit different usage environments (i.e., the context) in which the content will be consumed. DIA proved to be an opportunity to substantially extend that work and provide a set of tools that assist the adaptation of Digital Items, metadata and Resources.

The concept of DIA is best summarized by Figure 2.12. This figure illustrates the guiding principle that governed the development of the DIA Tools, that is, only the metadata tools to assist adaptation were standardized and the mechanisms of adaptation were left to proprietary implementation.

As mentioned previously, DIA is an extensive part of the MPEG-21 standard and the editors have ensured that two chapters on the topic (see Chapters 7 and 8) are included in this book. However, it is worth visiting briefly the primary sections of the DIA standard as a guide to the contents of the specification. These are also illustrated in Figure 2.13. The major sections of the standard are:

Usage environment description: Tools describing the environment in which the multimedia content will be transmitted, stored and consumed; they originate from the Users of Digital Items.

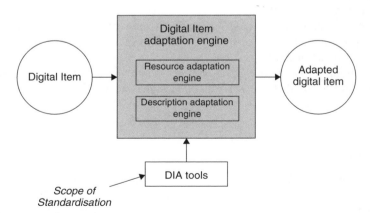

Figure 2.12 The Basic approach of DIA standardization [21]. Copyright ISO. Reproduced by permission

Figure 2.13   The DIA tools [21]. Copyright ISO. Reproduced by permission

Bitstream architecture and description: Tools providing for the specification of adaptation architectures and the description of bit streams. The description tools allow content agnostic transformation using simple editing-type operations.

Terminal and network quality of service: Tools providing a vital linkage to the behaviour of devices and transport media by describing the relationship between QoS constraints, the possible adaptation operations and the resulting media Resources. Using these tools, service providers can make trade-offs during delivery to constrained environments.

Universal constraints: Tools providing for the description of limitations and optimization constraints on adaptations.

Metadata adaptation: Tools providing hints as metadata that might be dropped or modified to reduce the handling complexity of Digital Items and XML in networks.

Session mobility: Tools providing an interoperable DID-based infrastructure for the transfer of a Digital Item session from one device to another or to storage.

DIA configuration: Tools allowing for the configuration of peers on the basis of DIA tools.

Finally, it is worth noting that a recent amendment to MPEG-21 DIA has provided for Rights Expressions governing adaptations in an interoperable framework. MPEG-21 DIA is explained extensively in Chapters 7 and 8 of this book and the interested reader is referred directly to that treatment [21].

2.8.8 MPEG-21 REFERENCE SOFTWARE

2.8.8.1 Motivation

As for previous MPEG standards, offering software that implements the MPEG-21 tools (for both normative and also some non-normative tools) is an important vehicle for faster understanding, deployment and acceptance of the standard. In fact, this software can be used for the implementation of compliant products as ISO waives the copyright of the code (of course, any patents on the techniques will still apply).

2.8.8.2 Objective

Part 8 should include the reference software implementing the normative parts of MPEG-21 as well as describe its functionality and operation. The reference software should be supplied in full in the package and should be also complemented by utility software that provides further insight into the parts of MPEG-21.

2.8.8.3 Summary

An advantage of MPEG standards is that they provide reference software that, if not always optimal, offers extra insight into complex standards text. The MPEG-21 Reference software will eventually cover all of the specification parts of the standard; however, at the time of writing it was restricted to Digital Item Declaration, Digital Item Identification, Rights Expression language, RDD and DIA.

One important consideration when using the reference and accompanying utility software is that they demonstrate the usage of the MPEG-21 standard from, usually, just one user's standpoint. Thus, it is often necessary to encapsulate and modify the software to meet any particular application's requirements [29].

2.8.9 MPEG-21 FILE FORMAT

2.8.9.1 Motivation

As MPEG-21 is an ISO standard, the ISO File Format (standardized jointly for JPEG2000 and MPEG-4) was extended to also be the stand-alone container for MPEG-21 Digital Items. Users of MPEG-21 were keen to see the ISO file format reused, as substantial infrastructure already exists for the handling and delivery of such files.

2.8.9.2 Objective

Part 9 should specify the MPEG-21 file format, in which an MPEG-21 DID and some, or all, of its referenced content can be placed in a single 'content package' file. This should enable the interchange, editing and 'playback' of MPEG-21 Digital Items without the necessity of network connectivity.

2.8.9.3 Summary

The MPEG-21 file format work was driven by a simple desire to have a mechanism by which a complete Digital Item could be packaged physically into a single object. Previously, MPEG had utilized the ISO file format [31], which can hold JPEG2000 images, and MPEG-4 content. The ISO base file format is derived from the Apple Quicktime file format and is an object-structured file format. The MPEG-21 file format is an extension of the ISO base file format with only minor modifications that allow for the fact that a DID is a mandatory part of a Digital Item. Hence, an ISO file with a brand indicator 'mp 21' contains certain mandatory structures that are tailored towards MPEG-21. ISO Files are organized as a collection of 'boxes' that can contain resources and metadata. The boxes can be nested so as to create logical collections of content and metadata.

At the top level of an MPEG-21 file, there will be an MPEG21Box, which is a particular type of MetaBox (the containers in the ISO base file format are referred to as boxes and certain types of boxes can contain certain types of data). This top-layer box contains the MPEG-21 DID usually in an XML box. There is also provision for a binary (compressed) version of the DID (encoded as per Part 16 of the standard) to be held in a 'bXML' box.

The format is also unusual in that it provides for files that can be 'multi-branded'. As an example, it is possible to have a single file that can be interpreted both as an MPEG-4 scene and as an MPEG-21 Digital Item. This highlights the potential overlap between the XML-enabled Internet framework standard that is MPEG-21 and the binary format, communications driven MPEG-4 coding format. This provides for interesting possibilities whereby a file and be read by either an MPEG-21 'player' or an MPEG-4 'player' or, alternatively, an MPEG-4 presentation maybe included such that it is only visible when an MPEG-21 player is invoked.

The MPEG-21 file format does not explicitly deal with streaming as, at that time, streaming was regarded as not being a general capability of Digital Items. Instead, the specification provides for streaming of resources that are part of the Digital Item but not the DID itself. At the time of writing, however, MPEG had embarked on standardization of approaches to streaming complete Digital Items and thus it is likely that the streaming of the file format will be revisited.

2.8.10 DIGITAL ITEM PROCESSING

2.8.10.1 Motivation

DIDs are just declarative and provide no instructions on usage. Digital Item Processing (DIP) was motivated by the need to make DIDs 'active' and allow methods of user interaction to be supplied with the DI. The work was motivated by user's asking 'What can I do with a Digital Item?'.

2.8.10.2 Objective

DIP should improve the processing of 'static' Digital Items by providing tools that allow a User to add specified functionality to a DID. The standardization of DIP (Part 10) should enable interoperability at the processing level.

2.8.10.3 Summary

During the development of the MPEG-21 standards, it became increasingly apparent that something was missing from a framework based purely around a static, structured digital object – the Digital Item. In particular, it became clear that there was no mechanism for a Digital Item author to provide suggested interaction mechanisms with a Digital Item. This is the role fulfilled by DIP. Effectively, it makes the 'static' declarative Digital Item a 'smart' Digital Item by providing executable code that provides interaction with the declared Digital Item content. The result is that a suitably-enabled terminal, on receiving a Digital Item including DIP mechanisms, can immediately present the Digital Item, its resources and metadata to a User on the basis of proposals made by the author. This may include, for example, playing an MP3 file and/or displaying metadata as a user message. While it was decided that a full presentation layer for MPEG-21 Digital Items was inappropriate, DIP goes some way to give an interoperable mechanism that makes a DI smart and hence allows users to gain immediate benefit from the information contained in the DI.

DIP specifies Digital Item Methods, which are based on an extended version of the ECMAScript language known as the Digital Item Method Language (DIML). The extension objects are provided by a normative set of base operations (DIBOs) that provide an interface between the DIM author, the execution environment of the MPEG-21 Digital Item and the User. DIP specifies only the semantics of the base operations and does not specify implementations – rather, this is an opportunity for proprietary implementations on the basis of interoperable standardized semantics and syntax.

Since some operations will be complex, DIP also defines a mechanism for Digital Item Extension Operations (DIxOs), beyond the scope of the base operations. Currently, the standard provides for Java implementations of DIxOs.

While the scope of DIP is carefully restricted, the usage possibilities of DIMs and DIBOs are many. One possibility is to deliver a Digital Item with a set of methods that can be listed as options to a User. It is also possible to link DIP functionality with the REL to express Rights of usage of certain methods. It is also easy to subconsciously restrict methods to 'end-Users'; in fact, methods could be used at any point in the delivery chain of a Digital Item. Other examples of 'Users' that might employ methods could be network proxy and cache nodes, adaptation nodes distributors, and aggregators.

A full explanation of DIP [32] and an investigation of the possibilities it enables is given in Chapter 9.

2.8.11 EVALUATION METHODS FOR PERSISTENT ASSOCIATION TECHNOLOGIES

2.8.11.1 Motivation

Digital Items contain resources that need to be bound permanently to metadata and the Digital Item itself. Hence, there is a need to employ technologies that are capable of achieving persistent association between resources and metadata. However, since there are many technologies available, users requested that MPEG provide a report with clear guidelines on how persistent association technologies should be selected for use with resources and Digital Items.

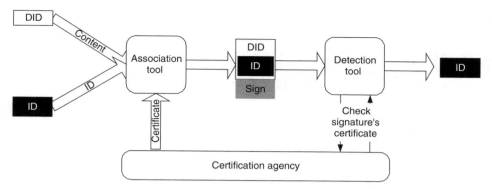

Figure 2.14 A typical scenario for persistent association in digital items (signed content) [33]. Copyright ISO. Reproduced by permission

2.8.11.2 Objective

In MPEG-21, coded representations of media resources will be juxtaposed with metadata descriptors and IPMP protection that apply to the content. This TR (Part 11) should examine methods to evaluate watermarking and fingerprinting techniques that offer persistent associations between audio resources and related metadata.

2.8.11.3 Summary

While MPEG has avoided standardization activities in areas such as watermarking and fingerprinting, a significant groundswell of support was found for the production of a TR on assessment of Persistent Association Technologies relevant to audio applications [33] that might find use within the MPEG-21 framework. One scenario (signing of content) is illustrated in Figure 2.14. This is an important requirement as Digital items contain metadata and Resources that must be intrinsically and persistently linked. One example might be that a Resource must always be associated with the DID of which it is a part. Two main families of persistent association technologies were identified (watermarking and fingerprinting) and this Technical Report gathers experiences of those technologies and considers the parameters for evaluation, assessment, issues and practical testing mechanisms. The key parameters for PAT are identified as fingerprint size, watermark payload, granularity, perceptibility, robustness, reliability and computational performance. Currently, this part only considers audio technologies; it is possible in the future that the part will be extended to video technologies.

2.8.12 TEST BED FOR MPEG-21 RESOURCE DELIVERY
2.8.12.1 Motivation

The multimedia framework is intended to fulfill the needs of universal access. This motivated the development of a test bed that would handle adaptation of scalable media content. Users have the opportunity to use this complete test bed to investigate MPEG-scalable technologies and the use of DIA description tools for adaptation of the content.[2]

---

[2] This test bed was adopted when MPEG-21 Part 13 was still dealing with the specification of a scalable video coding solution. With the move of this coding work to MPEG-4, the need for this test bed is less important to MPEG-21 but is still useful in the context of MPEG-scalable media standards.

2.8.12.2 Objective

The test bed provides a software framework, consisting of a streaming player, a media server, and an IP network emulator, for testing tools under development. The framework is provided in the form of an API. The report should describe the API of each component of the test bed and the platform should provide a flexible and uniform test environment, for example, for evaluating scalable media streaming technologies over IP networks.

2.8.12.3 Summary

Part 12 is in the form of a TR that provides a test environment aimed at the evaluation of scalable media streaming technologies for MPEG resources over IP networks. The work was originally aimed at providing a test bed for scalable video technology and is capable of simulating different channel and network characteristics. In summary, it is suitable for the evaluation of the following:

- Scalable codec (audio, video, scene composition) technologies
- Packetization methods and file formats
- Multimedia streaming rate control and error control mechanisms.

The TR is accompanied by a substantial API software suite (illustrated in the block diagram of Figure 2.15) that is aimed at providing software for resource delivery system development as well as demonstrating how MPEG technologies can be integrated into scalable audio/video streaming applications.

The test bed supports the following MPEG technologies:

- MPEG scalable audio and video codecs
- MPEG-4 on IP

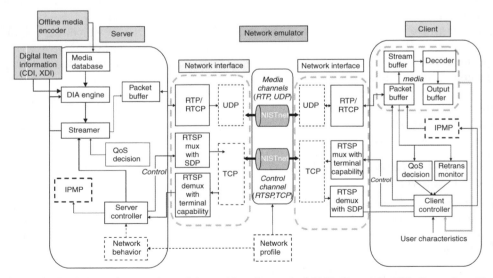

Figure 2.15 Schematic overview of the multimedia test bed [34]. Copyright ISO. Reproduced by permission

- A subset of MPEG-4 IPMP
- A subset of MPEG-21 DIA.

The architecture and the API are not tied to any particular media codecs but, to date, only the MPEG-4 FGS video and MPEG-4 BSAC audio are supported by the software [34].

The API is divided into three parts: the server component API, the client component API, and the common component API (Figure 2.15). The API is written primarily in C++ but employs a JAVA-based network emulator GUI.

2.8.13 PART 13: UNASSIGNED

This part was previously assigned to Scalable Video Coding. However, in January 2005, MPEG decided that the Scalable Video Coding activities underway at that time would be better placed within the MPEG-4 standards set. In fact, because the Scalable Video Coding solution is technologically speaking an extension of Advanced Video Coding (AVC) specified in MPEG-4 Part 10, it will be specified as an amendment to that part. Thus, Part 13 of MPEG-21, which might be construed as being unlucky in some countries, is currently not in use.

2.8.14 MPEG-21 CONFORMANCE

2.8.14.1 Motivation

All MPEG standards include conformance bitstreams or mechanisms. These address the needs of implementers to check and test implementations of the standard. This is vital as users want to be certain that MPEG-21 based products will interoperate and meet certain conformance criteria.

2.8.14.2 Objective

Part 14 should specify a series of conformance points that can be applied to MPEG-21 applications to test their conformance with the MPEG-21 standard. For each conformance point, it aims to provide test bitstreams and XML documents.

2.8.14.3 Summary

The conformance part of MPEG-21 is important as it allows implementers of the MPEG-21 standard to have confidence in the interoperability of their implementations. The conformance specification mirrors the reference software and provides, for example, test DIDs and DIA Bitstream definitions [35].

2.8.15 EVENT REPORTING

2.8.15.1 Motivation

The tracking of content usage has been a long held ambition for providers of digital media. Event Reporting (ER) is intended to provide such functionality for MPEG-21. Users should be able to request reports on the usage of 'their' content and the aim of the ER part of the framework is to create a standard mechanism for requests and the corresponding reports.

2.8.15.2 Objective

Event Reporting (Part 15) should standardize the way of expressing Event Report Requests (ER-R) that contain information about which Events to report, what information is to be reported and to whom; also it should allow expression of ERs created by an MPEG-21 Peer in response to an ER-R (when the conditions specified by that ER-R are met).

2.8.15.3 Summary

While Event Reporting was one of the seven elements identified in the early days of MPEG-21 standardization, the actual Event Reporting work item came much later, with the first Committee Draft of Part 15: Event Reporting only being published in early 2005 [23]. The final Event Reporting Standard is expected to be completed in January 2006.

Event Reporting provides a standardized framework in which 'reportable events' can be specified, detected and, on the basis of which, actions may be taken. Events can be related directly to Digital Item consumption or can be based around peer activity. Hence, the playing of a Digital Item Component or the connection of two peers are valid events.

ERs are generated as a result of an ER-R. This specifies the conditions that will be fulfilled before an ER is generated, the format of that ER, the recipient of the report and information regarding delivery timing, transport mechanisms, and so on. A general model for the generation of the ERs on the basis of an ER-R is shown in Figure 2.16.

2.8.16 MPEG-21 BINARY FORMAT

2.8.16.1 Motivation

XML is used for DIDs and the metadata that is utilized in many parts of MPEG-21. However, since XML is a textual format, it can be inefficient and it is important to provide

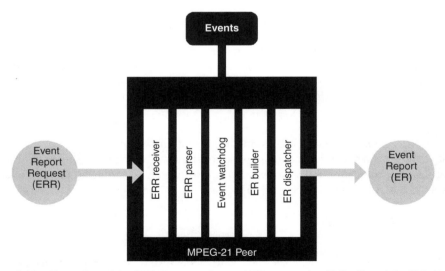

Figure 2.16 General model of ER-R processing and ER generation [23]. Copyright ISO. Reproduced by permission

users who have bandwidth or storage limitations with an alternative format option. The MPEG Binary Metadata (BiM) solution standardized in MPEG-7 was thus adapted to MPEG-21 usage.

2.8.16.2 Objective

Part 16 should provide a mechanism by which XML documents used in the MPEG-21 framework (e.g., DIDs, DIA metadata) can be transmitted in a binary, rather than textual, format. This should offer significant reductions in bit rate.

2.8.16.3 Summary

MPEG has recently shifted its core tools for the compression (binarization) of XML, known as BiM, to a new standard (known as MPEG-B Part 1). Part 16 of MPEG-21 makes use of the MPEG-B Part 1 tools to allow the compression of Digital Item Declarations, enclosed metadata and also for metadata that describes streams (the BSDL from MPEG-21 Part 7 DIA). MPEG has chosen to use a schema-based XML compression mechanism that utilizes the fact that XML is a hierarchical metadata approach and that XML Schema can be thought of as a finite state machine. This allows a binary sequence to be created on the basis of 'state' transitions between elements. In combination with encodings of element values, the latter results in a compressed binary form of an XML document instance.

The XML compression is also enhanced with specific codecs that cater to embedded content. The latter is particularly important as DIDs can have, for example, JPEG images embedded into descriptors and the Base-64 encoding of such content for carriage in XML is wasteful of bandwidth. The BiM tools thus allow specific coded content to be embedded in the binary format to avoid such wastage. Furthermore, one of the main features of BiM is that it provides efficient streaming capabilities for XML-based data; this offers significant improvements over plain text XML in, for example, mobile environments. As part of the efficient XML toolset, BiM features tools offering incremental delivery and selective updates of an XML document [36].

2.8.17 FRAGMENT IDENTIFIERS FOR MPEG RESOURCES
2.8.17.1 Motivation

The MPEG-21 DID provides Anchor and Fragment elements and this part was motivated by the need to provide users with a syntax and semantic for identification of those Fragments. Further, it was an opportunity for MPEG to standardize Fragment identifiers for MPEG media with registered MIME types.

2.8.17.2 Objective

Part 17 should specify a normative syntax for URI Fragment Identifiers to be used for addressing parts of any resource whose Internet Media Type is one of the following:

- audio/mpeg [RFC3003]
- video/mpeg [RFC2045,RFC2046]
- video/mp4 [draft-lim-mpeg4-mime-02]
- audio/mp4 [draft-lim-mpeg4-mime-02]

- application/mp4 [draft-lim-mpeg4-mime-02]
- video/MPEG4-visual [draft-lim-mpeg4-mime-02]
- application/mp21

MPEG URI Fragment Identifier schemes are intended to offer comprehensive and flexible mechanisms for addressing fragments of audio-visual content. Therefore, their use may potentially be extended to other audio-visual MIME types.

2.8.17.3 Summary

The MPEG-21 Digital Item Declaration language second edition provides a mechanism for the addressing of fragments (defined sections) of media resources using the Fragment element. This element was introduced in the second edition as the silence of the first edition on how to identify sections and points in media resources was felt to have caused substantial confusion. The original motivation had been to provide users with freedom of choice for fragments but the increased maturity of the URI Fragment Identifier specifications allowed the second edition DID to be complemented by a more substantial MPEG-defined identification mechanism. However, the introduction of Fragment as an element logically led to discussion of the URI structures available for the identification of fragments of MPEG media resources. It soon became apparent that over the years MPEG had not registered URI Fragment Identification schemes for its registered media types. Thus, Part 17 of MPEG-21 [37] provides a normative syntax for URI Fragment Identifiers to be used for the addressing of Resources with any of the following MPEG-registered MIME types:

- audio/mpeg
- video/mpeg
- video/mp4
- audio/mp4
- application/mp4
- video/MPEG4-visual
- application/mp21

The framework can also be used for more general audio-visual content, but MPEG can only mandate usage for the MIME types that it has registered.

The Fragment Identifiers are used after the # character of a URI reference and add the ability to address the following:

- Temporal, spatial and spatio-temporal locations
- Logical unit(s) of a resource according to a given Logical Model
- A byte range of a resource
- Items or Tracks of an ISO base media file

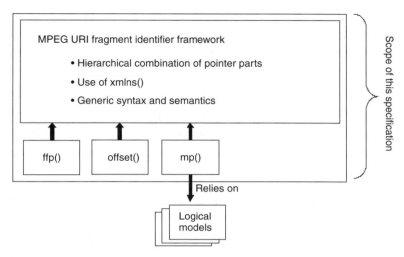

Figure 2.17   Scope of Part 17 of MPEG-21 [37]. Copyright ISO. Reproduced by permission

The scope of the specification can usefully be described using the diagram shown in Figure 2.17.

The three pointer schemes, ffp(), offset(), and mp() apply to ISO file formats, binary resources and streams, and multimedia resources, respectively. The mp() scheme is divided into a pointer scheme for, for example, audio and video files and then a second scheme that allows the addressing of fragments via a hierarchical Logical Model of a resource. The latter can be used to address parts of a track of a CD. Typical uses of the Fragment Identifiers might thus be

- a time point or a time range in an audio CD track;
- one, several or all the chapters of a DVD;
- a moving region in a video sequence;
- a volume of an object in a 3D image.

A simple example of the scheme that accesses a time point 50s into the track with the identifier 'track_ID' equal to '101' in an MPEG-21 file format could be:

```
http://example.com/myFile.mp21 #ffp(track_ID=101)*mp(/~time('npt','50'))
```

2.8.18 DIGITAL ITEM STREAMING

Digital Item Streaming was the newest part of MPEG-21 to be created before this book went to press [39]. It is considered in the final chapter of this book as it is very much a work in progress and the call for technologies and initial evaluation was only just complete at the time of writing.

2.9 MPEG-21 SCHEMAS

While the MPEG-21 schemas are not strictly a part of MPEG-21, they are treated here in the same way as the parts themselves for consistency.

2.9.1 MOTIVATION

Since XML Schemas need to be accessible online, MPEG-21 schemas are hosted on a public site by ISO. The Schemas give the syntax and base semantics of the metadata standardized in the foregoing descriptions of parts.

2.9.2 OBJECTIVE

Wherever metadata is used in the MPEG-21 standard, it is formally specified by an XML schema. This allows users to perform full validation of MPEG-21 conformant metadata documents.

2.9.3 Summary

As of May 2005, the MPEG-21 XML schemas are freely available and can be found at the address: http://standards.iso.org/ittf/PubliclyAvailableStandards/MPEG-21_schema_files/. The schemas are not explicitly a part of MPEG-21 and can be found in each part individually. However, they are collected at this single site for convenience and to allow online referencing. At this site, a hierarchy of schema files can be found and these are detailed in Table 2.2. When used in conjunction with suitable XML tools, the schemas allow users to create XML documents that are valid against the parts of MPEG-21. Further, the schemas

Table 2.2   The MPEG-21 schemas

Directory (MPEG-21 part)	Schemas available
DID (Part 2: Digital Item Declaration)	didl.xsd
	didlmodel.xsd
DII (Part 3: Digital Item Identification)	dii.xsd
REL (Part 5: Rights Expression Language)	rel-r.xsd
	rel-mx.xsd
	rel-sx.xsd
DIA (Part 7:Digital Item Adaptation)	AQoS.xsd
	ArgumentType.xsd
	BaseStackFunction.xsd
	BSDL-1.xsd
	BSDL2.xsd
	BSDLink.xsd
	DIA.xsd
	DIAConfiguration.xsd
	gBSDSchema.xsd
	MatrixDatatypes.xsd
	MetadataAdaptation.xsd
	SessionMobility.xsd
	SingleValueDatatypes.xsd
	UCD.xsd
	UED.xsd
	UnsignedInteger.xsd
	VectorDatatypes.xsd
DIP (Part 10: Digital Item Processing)	dip.xsd

can be used directly in software implementations to create valid documents and to validate incoming documents.

2.10 CONCLUSION

This chapter has introduced the overall vision and concepts that underpin MPEG-21. For such a broad framework, the exact way in which it will be utilized in given application spaces will always be unpredictable. However, MPEG-21 provides a set of standard parts that, in combination with other standards, especially MPEG standards, and proprietary mechanisms, can fulfil a substantial part of the vision of 'protected content anytime, anywhere'. The preceding chapters detail the parts of the framework and explain in detail how the standards work individually and interoperate. As this book was being completed, several application spaces had begun to examine MPEG-21 in more detail [42–44]. In each of these processes, there was a choice of the best tools from MPEG-21, other standards, and proprietary formats. The remaining chapters of this book are intended to inform those making future choices for their particular industry sector's usage of MPEG-21.

REFERENCES

[1] ISO/IEC 11172-1(2,3):1993. "Information Technology – Coding of Moving Pictures and Associated Audio for Digital Storage Media at up to about 1.5 Mbit/s – Part 1: Systems, Part 2: Video, Part 3: Audio", 1993.
[2] ISO/IEC 13818-1(2,3):2000. "Information Technology – Generic Coding of Moving Pictures and Associated Audio Information – Part 1: Systems, Part 2: Video, Part 3: Audio", 2000.
[3] ISO/IEC 14496-1(2,3): 2001, "Coding of Audio-Visual Objects – Part 1: Systems, Part 2: Visual, Part 3: Audio", 2001.
[4] ISO/IEC 15938-1(2,3,4,5): 2001, "Multimedia Content Description Interface – Part 1: Systems, Part 2: Description Definition Language, Part 3: Visual, Part 4: Audio, Part 5: Multimedia Description Schemes", 2001.
[5] See http://www.live8live.com/ for details, accessed in August 2005.
[6] See http://www.apple.com/itunes/ for details, accessed in August 2005.
[7] ISO/IEC TR 21000-1: 2002. "Information technology – Multimedia framework (MPEG-21) – Part 1: Vision, Technologies and Strategy", 2002.
[8] L. Chiariglione, "(Convenor, MPEG) at MPEG-21 Workshop held in Noordwijkerhout", The Netherlands, March 2000.
[9] MPEG Convenor. "Request for New Project MPEG-21, ISO/MPEG N3200", 50th MPEG meeting, December, Hawaii, USA, 1999.
[10] MPEG Multimedia Description Schemes (MDS). "Text of ISO/IEC 21000-2 Digital Item Declaration", (2nd Edition), ISO/MPEG N6927, 71st MPEG Meeting, Hong Kong, China, January, 2005.
[11] ISO 2108:2005. "Information and documentation – International Standard Book Number (ISBN)", 2005.
[12] ISO 3901:2005. "Information and documentation – International Standard Recording Code (ISRC)", 2005.
[13] ISO 15707:2001. "Information and documentation – International Standard Musical Work Code (ISWC)", 2001.
[14] ISO 15706:2002. "Information and documentation – International Standard Audiovisual Number (ISAN)", 2002.
[15] See http://www.doi.org/index.html, "The Digital Object Identifier System", for details, accessed in August 2005.
[16] See http://www.cidf.org/ for details, accessed in August 2005.
[17] See www.sun.com for details, accessed in August 2005.
[18] See http://www.coral-interop.org/ for details, accessed in August 2005.
[19] See http://www.dmpf.org/ for details, accessed in August 2005.

[20] MPEG Multimedia Description Schemes (MDS). "FCD MPEG-21 IPMP Components, ISO/MPEG N7196", 72nd MPEG Meeting, Busan, Korea, April, 2005.

[21] ISO/IEC TR 21000-7:2004. "Information technology – Multimedia framework (MPEG-21) – Part 7: Digital Item Adaptation", 2004.

[22] B.S. Majunath, P. Salembier, T. Sikora (eds), "Introduction to MPEG-7 Multimedia Content Description Interface", John Wiley & Sons Ltd., Chichester, 2002.

[23] MPEG Multimedia Description Schemes (MDS). "FCD MPEG-21 Event Reporting, ISO/MPEG N7444", 73rd MPEG Meeting, Poznan, Poland, July, 2005.

[24] ISO/IEC TR 21000-1: 2004. "Information technology – Multimedia framework (MPEG-21) – Part 1: Vision, Technologies and Strategy", 2004.

[25] ISO/IEC 21000-2:2003. "Information technology – Multimedia framework (MPEG-21) – Part 2: Digital Item Declaration", 2003.

[26] ISO/IEC 21000-3:2003. "Information technology – Multimedia framework (MPEG-21) – Part 3: Digital Item Identification", 2003.

[27] ISO/IEC 21000-5:2004. "Information technology – Multimedia framework (MPEG-21) – Part 5: Rights Expression Language", 2004.

[28] ISO/IEC 21000-6:2004. "Information technology – Multimedia framework (MPEG-21) – Part 6: Rights Data Dictionary", 2004.

[29] MPEG Multimedia Description Schemes (MDS). "Text of ISO/IEC 21000-8 Reference Software, ISO/MPEG N7206", 72nd MPEG Meeting, Busan, Korea, April, 2005.

[30] ISO/IEC 21000-9:2004. "Information technology – Multimedia framework (MPEG-21) – Part 9: File Format", 2004.

[31] ISO/IEC 14496-12:2005. "Information technology – Coding of audio-visual objects – Part 12: ISO base media file format", 2005.

[32] MPEG Multimedia Description Schemes (MDS). "Text of ISO/IEC 21000-10 Digital Item Processing, ISO/MPEG N7208, 72nd MPEG Meeting, Busan, Korea, April, 2005.

[33] ISO/IEC 21000-11:2004. "Information technology – Multimedia framework (MPEG-21) – Part 11: Evaluation Tools for Persistent Association Technologies", 2004.

[34] ISO/IEC 21000-12:2005. "Information technology – Multimedia framework (MPEG-21) – Part 12: Test Bed for MPEG-21 Resource Delivery", 2005.

[35] MPEG Multimedia Description Schemes (MDS). "ISO/IEC 21000-14 MPEG-21 Conformance CD, ISO/MPEG N7213", 72nd MPEG Meeting, Busan, Korea, April, 2005.

[36] MPEG Systems, "Text of ISO/IEC 21000-9/FDIS, ISO/MPEG N6975", 71st MPEG Meeting, Hong Kong, China, January, 2005.

[37] MPEG Multimedia Description Schemes (MDS). "ISO/IEC 21000-17 FCD MPEG-21 Fragment Identifiers for MPEG Resources, ISO/MPEG N7446", 73rd MPEG Meeting, Poznan, Poland, July 2005.

[38] MPEG Requirements. "ISO/IEC 21000-14 DI Streaming Call for Proposals, ISO/MPEG N7066", 72nd MPEG Meeting, Busan, Korea, April, 2005.

[39] See http://www.indecs.org/ for details and documents, accessed in August 2005.

[40] J.F. Sowa, "Knowledge Representation: Logical, Philosophical, and Computational Foundations", Brooks Cole Publishing Co., Pacific Grove, CA, 2000.

[41] G.A. Brox, "MPEG-21 as an access control tool for the National Health Service Care Records Service Health Journal of Telemedicine and Telecare", vol. 11, Supplement 1, Royal Society of Medicine Press, July 2005, pp. 23–25.

[42] TV Anytime Phase 2 Specification S-3-3, version 2, "Extended Metadata Schema", May 2005.

[43] J. Bekaert, L. Balakireva, P. Hochstenbach and H. Van de Sompel. "Using MPEG-21 and NISO OpenURL for dynamic dissemination of complex digital objects in the Los Alamos National Laboratory digital library", D-Lib Magazine, vol. 10, no. 2, February 2004.

3

Digital Item Declaration and Identification

Frederik De Keukelaere and Rik Van de Walle

3.1 INTRODUCTION

Within any system that proposes to facilitate a wide range of actions involving Digital Item (DI), there is a need for a very precise definition of what exactly constitutes such a 'Digital Item'. Currently, there are many types of content, and probably just as many possible ways of describing them and the context in which they can be used.

This results in a strong challenge to design a powerful and flexible model for DIs. This model must be able to accommodate the variety of forms that content can take and also provide support for new forms of content that will be developed in the future. Such a model will be truly useful only if it results in a format that can be used to unambiguously represent any DI defined within the model. The success of such a model depends on its capability to express the DIs, including the additional information, generally called *metadata*, in an interoperable manner.

Besides the requirement of a basic definition of a model for declaring DIs, a truly interoperable framework requires a technology for identifying, locating and assigning types to DIs. The identification allows DIs to contain unique markers. On the basis of such markers, the DIs can be located and tracked within the framework. Because DIs are a very flexible concept, there will be a variety of types of DIs, for example, music albums, movie collections, e-tickets and so on. To increase interoperability within a framework, having a mechanism that allows typing of those DIs will be required.

In this chapter, the Digital Item Declaration (DID) specification and the Digital Item Identification (DII) specification are discussed; these specifications correspond to Parts 2 and 3 of the MPEG-21 standard. The DID, which will be discussed in the first part of this chapter, describes a model for declaring DIs [1, 2]. The DII, which will be discussed in the last part of this chapter, specifies a technology that allows unique identification and typing of DIs [3].

The MPEG-21 Book   Ian S Burnett, Fernando Pereira, Rik Van de Walle, Rob Koenen

3.2 DIGITAL ITEM DECLARATION

In MPEG-21 (formally called *Multimedia Framework* and numbered ISO/IEC 21000), the concept of a 'Digital Item' is the key to the whole framework; every transaction, every message, every form of communication is performed using a DI [4]. Within ISO/IEC 21000–1, also called *MPEG-21 Part 1*, DIs are defined as 'structured digital objects, with a standard representation, identification and *metadata* within the MPEG-21 framework' [5].

ISO/IEC 21000–2 [1], better known as the *Digital Item Declaration*, consists of three parts:

- *Abstract model*: The DID Model defines a set of abstract elements and concepts that form a useful model for declaring DIs. Within this model, a DI is the digital representation of 'an asset' (see [5]), and as such, it is the entity that is acted upon (managed, described, exchanged, collected and so on.) within the model.
- *Representation of the model*: Normative description of the syntax and semantics of each of the DID elements, as represented in the Extensible Markup Language (XML) [6]. This section of the specification also contains some examples for illustrative purposes.
- *Corresponding MPEG-21 DIDL schemas*: Informative XML Schemas [7] comprising the entire grammar of the DID model and Digital Item Declaration Language (DIDL) in XML.

In Section 3.2, each of these parts and their functionality within the DID specification are discussed. The first part of the DID specification is the Abstract Model. As stated above, this model contains a set of abstract terms and concepts that can be used to declare DIs. Figure 3.1 is an example of a declaration of such a DI. At the left side, building blocks of the Abstract Model are displayed. At the right side, a music collection is displayed. Throughout the discussion of this Abstract Model, this music collection example is used to illustrate the mapping of the abstract terms and definitions to a real-world example DI.

Because the Abstract Model is merely a model, and not a concrete implementation, there are several possible ways to implement it. One possible implementation, called the *DIDL*, is discussed in Section 3.2.2. Starting from a description of the Abstract Model, MPEG issued a call for proposals for a DIDL. On the basis of the responses to this call MPEG developed the second part of ISO/IEC 21000–2, the DIDL.

The DIDL is an XML representation of the abstract terms and concepts that are put forward in the Abstract Model. It forms the basic language that allows the unambiguous declaration of DIs.

3.2.1 THE ABSTRACT MODEL

The Abstract Model of the DID defines a set of abstract terms and concepts that represent a useful model for defining DIs. Within this model, a DI is the digital representation of 'an asset', and as such, it is the entity that is acted upon (managed, described, exchanged, collected and so on) within the model. This model has been designed to be as flexible and general as possible, providing for the 'hooks' that enable higher-level functionality and interoperability. This allows the model to serve as a key foundation in the building

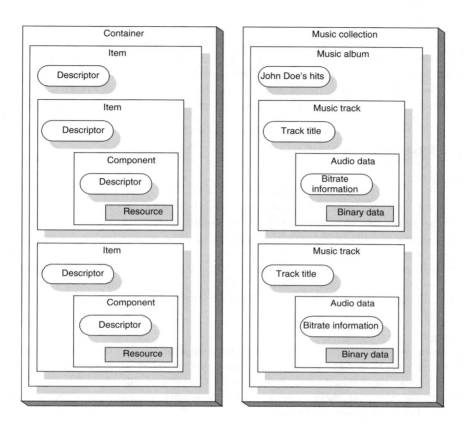

Figure 3.1   Building blocks of the Digital Item Declaration

of higher-level models in other MPEG-21 parts like the DII, which will be discussed later in this chapter. As described earlier, this model does not define a language in and by itself. Instead, it provides a common set of abstract concepts and terms that can be used to define a language, or to perform mappings between existing languages capable of DID.

3.2.1.1 An Example MPEG-21 DID

Currently, music albums typically consist of a set of audio tracks. This media content is distributed on a CD disk or downloaded from an on-line music store. Apart from the fact that audio tracks can be played on audio equipment, there is not much functionality included in such music albums. At best, CDs contain some 'bonus material' – some additional movie clips, or some previews of related merchandize. However, the possibility of including additional features with music albums is very limited. For example, there is no possibility to express the relation between the extra content and the music album; in most cases, special equipment, such as a personal computer with a certain type of player, is necessary to be able to consume the additional content.

Within MPEG-21, a music album is no longer limited to a set of music tracks. A 'Music DI' can contain a variety of resources and *metadata*, text, images, audio, video and descriptive information (for example, for indexing or searching) can all be included in a DI. A step-by-step approach is followed below in creating a DID that expresses a music album. An overview of the terms and definitions of the DID Abstract Model is given, and it is demonstrated how they can be used to create a complex 'Music DI'.

3.2.1.2 Terms and Definitions

Resource
A *resource* is an individually identifiable asset such as a video or audio clip, an image or a textual asset. A *resource* may also potentially be a physical object (e.g., a painting). All *resources* must be locatable via an unambiguous address.

In the music collection example, the *resource* can be associated to the actual audio content of the music album. This audio content, for example, encoded with MPEG-1 [8], can be seen as the part of the music album that has the most value to the author of the DID, therefore it is definitely an asset. Similar to the audio content, the cover of the album also can be a *resource*. This image is an asset that represents the cover of the music album.

Statement
A *statement* is a literal textual value that contains information, but not an asset. Examples of likely *statements* include descriptive, control, revision tracking or identifying information (such as an identifier as described in ISO/IEC 21000–3 DII [3]).

For a music album, a likely *statement* could be a copyright statement. The text of this copyright statement should be kept as it is during the lifetime of the digital music album, and therefore it can be considered as a literal textual value. Other examples of *statements* could be the title of the music track, the title of the music album, a unique identifier such as a barcode number and so on. Note that not all textual values are necessarily *statements*, for example, the lyrics of a song are a textual asset, and therefore not a *statement*, as opposed to a copyright notice.

Descriptor
A *descriptor* associates information with the enclosing element (for example, an enclosing *item* or *component*). This information may be a *component* (such as a thumbnail of an image or a text *component*), or a textual *statement*. A *descriptor* can be formally defined by using the Extended Backus–Naur Form (EBNF) [9] (note: the *item* and *component* elements will be defined later):

```
descriptor ::= condition* descriptor* (component | statement)
```

A *descriptor* gives information about its parent. In Figure 3.1, the *descriptor* of the music track contains the title of the music track, which is expressed using an *item*. This *descriptor* is an example of a description expressed by a literal text value and therefore this *descriptor* will contain a *statement*. In addition, it would have been possible to add an extra *descriptor*, for example, containing MPEG-7 [10] data describing the track.

Component

A *component* is the binding of a *resource* to a set of *descriptors*. These *descriptors* are information related to all or part of the specific *resource* instance. Such *descriptors* will typically contain control or structural information about the *resource* (such as bitrate, character set, start points or encryption information) but not information describing the 'content' within. It should be noted that a *component* itself is not an *item; components* are *building blocks* of *items*. The EBNF definition of a *component* is as follows:

```
component ::= condition* descriptor* resource anchor*
```

In the music album example, the *component* binds the *descriptor* to the *resource*. More precisely, the *component* contains the audio-related data (in a *resource*), such as the binary data of the audio track, and *metadata* (in a *descriptor*) such as the corresponding bitrate information. *Descriptors* that are grouped together with *resources* typically contain technical information about the *resource*, not the semantic description of the information that is contained within that *resource*. For example, the *descriptors* in *components* may contain bitrate information about the *resource*. They do not contain the title of the track that is represented by the audio data. This information is contained one level higher, as a *descriptor* of the parent *item* (see Figure 3.1).

Item

Items are a grouping of *components* bound to a set of relevant *descriptors* containing information about the *item* as a representation of an asset. *Items* may contain *choices*, which allow them to be customized or configured. *Items* may be conditional (on *predicates* asserted by *selections* defined in the *choices*). *Items* may also contain *annotations* to their subparts. The EBNF definition of an *item* is given below (note: *choices, predicates, selections and annotations* will be defined later):

```
item ::= condition* descriptor* choice* component* annotation*
```

However, *items* can also have a nested structure, they can also contain *subitems*, that is, an *item* within an *item*. An *item* that contains no *subitems* can be considered an entity, a logically indivisible asset. An *item* that does contain *subitems* can be considered a compilation of *items*, an asset composed of potentially independent subparts.

This leads to the final EBNF definition of an *item*:

```
item ::= condition* descriptor* choice* (item | component)* annotation*
```

The music collection example contains different types of *items*. The first type is the *item* without *subitems*. The music tracks are examples of such *items*. Those *items* are logically indivisible assets. Each track is considered as one entity. This entity is constructed using the basic building blocks, *items* and *components*. In Figure 3.1 there is only one *component*, but it is possible to add additional *components* with different formats for the audio data, for example, an uncompressed version and a compressed version of the audio data. The second type of *item* contains *subitems*. In Figure 3.1, the music album is a compilation of independent subparts, comprising the music tracks.

Container

A *container* is a structure that allows *items* and/or *containers* to be grouped. These groupings of *items* and/or *containers* can be used to form logical packages (for transport or exchange) or logical shelves (for organization). *Descriptors* allow for the 'labeling' of *containers* with information that is appropriate for the purpose of grouping (for example, delivery instructions for a package or category information for a shelf).

It should be noted that a *container* itself is not an *item; containers* are groupings of *items* and/or *containers*. The EBNF definition of a *container* is as follows:

```
container ::= descriptor* container* item*
```

In a music collection example, the whole collection can be seen as a *container*. This *container* contains a set of music albums, which are represented by *items*. The *container* representing the music collection can be seen as the digital shelf on which digital music albums are stored. This *container* could have contained a *descriptor* describing the type of the music collection. For example, a *descriptor* that contains 'this is my rock collection' could have been placed within the *container*.

Predicate – Condition – Selection – Choice

The combination of the *predicate, condition, selection* and *choice* elements make the DID model truly unique because they introduce a choice mechanism to the static declaration of the DIs, that is, they make DIs configurable. These building blocks allow for the interoperable expression of the fact that parts of DIs are only conditionally available. They allow for run-time configuration of DIs. Before starting the discussion on these building blocks of the Abstract Model, the definition of each of these building blocks is given.

A *predicate* is an unambiguously identifiable declaration that can be true, false or undecided. A *predicate* can be compared with a variable in a programming language that has three possible values: true, false and undecided. A *predicate* is declared in a *selection*. Sets of selections are grouped together into *choices*. An example of a possible *choice* could be 'Do you want to pay for the high-resolution version'? This *choice* will have two *selections*, one with a *predicate* representing 'yes', and the other with a *predicate* representing 'no'. If the MPEG-21 User decides to pay for the high-resolution version, the payment mechanism starts working and the high-resolution version becomes available.

A *condition* describes the enclosing element as being optional, and links it to the *selection(s)* that affects its inclusion. Multiple *predicates* within a *condition* are combined as a conjunction (an AND relationship). Any *predicate* can be negated within a *condition*. Multiple *conditions* associated with a given element are combined as a disjunction (an OR relationship) when determining whether to include the element. The EBNF specification of a *condition* is given as follows:

```
condition ::= predicate+
```

A *selection* describes a specific decision that will affect one or more *conditions* somewhere within an *item*. If the *selection* is chosen, its predicate becomes true; if it is not chosen, its *predicate* becomes false; if it is left unresolved, its *predicate* is undecided. The *descriptor* element in the *selection* gives the User additional information when choosing *predicates*. The EBNF definition of the *selection* element is as follows:

```
selection ::= condition* descriptor* predicate
```

A *choice* describes a set of related *selections* that can affect the configuration of an *item*. The *selections* within a *choice* are either exclusive (choose exactly one) or inclusive (choose any number, including all or none). The EBNF definition of the *choice* element is given by the following:

```
choice ::= condition* descriptor* selection+
```

Throughout this chapter, it is important to keep in mind that there is a clear distinction between the terms 'available' and 'accessible'. When the word 'available' is used in the context of the choice mechanism, it means that it is possible for the MPEG-21 User to consume the content from the DID point of view. This means that all the *conditions* that make a certain *component* conditionally available are satisfied.

This, however, does not mean that the *resource* contained within the component is 'accessible'. Accessible in this context means that it is possible for the MPEG-21 User to get access to the content. For example, it could mean that he can stream a music track to his terminal and play the track. To have access to the content, it requires that the rights for accessing the content be satisfied.

The choice mechanism can be used to configure the DI at run-time, that is, at the time the DI will be consumed. *Choices* can be related to different types of questions, going from questions about payment to more technical questions about bitrate. *Choices*, can be resolved by the User, therefore it is possible that *choices*, are also resolved by either hardware or software in addition to a human end-user. For example, *choices* with respect to the bitrate of a movie could be resolved by the terminal, on the basis of the information about the available bandwidth of the network.

Assertion

The *assertion* is relevant to the *choice* mechanism because it allows the storage of the configuration state of a *choice*. In other words, an *assertion* can be placed within a DI to store the configuration state of a certain *choice*. For example, an *assertion* can be used when a DI has been configured into a certain state but the consumption of the DI needs to be paused. If the MPEG-21 User wants to resume the consumption of the multimedia content afterwards, he/she does not need to reconfigure the DI. The terminal can do this for him/her using the information stored in the *assertion*. A more formal definition of the *assertion* is as follows: An *assertion* defines a full or partially configured state of a *choice* by asserting true, false or undecided values for some number of *predicates* associated with the *selections* for that *choice*. In EBNF notation, the *assertion* element is defined by the following:

```
assertion ::= predicate*
```

Annotation

An *annotation* contains information about another identified element of the model without altering or adding to that element. The information can take the form of *assertions, descriptors* and *anchors*. In EBNF, *annotations* are defined by the following (note: the *anchors* elements will be defined later):

```
annotation ::= assertion* descriptor* anchor*
```

The *annotation* allows MPEG-21 Users to add additional information to their DIs. This additional information can be contained within *assertions, descriptors* and *anchors*. In the music album example, the additional information could be a textual description such as 'this is a cool track'. *Annotations* can also be application-specific control information, such as a bookmark of the last song that the MPEG-21 User listened to.

When DIs are transferred in a secure way, Users may want to calculate digital signatures of their DIs (e.g., using XML Signature [11]), for example, to see whether there has been any tampering with the DI during transmission. When the digital signature of the DI needs to be calculated, the *annotations* and the information contained within the *annotations* are ignored. This information is indeed additional information and not part of the actual content; it is up to the terminal (or the end-user) to display, or process that information or not.

Fragment – Anchor

A *fragment* unambiguously designates a specific point or range within a *resource*. *Fragments* may be *resource* type specific.

An *anchor* binds *descriptors* to a *fragment* that corresponds to a specific location or part of a *resource*. In EBNF notation, the *anchor* element is defined by:

```
anchor ::= condition* descriptor* fragment
```

While *annotations* give User the possibility to add information to the building blocks of the model, the *fragment anchor* construction allows the User to add information to the DI about a specific location in a *resource*. For example, it allows the User to add the *descriptor* 'this is John's greatest drum solo' to a specific point in time of the audio track. To realize this, the *fragment* identifies the point in time to which the *descriptor* applies, and the *anchor* binds that *fragment* to the *descriptor*. It should be noted that *anchors* are considered as a part of the DI (as opposed to *annotations*) and therefore they are included when, for example, digital signatures need to be calculated. *Anchors* that are not supposed to be considered as part of the DI need to be included in *annotations*.

3.2.2 DIGITAL ITEM DECLARATION LANGUAGE

The DIDL, which is an implementation of the Abstract Model of Section 3.2.1 [1], is a language that has been developed by MPEG. The DIDL is an XML-based language allowing authors of DIs to express the structure of their multimedia content within a standardized XML framework. The DIDL is basically a set of XML building blocks corresponding to the entities in the Abstract Model. With the DIDL, it is possible to declare DIs in an unambiguous way. Such declarations, called *Digital Item Declarations (DIDs)*, which are declared in DID documents, express the structure of a DI, and the relationship between the different entities of which the DI is constructed.

In the following subsections, a step-by-step approach is taken in creating a DID that represents a music album. During this process an overview of the XML building blocks of the DIDL is given, and there is a demonstration on how they can be used to create a 'Music DI' using MPEG-21 technology. This demonstrates how a music album using MPEG-21 technology can be richer than a standard CD, for example, by containing video clips, animations, interviews with the composer and the artists and so on. By using other MPEG-21 parts, it will be possible to protect this Music DI by allowing distribution and consumption of rights-protected content in a transparent and governed way.

3.2.2.1 XML Building Blocks

Resource

The `Resource` element in the DIDL is the XML representation of the *resource* from the Abstract Model. It defines an individual identifiable asset, such as an image, an audio or video clip and so on. This video clip could be the audio, video, multiplexing and synchronization combined into one resource or each of them in a separate resource. In the latter, the synchronization and multiplexing would be done by the player that consumes the resources. In other words, a `Resource` element typically contains the multimedia content (or part of it) of the DI.

A `Resource` element has the possibility to include a variety of content with a variety of data types. Identifying the data type of the resources can be done using the `mimeType` attribute. The `mimeType` attribute is a concatenation of MIME (Multipurpose Internet Mail Extensions) media type and the subtype indicating the data type of a `Resource`, as defined in RFC 2045 [12]. For example, the `mimeType` attribute of a `Resource` containing an mp3 audio file [8] is 'audio/mpeg'. The first part of the MIME type identifies that the data of the `Resource` is audio data, and the second part identifies that this audio data is encoded as audio frames, as defined in RFC 3003 [13]. By using this information, it is possible, for example, for an MPEG-21 media player, to unambiguously determine the data type of the `Resource` and take the appropriate action when consuming it (displaying the image, playing the audio or video clip, starting a text editor and so on).

Besides the `mimeType` attribute, the `Resource` element has a `contentEncoding` attribute providing information about content encoding as defined in RFC 2045 [12]. When present, the value of the `contentEncoding` attribute indicates what additional content encodings have been applied to the data, and thus what decoding mechanisms to apply. Content encoding is primarily used to allow a document to be compressed without losing the identity of its underlying MIME media type. A possible example would be to apply 'gzip' compression (as per RFC 1952 [14]) to a textual resource before using it in a DID, the value of the `contentEncoding` attribute would then be 'gzip'.

A `Resource` element can be defined both by reference and by value. When defining a `Resource` by value, it can contain either textual data or base64 [12] encoded data. When a `Resource` is base64 encoded, the `encoding` attribute has the value 'base64'. Having this attribute allows for MPEG-21 terminals to determine whether or not the `resource` is base64 encoded. In Example 3.1, the `Resource` contains textual data, a small fragment of text saying 'This is a plain text resource'.

As an alternative to defining by value, a `Resource` element can be defined by reference. In this case, the `Resource` has a `ref` attribute containing a Universal Resource Identifier (URI) [15] that points to the actual data. In Example 3.2, the `Resource` contains a link to an mp3 audio track that is stored somewhere on a web server.

Having the ability to either include a `Resource` by reference or by value gives the author of the DID the possibility to optimize the DID for his/her purpose. For example, when dealing with a large set of small `Resources`, it may be more efficient when the `Resources` are declared by value. Otherwise, processing all the `Resources` would

```
<Resource mimeType="text/plain">
   This is a plain text resource.
</Resource>
```

Example 3.1   A `Resource` declared by value

```
<Resource mimeType="audio/mpeg" ref="http://server/track01.mp3"/>
```

Example 3.2   A `Resource` declared by reference

introduce the overhead of resolving all the URIs pointing at the actual data. On the other hand, when a DID contains a lot of binary data, encoding all the `Resources` in base64 and including them in the DID, will result in a large and bloated DID.

Statement

The `Statement` element in the DIDL is the XML representation of the *statement* from the Abstract Model. It contains a piece of information that can be attached to a parent element.

A `Statement` can contain any data format, including plain text and various machine-readable formats such as well formed XML. The same attributes that are available on `Resources` are also available on `Statements`.

`Resources` and statements look very similar at first sight, they both contain a variety of data types, they both support the base64 encoding for inline data and so on. However, there are several differences between a `Resource` and a `Statement`. The first difference between a `Resource` and a `Statement` is the functionality of the data they contain. A `Resource` will typically contain the actual content of a DI, that is, the audio, video, text and so on. Opposed to that, a `Statement` typically contains information that will be attached to another element (for example, to a `Resource`). This information is typically descriptive, control, revision tracking or identifying information.

Another difference between a `Resource` and a `Statement` is the semantic meaning of the data. The data in a `Resource` is considered as an individually identifiable asset, while the data in a `Statement` is considered as information, but not as an asset. For example, a picture, which has real value to the author of the DI, is typically included in the DID as a `Resource`, while a copyright statement that has no real value on its own (although being very important, of course) will typically be included as a `Statement`.

A `Statement` containing plain text can be used to associate a human-readable description with the parent element, for example, a title of a music album as given in Example 3.3.

As an alternative, a `Statement` can contain data in some machine-readable format, such as well formed XML. Those `Statements` can be used to associate application-specific *metadata* with a parent element. This machine-readable data can be information about the bitrate of the parent element, identifying information and so on. The grammar that can be used for `Statements` containing well formed XML is identified by the namespace of the fragment. In Example 3.4, the `Statement` contains the description of the bitrate in MPEG-7 format [16]. The namespace `urn:mpeg:mpeg7:schema:2001` identifies that the `Statement` contains MPEG-7 compliant data.

The `Statement` element provides the link between DID and the other parts of MPEG-21. For example, DII information, which will be discussed at the end of this chapter, is normatively included in `Statement` elements.

Descriptor

The `Descriptor` element in the DIDL is the XML representation of the *descriptor* from the Abstract Model. Typically, a `Descriptor` is used to associate descriptive data with a parent element. Descriptive data can take the form of a `Statement` or `Resource`

```
<Statement mimeType="text/plain">
   All time greatest hits of John Doe
</Statement>
```

Example 3.3 A `Statement` declared by reference

```
<Statement mimeType="text/xml">
  <Mpeg7 xmlns="urn:mpeg:mpeg7:schema:2001">
    <Description xsi:type="MediaDescriptionType">
      <MediaInformation>
        <MediaProfile>
          <MediaFormat>
            <Content href="http://server/track01.mp3"/>
            <BitRate>131072</BitRate>
          </MediaFormat>
        </MediaProfile>
      </MediaInformation>
    </Description>
  </Mpeg7>
</Statement>
```

Example 3.4   A `Statement` containing MPEG-7 data

```
<Descriptor>
  <Component>
    <Resource mimeType="image/jpeg" ref="thumbnail.jpg"/>
  </Component>
</Descriptor>
```

Example 3.5   A `Descriptor` containing an image

(contained in a `Component`) (note: the `Component` element will be defined later). An example of a `Resource` containing descriptive data is found in Example 3.5. This `Resource` contains a thumbnail version of an image. It is enclosed within a `Descriptor`, which is why this image is to be interpreted as descriptive data.

An example of a `Statement` containing descriptive data is that of a simple textual description, such as the title and author of a work.

It should be noted that it is possible to attach multiple `Descriptors` to a parent element. Attaching multiple `Descriptor` elements to a parent element has the advantage that it is possible to include different versions or different formats of the description to the parent. For example, including a thumbnail in JPEG format [17] and in TIFF [18] in a DI has the advantage that terminals supporting only one of those formats have the possibility to render that specific format and will ignore the other. Furthermore, a terminal that supports none of those formats, and is therefore not capable of rendering the image, will still be able to know that this information is descriptive information about a parent element.

The *descriptor* from the Abstract Model has a second representation in the DIDL. DIDL elements that can have a `Descriptor` element as a child element allow attributes from any namespace to be included in that element. Those attributes have the same semantic

```
<Descriptor>
  <Statement mimeType="text/plain">
    All time greatest hits of John Doe
  </Statement>
</Descriptor>
```

Example 3.6   A `Descriptor` containing a plain text `Statement`

and functionality as the `Descriptor` element but are represented as attributes. The elements to which the attribute *descriptors* can be added are the `Anchor`, `Annotation`, `Choice`, `Component`, `Container`, `Descriptor`, `Item` and `Selection`.

Component

The `Component` element in the DIDL is the XML representation of the *component* from the Abstract Model. A `Component` is the DIDL element that groups `Resource` elements with `Descriptor` elements. As such, it attaches descriptive information to resources. Because a `Component` logically groups resources and descriptors, it can be seen as the basic building block of an `Item`. In Example 3.7, an MP3 resource and an MPEG-7 description are grouped together with a `Component` element. In this example, the `Component` element has the functionality to group the `Resource` element with its technical description. The latter, expressed as a `Descriptor` element, describes that the bitrate of the resource is 128 kbps.

It is possible to include multiple `Resource` elements in a `Component` element. In this case, the `Resource` elements are considered to be bit equivalent. This functionality can be very useful when an author wants to provide multiple versions of the same resource. In Example 3.8, the author included several versions of a resource, which were stored on different mirrored servers. When an MPEG-21 compliant terminal receives this `Component`, it can choose the version of the resource to which it has the fastest access, the cheapest access and so on.

Additionally, a `Component` element also binds a `Resource` to an `Anchor`. Later in this chapter, the functionality of the `Anchor` element is fully explored. For now, it is

```
<Component>
  <Descriptor>
    <Statement mimeType="text/xml">
      <Mpeg7 xmlns="urn:mpeg:mpeg7:schema:2001">
        <Description xsi:type="MediaDescriptionType">
          <MediaInformation>
            <MediaProfile>
              <MediaFormat>
                <Content href="http://server/track01.mp3"/>
                <BitRate>131072</BitRate>
              </MediaFormat>
            </MediaProfile>
          </MediaInformation>
        </Description>
      </Mpeg7>
    </Statement>
  </Descriptor>
  <Resource mimeType="audio/mpeg" ref="http://server/track01.mp3"/>
</Component>
```

Example 3.7   `Components` bind `Resources` to relevant `Descriptors`

```
<Component>
  <Resource mimeType="audio/mpeg" ref="http://fast.server/track01.mp3"/>
  <Resource mimeType="audio/mpeg" ref="http://slow.server/track01.mp3"/>
</Component>
```

Example 3.8   Multiple `Resources` in one `Component`

sufficient to know that an Anchor binds a Descriptor to a part of a Resource, for example, to the first 5 s, as opposed to a Component, which binds a Descriptor to the Resource as a whole.

Item

In the next step of the construction of a music album DI, the Components, Descriptors and so on are grouped together until they form a logically indivisible asset. Such grouping can be realized by means of an Item element, which is the DIDL representation of the *item* from the Abstract Model.

According to the Abstract Model, an Item is a grouping of sub-Items, Components, and relevant Descriptors. The Descriptors typically contain descriptive information about the content of the Item itself. The Components typically contain the actual multimedia data.

In Example 3.9, there are two Descriptors. The first Descriptor is a plain text description of the music album, which contains the title of the album. The second Descriptor is an example of an image Descriptor. This image, representing the cover of the music album, can be used as information about the album when browsing a collection of music albums.

Immediately after the description of the album, there is a collection of Components that contain the actual multimedia content. In this case, two different tracks are included by reference to a multimedia content server. In reality, a typical music album would be more complex than the one that is displayed below. Such a music album would probably contain additional *metadata*, for example, additional Descriptors with short preview tracks for each Component. Other additional data could be technical *metadata* about the encoding of the different tracks, such as the bitrate.

To really use the power of the DIDL, one could use sub-Items in the music album Item. Those sub-Items could extend the music album beyond the basic functionality. In a sub-Item, there could be all sorts of information that is related to the original music album. For example, additional video clips, advertisement pointing to related music albums, related TV shows, portable music players and so on.

```
<Item>
  <Descriptor>
    <Statement mimeType="text/plain">
      All time greatest hits of John Doe
    </Statement>
  </Descriptor>
  <Descriptor>
    <Component>
      <Resource mimeType="image/jpeg" ref="cover.jpg"/>
    </Component>
  </Descriptor>
  <Component>
    <Resource mimeType="audio/mpeg" ref="http://server/track01.mp3"/>
  </Component>
  <Component>
    <Resource mimeType="audio/wav" ref="http://server/track01.wav"/>
  </Component>
</Item>
```

Example 3.9 An Item as a logically indivisible asset

Using sub-Items for this additional content has the advantage that the same MPEG-21 technology can be used for the additional content. Nowadays, when buying a CD with additional content, a specific set of additional players/hardware is required for utilizing that additional content. Using MPEG-21 throughout the whole music album, even for the additional content, has the advantage that interoperability will be improved.

DIDL

The DIDL element is the root element of an XML document, defining a DI according to the DIDL grammar. The DIDL element must include a namespace declaration that states the DIDL namespace for the root DIDL element and its contents. This is required so that applications will recognize the document as a DIDL document. The original DIDL namespace URI was 'urn:mpeg:mpeg21:2002:01-DIDL-NS'. In this namespace the ' 01' represents a serial number that changes as the MPEG-21 DIDL Schema evolves along with the rest of ISO/IEC 21000. Therefore, the second edition, on which this chapter is based, requires the URI 'urn:mpeg:mpeg 21:2002:02-DIDL-NS' as its namespace. More information about the different editions of the DID can be found in Section 3.2.5.

At this point in the discussion about the different elements in the DIDL specification, it is possible to construct a valid DIDL document. What is actually meant by 'valid' for DID will be explained in Section 3.2.2.2, for now it is sufficient to think of valid as 'according to the spirit of the DIDL specification'. To create such a document that starts from the previous example, it is required to add the XML processing directive and the DIDL root element with the necessary namespace declaration at the top of this document. This directive implies that the following document is an XML document, and it identifies the encoding of that document. As a result, Example 3.10 is a first valid MPEG-21 DI.

It is possible to add a DIDLDocumentID attribute to the DIDL element. This attribute is used to convey the identifier of the DIDL document. It should not be used to convey

```
<?xml version="1.0" encoding="UTF-8"?>
<DIDL xmlns="urn:mpeg:mpeg21:2002:02-DIDL-NS">
  <Item>
    <Descriptor>
      <Statement mimeType="text/plain">
        All time greatest hits of John Doe
      </Statement>
    </Descriptor>
    <Descriptor>
      <Component>
        <Resource mimeType="image/jpeg" ref="cover.jpg"/>
      </Component>
    </Descriptor>
    <Component>
      <Resource mimeType="audio/mpeg" ref="http://server/track01.mp3"/>
    </Component>
    <Component>
      <Resource mimeType="audio/wav" ref="http://server/track01.wav"/>
    </Component>
  </Item>
</DIDL>
```

Example 3.10   A first valid Digital Item

```
<?xml version="1.0" encoding="UTF-8"?>
<DIDL xmlns="urn:mpeg:mpeg21:2002:02-DIDL-NS"
      xmlns:app="urn:applicationdata:NS"
      app:attr="my_application_specific_attribute">
  <DIDLInfo>
    <app:data>my application specific data</app:data>
  </DIDLInfo>
  ...
</DIDL>
```

Example 3.11   The use of application-specific data in DIDL

identifiers of the DIs declared within the DIDL document only for identifying XML document itself.

Besides the `DIDLDocumentID` attribute, attributes from any namespace can be added to the DIDL element. Since the DIDL element is not a representation of any of the concepts in the DID model itself, it is simply a root element for the XML document, the attributes that appear in the DIDL element are allowed as application-specific attributes. They will not contain any information about the DI represented by the DIDL document.

DIDLInfo

Since it might not always be possible, or convenient, to convey all application-specific information about the document in attributes, the `DIDLInfo` element was created. This element has the same functionality of the attributes from any namespace that can be added to the DIDL element. Example 3.11 illustrates the use of application-specific data in DIDL, both with an application-specific attribute and the `DIDLInfo` element.

Container

The `Container` element in the DIDL is the XML representation of the *container* from the Abstract Model. The `Container` is a grouping of `Items` and possibly other `Containers` bound to a set of `Descriptors` that contain information about the `Container`. Starting from the definition above, a `Container` and an `Item` seem to be similar regarding their functionality. As explained previously, an `Item` is generally considered as a logically indivisible asset. A `Container`, however, is a grouping of such assets. Therefore, grouping different `Items` together into one `Item` with sub-`Items` is semantically different than grouping different `Items` together into a `Container`. In the former, the top-level Item is considered as a logically indivisible asset. In the latter, the different Items in the Container are each considered as a logically indivisible asset.

Consider a real-world scenario that clearly explains the difference between a `Container` and an `Item`. Nowadays, most people have a collection of audio CDs that contain a small set of audio tracks. This collection of audio CDs is typically stored somewhere on a shelf. When moving this scenario into the digital world, a large set of audio files will exist. Those audio files can be structured as music albums using the `Item` elements, as in Example 3.10. Those music album `Items` correspond to the audio CDs. To group the whole collection of music albums together, a `Container` element can be used. The `Container` element expresses that the music collection is a grouping of different music albums. The `Item` element expresses, that the music albums can be considered as a single entity within the collection. Example 3.12 illustrates how a music collection can be stored within one DIDL document by using the `Container` element.

```
<?xml version="1.0" encoding="UTF-8"?>
<DIDL xmlns="urn:mpeg:mpeg21:2002:02-DIDL-NS">
  <Container>
    <Descriptor>
      <Statement mimeType="text/plain">
        My favorite music albums
      </Statement>
    </Descriptor>
    <Item>
      <Descriptor>
        <Statement mimeType="text/plain">
          All time greatest hits of John Doe
        </Statement>
      </Descriptor>
      ...
    </Item>
    <Item>
      ...
    </Item>
    ...
  </Container>
</DIDL>
```

Example 3.12 A `Container` as a digital shelf

Condition

The next three DIDL elements are closely related to each other. The `Condition` element, the `Selection` element and the `Choice` element form a group of DIDL elements allowing the inclusion of choices in DIs. Using those choices, DIs can be configured in different ways, and behave differently according to the different configurations that have been created.

The choice mechanism is based upon *predicates*. By definition, a *predicate* is an unambiguously identifiable declaration that can be true, false or undecided. Note that this implies that undecided is not equal to true and not equal to false; it is a separate value. The value undecided can be compared with the following expression: 'I do not know (yet) whether the answer to this question will be true or false'.

Now that *predicates* have been introduced, consider the `Condition` element. The `Condition` element in the DIDL is the XML representation of the *condition* from the Abstract Model. As such, it is an element describing that the enclosing element is conditioned on a certain set of *predicates*.

The `Condition` element can have two attributes, a `require` attribute and an `except` attribute. A `Condition` with a `require` attribute requires that the *predicate* associated with that attribute is true. In Example 3.13, this means that the *predicate Predicate1* needs to be true before the `Component` becomes 'available' (see earlier definition of 'available' in 3.2.1.2) to the User. Note that if *Predicate1* has the value 'undecided' or 'false', the `Component` will not be made available to the User.

The behavior of the `except` attribute is similar to that of the `require` attribute. The enclosing element of the `Condition` element becomes available to the User if the *predicate* identified by the `except` attribute is false. Note that if that predicate has the value 'undecided' or 'true', the `Component` will not be made available to the User.

```
<Component>
  <Condition require="Predicate1"/>
  <Resource mimeType="audio/mpeg" ref="http://server/track01.mp3"/>
</Component>
```

Example 3.13   `Condition`s and *predicates*

Different *predicates* can be combined in one `require` or `except` attribute. This can be realized by putting a list of *predicates* into one `require` or `except` attribute. In that case, they are combined as a conjunction. From a programmer's point of view, this means that they are combined with an AND relationship.

It is also possible to put several separate `Condition` elements into a single enclosing element. In that case, the different `Condition` elements are combined as a disjunction. From a programmer's point of view, this means that they are combined with an OR relationship.

Consider an example that combines the different *predicate* combinations into one example. In Example 3.14, the `Component` becomes available if the following expression is satisfied.

(P1 has the value 'true' AND P2 has the value 'true' AND P3 has the value 'False') OR (P4 has the value 'true').

The following DIDL elements can be made conditionally available using the `Condition` element: `Anchor`, `Choice`, `Component`, `Descriptor`, `Item`, and `Selection`.

Selection

In the previous paragraphs, it was demonstrated how to combine several *predicates* into `Condition`s and how those `Condition`s can be combined into one large Boolean expression. Until now it was quietly assumed that the *predicates* were declared somewhere in the DID. Using the `Selection` element, which corresponds to the *selection* in the Abstract Model, it is possible to declare *predicates*. The declaration of the predicates is done in the `select_id` attribute of the `Selection`.

Example 3.15 declares the *predicate* with identifier 'Predicate1'. This example illustrates how the predicates are defined in the `select_id` attribute of the `Selection` element.

The `Selection` element is used in the `Choice` element. This allows DID authors to make their DIDs more flexible and configurable. This will be illustrated later in this chapter.

```
<Component>
  <Condition require="P1 P2" except="P3"/>
  <Condition require="P4"/>
  <Resource mimeType="audio/mpeg" ref="http://server/track01.mp3"/>
</Component>
```

Example 3.14   Combining `Condition`s to a Boolean expression

```
<Selection select_id="Predicate1"/>
```

Example 3.15   Declaring *predicates* in `Selection` elements

```xml
<?xml version="1.0" encoding="UTF-8"?>
<DIDL xmlns="urn:mpeg:mpeg21:2002:02-DIDL-NS">
  <Item>
    <Choice>
      <Descriptor>
        <Statement mimeType="text/plain">
          What format do you want?
        </Statement>
      </Descriptor>
      <Selection select_id="MP3_FORMAT">
        <Descriptor>
          <Statement mimeType="text/plain">MP3 Format</Statement>
        </Descriptor>
      </Selection>
      <Selection select_id="WMA_FORMAT">
        <Descriptor>
          <Statement mimeType="text/plain">WMA Format</Statement>
        </Descriptor>
      </Selection>
    </Choice>
    <Component>
      <Condition require="MP3_FORMAT"/>
      <Resource mimeType="audio/mpeg" ref="http://server/track01.mp3"/>
    </Component>
    <Component>
      <Condition require="WMA_FORMAT"/>
      <Resource mimeType="audio/wma" ref="http://server/track01.wma"/>
    </Component>
  </Item>

</DIDL>
```

Example 3.16   Choices, a complete example

Choice

The final DIDL element that is needed to create an interoperable choice mechanism is the Choice element. The Choice element in the DIDL is the XML representation of the *choice* from the Abstract Model. This means that the Choice element is the grouping of a set of Selection elements into one choice.

The result of having a choice mechanism is that a DIDL document is no longer a completely static XML document. It becomes an XML document together with a collection of states for the predicates. This allows run-time configuration of the DID. This is illustrated in Example 3.16. In this example, a Choice containing two Selections is included. Within the Choice there is a plain text Descriptor containing the human-readable question 'What format do you want'? This human-readable information can be used at run-time when the Choice is presented to a human User. Using Descriptors has the advantage that the User gets additional information about the choice he/she has to make.

The same applies to the Selections. Within the example, the Selections are plain text Descriptors that contain human-readable information about those Selections. The Selections themselves have a select_id with the *predicates*. In the example, there are two *predicates*, the MP3_FORMAT *predicate* and the WMA_FORMAT *predicate*. Before the Choice has been configured, both *predicates* have the value undecided.

The third part in the choice mechanism is the Condition element. In the example, there are two Condition elements. Each of them makes a certain Component conditionally available on the basis of the earlier discussed Choice.

This example is concluded with a scenario in which this music album DI can be used. Suppose that an MPEG-21 User wants to listen to some music. To do this, he/she walks to his MPEG-21 compliant audio set and loads his/her music collection. The hi-fi device displays the question: 'Which album do you want to hear'? This question is included in the music collection DI, which is in fact a Container of different music album Items. After answering this question, the chosen music album Item becomes available. Because there are some other relevant questions to ask with respect to this music album, those questions are better resolved before the actual playing is done. For example, the chosen music album item contains a Choice with two possible selections, namely the WMA and MP3 versions. Because this is a fairly technical matter, the audio set does not bother the User with the selection of the most appropriate format; it configures the music album Item by itself. The decision that the audio set makes can be on the basis of availability of codecs, or on user preferences that the audio set collected during its installation, configuration or usage. When all relevant questions are resolved, the audio set builds a list of available tracks and starts to play them. Such a list can be built using the Condition elements included with the Components that contain the music tracks.

From the previous example, it becomes clear that the Choice, Selection, and Condition elements allow flexible run-time configuration of DIs. To extend the possibilities of what was discussed so far, it is possible to include Condition elements in both the Choice and the Selection elements. This allows the creation of complex decision trees in which certain Selections may make certain subsequent Choices or Selections redundant. To come back to the music album example, a music album DI could contain both an audio-only and a video version of a certain music clip. If the User chooses to play the audio-only version of the music track, all questions related to the encoding of the video become irrelevant, and therefore, they no longer need to be resolved. Such a scenario can be implemented by conditioning the Selections about the encoding of the video on the Selections that select the video clips. In Example 3.17, the User can configure the second Choice based upon the Selection made for the first Choice. If the User has set the VIDEO *predicate* to true, he has the possibility to choose between MPEG-4 format and Real Video format. Alternatively, if he/she had set the AUDIO *predicate* to true, he/she would have had the possibility to choose between MPEG-1 Audio layer 3 format and Windows Media Audio format.

To conclude this section about the Choice element, the set of attributes that allow putting various constraints on the Selections that can be made within a Choice are discussed. The minSelections and maxSelections attributes allow the author of the DID to constrain the number of Selections that have to be chosen in a DID. In this context, a Selection element is chosen if the value of the *predicate* associated with the Selection element is set to true. The minSelections attribute defines the minimum number of Selections that must be chosen. If this attribute is not present, there is no minimum number. The DID specification leaves it up to the author of the DI to decide whether this is meaningful or not. The maxSelections attribute defines the maximum number of Selections that can be chosen. If this attribute is not present, there is no maximum number.

```
<?xml version="1.0" encoding="UTF-8"?>
<DIDL xmlns="urn:mpeg:mpeg21:2002:02-DIDL-NS">
  <Item>
    <Choice>
      <Selection select_id="AUDIO"/>
      <Selection select_id="VIDEO"/>
    </Choice>
    <Choice>
      <Selection select_id="MP4_FORMAT">
        <Condition require="VIDEO"/>
      </Selection>
      <Selection select_id="REAL_FORMAT">
        <Condition require="VIDEO"/>
      </Selection>
      <Selection select_id="MP3_FORMAT">
        <Condition require="AUDIO"/>
      </Selection>
      <Selection select_id="WMA_FORMAT">
        <Condition require="AUDIO"/>
      </Selection>
    </Choice>
  </Item>
</DIDL>
```

Example 3.17   Conditioning `Choices`, and `Selections`

Finally, there is also a `default` attribute. This attribute allows the author to preconfigure the DID at the time of its design. The `default` attribute contains a list of *predicates* that are chosen by default. The *predicates* contained in the `default` attribute are set to the value true before any run-time configuration is done. During run-time configuration, the values of those predicates can still be changed to some other value. Therefore, the default configuration serves the purpose of creating a starting point for configuration.

In Example 3.18, there is a `Choice` element with a `minSelections` attribute with value 1 and a `maxSelections` attribute with value 2. To resolve the following `Choice` correctly, at least one `Selection` must be chosen and at most two `Selections` may be chosen.

When the configuration of the `Choice` starts at run-time, the `Choice` has the state that *predicate* AUDIO has the value 'true' and *predicate* VIDEO has the value 'undecided'. This state corresponds to state 1 in Table 3.1. At that point, the User can change all the values of the *predicates* to any of the combinations that are displayed in this

```
<?xml version="1.0" encoding="UTF-8"?>
<DIDL xmlns="urn:mpeg:mpeg21:2002:02-DIDL-NS">
  <Item>
    <Choice minSelections="1" maxSelections="2" default="AUDIO">
      <Selection select_id="AUDIO"/>
      <Selection select_id="VIDEO"/>
    </Choice>
  </Item>
</DIDL>
```

Example 3.18   The minSelections and maxSelections attributes

Table 3.1 Possible *predicate* configurations for the example `Choice`

	Audio	Video
1	True	Undecided
2	True	True
3	True	False
4	False	Undecided
5	False	True
6	False	False
7	Undecided	Undecided
8	Undecided	True
9	Undecided	False

table. Some of those combinations are valid, which means that at least one predicate has value 'true' and at most two predicates have value 'true'. The other combinations will be invalid. In Table 3.1, the valid states are state 1, state 2, state 3, state 5 and state 8; the other states are invalid. Note that it was in fact not necessary to add the `maxSelections` attribute with value 2 to this example because there were only two `Selections` in the example. Therefore, it would not have been possible to choose more than two `Selections` anyway.

Annotation
The next DIDL element in the discussion of the DIDL is the `Annotation` element. The `Annotation` element in the DIDL is the XML representation of the *annotation* from the Abstract Model. The purpose of this element is to attach additional information to other elements without affecting the original content of the DI. For example, this allows a User to associate bookmarks and commentary to a digitally signed work, without invalidating the signature of that work.

To support the unique identification of DIDL elements, the DIDL has a mechanism allowing identifying elements within DIDs using the `id` attribute. This `id` attribute, which has the XML data type ID [7], gives the author the possibility to add an identifier to a certain element. This identifier can be used by other elements, for example, by the `Annotation` element. The following elements have an `id` attribute: `Container`, `Item`, `Descriptor`, `Component`, `Anchor`, and `Annotation`.

Other than the `id` attribute, there are two other attributes that have similar functionality within DIDL. The `choice_id` attribute allows the unique identification of the `Choice` elements within a DID, and the `select_id` attribute creates a unique coupling between the `Selection` element and the *predicate* that the `select_id` represents. Both the attributes are of the XML ID data type.

The `Annotation` element can use those attributes when annotating DIDL elements. The `target` attribute of the `Annotation` element contains a URI identifying the elements that are targeted by that `Annotation`. Because the data type of the `target` attribute is a URI, the `target` attribute supports a wide variety of ways of targeting elements. In Example 3.19, the `Annotation` element contains a relative URL (Uniform Resource Locator), which is in fact a special type of URI, which targets the `Item`

```
<?xml version="1.0" encoding="UTF-8"?>
<DIDL xmlns="urn:mpeg:mpeg21:2002:02-DIDL-NS">
  <Item id="my_music_album">
    <Component>
      <Resource mimeType="audio/mpeg" ref="http://server/track01.mp3"/>
    </Component>
    <Component>
      <Resource mimeType="audio/mpeg" ref="http://server/track02.mp3"/>
    </Component>
    <Annotation target="#my_music_album">
      <Descriptor>
        <Statement mimeType="text/plain">
          This music album is awesome!
        </Statement>
      </Descriptor>
    </Annotation>
  </Item>
</DIDL>
```

Example 3.19   Annotations allow adding custom information

with the ID 'my_music_album'. In this example, the value of the `target` attribute, '#my_music_album', is constructed as the '#' character + the `id` attribute of the element that it is targeting. Other possible ways of targeting DIDL elements are the use of XPointer expressions [19], or any pointer schema that is based on the URI data type.

An `Annotation` element can contain various other elements, as depicted in the example, one of those being the `Descriptor` element. Within an `Annotation`, the `Descriptor` element can be used to add additional information to a certain DIDL element. For example, a `Descriptor` can be used to add bookmarks to web page expressed as a DI, or even to add additional comments to a family photo album. In Example 3.19, the `Annotation` is used to add a text note to the music album.

There are two other elements that can be contained within an `Annotation` element, the `Assertion` element, and the `Anchor` element. These elements are discussed in the following sections.

Assertion

The `Assertion` element in the DIDL is the XML representation of the *assertion* from the Abstract Model. The `Assertion` element allows the configuration of a set of *predicates* from a `Choice`. It can be used to set those *predicates* to true or false. This DIDL element, which can only be used from within an `Annotation` element, allows capturing the state of a Choice and its associated *predicates* without modifying the original DID. This is possible because `Annotations` do not modify the original DID. The `Assertion` element can have several attributes: the `target` attribute, the `true` attribute and the `false` attribute. The `target` attribute identifies the `Choice` that is asserted by the `Assertion`. It is a URI that uniquely identifies the targeted `Choice`. The `true` attribute contains a list of *predicates* of the targeted `Choice` that are asserted to true by the `Assertion` element. The `false` attribute contains a list of *predicates* of the targeted `Choice` that are asserted to false by the `Assertion` element.

`Assertions` may either fully or partially resolve the given `Choice`. The *predicates* represented by the missing `select_id` values are left undecided. It is possible to continue resolving a partially resolved Choice by simply assigning true or false values to some or all of the undecided *predicates* to arrive at a new Assertion. It is important that, when continuing resolving a partially resolved Choice, only those predicates that are undecided are further resolved. Once a predicate has been asserted, further assertions of that predicate should be ignored. The order in which Assertions are processed in DIDL documents is in document order, as defined in [20], starting at the root DIDL element.

Another important aspect to Assertions is that the Assertions always need to satisfy the requirements imposed by the minSelections and the maxSelections attributes. This requirement needs to be satisfied throughout the whole resolving chain, even when an Assertion is only partially resolving a Choice. This will be further discussed in Section 3.2.3.

In Example 3.20, a DID with an Item with id 'my_item' is presented. This Item contains a Choice, with choice_id 'my_choice' that is configured using Annotation and Assertion elements. Before any Assertions have been processed, the predicates have the following state: P1 is true because of the default attribute of the Choice; P2, P3, and P4 are undecided.

The first Annotation targets the Item with id 'my_item'. Within the Annotation there is an Assertion element that targets a child of that Item, the Choice element with choice_id 'my_choice'. This Assertion sets the value of the predicate P2 to false. At this point, predicates P3 and P4 are undecided. The second Annotation targets the same Item and also contains an Assertion. That Assertion further resolves the Choice and sets the predicate P3 to false and P4 to true. At this point, the Choice is fully resolved. It is up to the application that uses the DI to decide if it is necessary to again ask the questions previously asked whenever the Choice is used by the application. For example, some DIDL applications will configure a set of Choices once and then use the configured values every time when accessing the Choice (e.g., for configuring user preferences). Other DIDL applications will ask to resolve the Choice

```
<?xml version="1.0" encoding="UTF-8"?>
<DIDL xmlns="urn:mpeg:mpeg21:2002:02-DIDL-NS">
  <Item id="my_item">
    <Choice choice_id="my_choice"
            minSelections="1" maxSelections="2" default="P1">
      <Selection select_id="P1"/>
      <Selection select_id="P2"/>
      <Selection select_id="P3"/>
      <Selection select_id="P4"/>
    </Choice>
    <Annotation target="#my_item">
      <Assertion target="#my_choice" false="P2"/>
    </Annotation>
    <Annotation target="#my_item">
      <Assertion target="#my_choice" false="P3" true="P4"/>
    </Annotation>
  </Item>
</DIDL>
```

Example 3.20 Using Assertions for storing the configuration state of Choices whenever it is accessed (e.g., choosing a song for playback from a music album). And other programs might do a combination of both.

Fragment

The Fragment element in the DIDL is the XML representation of the *fragment* from the Abstract Model. Therefore, the Fragment element can be used to identify a specific location or a range in a *resource*. To realize this, the Fragment element can contain any type of *metadata*.

The *metadata* for identifying the location or range can be expressed in several ways. One possibility is the use of children of the Fragment element. In Example 3.21, MPEG-7 is used to identify a specific location on an image.

Alternatively, the fragmentId attribute of the Fragment element can be used. This attribute contains a string specifying a fragment identifier as defined in RFC 3986 [15]. Example 3.22 demonstrates how this fragmentId attribute can be used to identify a point in time, being 17 min and 30 s past the start, in a *resource* based on MPEG-21 part 17 Fragment Identification of MPEG Resources [21].

Finally, it is also possible that the Fragment element contains both the fragmentId attribute and children. In that case, the *metadata* in the children is used as a sublocation or subpart of the location identified by the URI.

Anchor

The Anchor element in the DIDL is the XML representation of the *anchor* from the Abstract Model. It binds Descriptors to a specific location or range in a Resource. This location or range is defined using the Fragment element as a child of the Anchor element.

In Example 3.23, there is a Component containing an Anchor. This Anchor adds additional information to a specific point in the Resource contained within that Component. It adds the plain text description 'John's longest trumpet solo' to the time point '1 min and 10 s.' The location in the actual media resource can be identified by the following URI fragment 'mp(/~time('npt','70')'.

```
<Fragment fragmentId="ConferenceRoom">
  <mpeg7:Mpeg7>
    <mpeg7:Description xsi:type="mpeg7:ContentEntityType">
      <mpeg7:MultimediaContent xsi:type="mpeg7:ImageType">
        <mpeg7:Image>
          <mpeg7:SpatialLocator>
            <mpeg7:Box mpeg7:dim="1 4"> 12 16 30 40 </mpeg7:Box>
          </mpeg7:SpatialLocator>
        </mpeg7:Image>
      </mpeg7:MultimediaContent>
    </mpeg7:Description>
  </mpeg7:Mpeg7>
</Fragment>
```

Example 3.21   Using an MPEG-7 Fragment to identify a specific location in an image.

```
<Fragment fragmentId="mp(/~time('npt','1050')"/>
```

Example 3.22   Using URI-based Fragment Identification to identify a point in time.

```xml
<?xml version="1.0" encoding="UTF-8"?>
<DIDL xmlns="urn:mpeg:mpeg21:2002:02-DIDL-NS">
  <Item>
    <Component>
      <Resource mimeType="audio/mpeg" ref="http://server/track01.mp3"/>
      <Anchor>
        <Descriptor>
          <Statement mimeType="text/plain">
            John's longest trumpet solo
          </Statement>
        </Descriptor>
        <Fragment fragmentId="mp(/~time('npt','70')"/>
      </Anchor>
    </Component>
  </Item>
</DIDL>
```

Example 3.23   Anchors allow adding custom information to specific parts of `Resources`

```xml
<?xml version="1.0" encoding="UTF-8"?>
<DIDL xmlns="urn:mpeg:mpeg21:2002:02-DIDL-NS">
  <Item>
    <Component id="my_component">
      <Resource mimeType="audio/mpeg" ref="http://server/track01.mp3"/>
    </Component>
    <Annotation target="#my_component">
      <Anchor>
        <Descriptor>
          <Statement mimeType="text/plain">
            John's longest trumpet solo
          </Statement>
        </Descriptor>
        <Fragment fragmentId="mp(/~time('npt',''70')"/>
      </Anchor>
    </Annotation>
  </Item>
</DIDL>
```

Example 3.24   Anchors can be used within `Annotations`

At first, the `Anchor` and the `Annotation` element look similar regarding their functionality. Both DIDL elements add additional information to DIDs. However, there are some clear differences between both elements. The main difference between an `Anchor` and an `Annotation` is the type of data to which they are adding information. An `Annotation` adds information to a DIDL element; an `Anchor` adds information to an actual multimedia resource, or to a part of it.

Another difference is the fact that `Annotations` are ignored when calculating a digital signature for a DID, opposed to what happens for `Anchors`.

Those differences become clearer when using `Anchors` within `Annotations`. In Example 3.24, an `Anchor` is contained within an `Annotation`. In this example, it is clear that their functionality is situated at a different level. The `Annotation` targets the `Component` element with the `id` 'my_component'. On the other hand, the `Anchor`'s `fragment` targets a certain position within a media stream – the time point '1 min and 10 s'. In most cases, from an end-user's point of view, the result of both examples will be the same. However, when calculating the digital signature of the first example, the `Anchor` element is taken into account, as opposed to the second example, where the `Anchor` is encapsulated in an `Annotation` element and therefore left out of the signature.

Finally, an `Anchor` element can have a `precedence` attribute. This `precedence` attribute is a nonnegative integer value indicating the relative ranking of the `Anchor` among other `Anchors` in an `Item`. The `Anchor` with the highest precedence value is the default `Anchor` for the `Item`. If the precedence attribute is not present, the precedence of the `Anchor` is zero. This ranking could be used to create a hierarchy of `Anchors`. For example, consider a set of `Anchors` containing `Descriptors` with precedence ranging from 0–100. If the DID author assigns precedence 0 to the most detailed `Descriptor` and precedence 100 to the least detailed `Descriptor`, it is possible for an application to show more or less detailed `Anchor` data based on a threshold being the `precedence` value.

Declarations

The final element in the discussion about the DIDL elements is the `Declarations` element. The `Declarations` element is used to define a set of DIDL elements, without instantiating them, for later use in that document. Instantiating the declared elements can be realized using XML Inclusions (XInclude) [22], which will be discussed in Section 3.2.2.2.

The `Declarations` element can contain `Items`, `Descriptors`, `Components`, `Annotations` and `Anchors`. It is important to note that the `Declarations` element does not instantiate the elements it declares; the actual instantiation is done at the time the elements are referenced using XInclude. The differentiation between instantiating the elements and not instantiating them is important when validating the DIDs. Elements that are not instantiated do not need to be checked when validating DIDs. Validation of DIDs will be discussed in detail in Section 3.2.3.

In Example 3.25, the `Declarations` element contains a copyright statement for the `Items` contained in the DID. That copyright statement is instantiated two times within the `Items` using XInclude technology. Note that 'copyright/1' is an XPointer expression pointing to the first child of the element with id 'copyright'.

At this point in this chapter it is possible to use the full potential of the DIDL. The concept of the DI was introduced and a step-by-step approach was used when introducing the different building blocks that conceive the DIDL. It is now possible to declare DIs for various application spaces using those building blocks. In the following sections about the DIDL, it is demonstrated how modularity can be realized and how DIs need to be validated according to MPEG-21 DID. As a final conclusion about DIDs, a real-world example of a DID is given.

3.2.2.2 Document Modularity using XML Inclusions

The XML Inclusions technology can be used in the DIDL to realize modularity of DIDL documents. XInclude can be used to include elements within the same document, called an *internal inclusion*, or to include elements from another document, called an *external inclusion*. Before actually consuming the DID, the XInclude elements in the DID need to be processed using an XInclude processor.

```
<?xml version="1.0" encoding="UTF-8"?>
<DIDL xmlns="urn:mpeg:mpeg21:2002:02-DIDL-NS"
      xmlns:xi="http://www.w3.org/2001/XInclude">
  <Declarations>
    <Descriptor id="copyright">
      <Statement mimeType="text/plain">
        Copyright MyCompany.  All rights reserved.
      </Statement>
    </Descriptor>
  </Declarations>
  <Container>
    <Item id="pictures_2004">
      <Descriptor>
        <xi:include xpointer="copyright/1"/>
      </Descriptor>
      <Component>...</Component>
    </Item>
    <Item id="pictures_2005">
      <Descriptor>
        <xi:include xpointer="copyright/1"/>
      </Descriptor>
      <Component>...</Component>
    </Item>
  </Container>
</DIDL>
```

Example 3.25   Using a `Declaration` element to allow reuse of DIDL elements

Example 3.25 shows the use of internal inclusions. Internal inclusions can be used to allow the author to maintain a single source of an element that occurs in more than one place within a single DIDL document. In the example, it is used to maintain a single source of the copyright statement. If the copyright statement changes (for example, there are new company policies regarding copyright), these changes become automatically available in the two instantiations, being the `Descriptors` for the `Items` containing the pictures for 2004 and 2005.

XInclude can also be used for external inclusions. In that case, the target element is not a DIDL element from within the same DID. Instead, it is a DIDL element from another DID. External inclusions allow lengthy or complex DIDL documents to be split up into multiple separate documents. This can be used, for example, for load balancing to make more efficient use of computing/network resources. Another case of usage in which external inclusions are very useful is the use of copyright statements. Suppose that a company wants to add copyright statements to every DID they produce. One possible way to realize this would be to add that statement inline in every DID. However, if this copyright statement needs to change, this requires a change of every DID that contains that statement. By using external references, it becomes possible to store that copyright statement in a separate DID and refer to that DID from within every DID that requires the copyright statement. When the copyright statement changes, it only needs to be changed in one DID; all referencing DIDs will automatically refer to the updated copyright statement.

In Example 3.26, it is demonstrated how external inclusions can be used to create DIDs that exist out of multiple files. In this case, there are three files, album.xml, pictures_2004.xml and pictures_2005.xml. The files pictures_2004.xml and pictures_2005.xml contain a collection of pictures taken respectively in the years 2004 and 2005. The file album.xml contains an overview of the pictures of the last two years, in this case of the years 2004 and 2005. Note that in this example, in the year 2005, the descriptions 'Last year's pictures' and 'This year's pictures' would not make any sense in the files pictures_2004.xml and pictures_2005.xml, because next year, in 2006, pictures_2006.xml will be created and the references in the album.xml file will be updated accordingly.

album.xml

```
<?xml version="1.0" encoding="UTF-8"?>
<DIDL xmlns="urn:mpeg:mpeg21:2002:02-DIDL-NS"
      xmlns:xi="http://www.w3.org/2001/XInclude">
  <Container>
    <Item>
      <Descriptor>
        <Statement mimeType="text/plain">
          Last year's pictures
        </Statement>
      </Descriptor>
      <xi:include ref="pictures_2004.xml" xpointer="the_pictures/1"/>
    </Item>
    <Item>
      <Descriptor>
        <Statement mimeType="text/plain">
          This year's pictures
        </Statement>
      </Descriptor>
      <xi:include ref="pictures_2005.xml" xpointer="the_pictures/1"/>
    </Item>
  </Container>
</DIDL>
```

pictures_2004.xml

```
<?xml version="1.0" encoding="UTF-8"?>
<DIDL xmlns="urn:mpeg:mpeg21:2002:02-DIDL-NS">
  <Item id="the_pictures">
    <Component>...</Component>
  </Item>
</DIDL>
```

pictures_2005.xml

```
<?xml version="1.0" encoding="UTF-8"?>
<DIDL xmlns="urn:mpeg:mpeg21:2002:02-DIDL-NS">
  <Item id="the_pictures">
    <Component>...</Component>
  </Item>
</DIDL>
```

Example 3.26   Including from different XML documents

3.2.3 VALIDATION OF MPEG-21 DIDs

To make sure that a framework such as MPEG-21 will be interoperable, MPEG decided to use the DID as the base for the whole framework. Every transaction, every message, every form of communication is performed using a DI. In MPEG-21, such DIs are always declared in the DIDL.

This general and flexible language, described in the previous sections, allows the authors of DIs to embed any type of content within DIs. To make sure that interoperability is preserved, it is necessary that a set of strict rules with respect to DIDs are met. In other words, to ensure interoperability, DIDs used within the MPEG-21 framework need to be valid. In the following sections, validation of MPEG-21 DIDs is discussed in detail.

3.2.3.1 XML Validation

When validating a DID, the XML document of the DID needs to be validated first. XML is a text-based language that allows the creation of new grammars for a specific application domain. For example, the DIDL is a grammar for declaring DIs. To check whether a DID contains valid DID XML code, there are two steps that need to be taken.

During the first step, it is checked whether the DID is well formed XML code. This checking can be done by standard XML parsers such as Xerces2 [23] and MSXML [24]. Well formed XML code is basically XML code that is compliant with the XML syntax. For example, well formed XML code does not nest XML tags, escapes certain characters and so on. For a complete overview of the XML syntax, see the W3C XML specification [6].

In the second step, a check against the grammar of the DIDL will be performed. This grammar is expressed in an XML Schema. Generally speaking, XML Schemas express shared vocabularies; in addition, they provide a means for defining the structure, content and semantics of XML documents. ISO/IEC 21000–2, the DIDL specification, contains an XML Schema for the DIDL [1].

To check whether a DID is conformant to the MPEG-21 DIDL XML Schema, validating XML parsers can be used. Both the Xerces2 and the MSXML parser are examples of validating XML parsers. If a DID is conformant to that MPEG-21 DIDL Schema, the DID XML code can be considered as valid XML.

3.2.3.2 Validation Rules

Not every possible (validation) rule can be (easily) expressed within XML Schema, so the DIDL specification contains sets of validation rules for certain DIDL elements. Those rules cannot be checked with standard XML parsers. Therefore, additional software is required for checking those rules. In the following sections, all DIDL elements that require validation against such rules are discussed. Some examples of possible violations of those rules are given as well.

Resource

The Resource element has three validation rules:

- If the ref attribute is specified, then the Resource shall not be defined by value, and vice versa.
- If the Resource is not included by value, the encoding attribute shall be omitted.
- If the encoding attribute is present, the value shall be set to 'base64'.

```
<?xml version="1.0" encoding="UTF-8"?>
<DIDL xmlns="urn:mpeg:mpeg21:2002:02-DIDL-NS">
  <Item>
    <Component>
      <Resource mimeType="text/plain" ref="http://server/mytext.txt">
        My favorite text ...
      </Resource>
    </Component>
  </Item>
</DIDL>
```

Example 3.27   Violation of the first validation rule of the `Resource` element

The first validation rule makes sure that the `Resource` does not contain its data both by reference and by value at the same time. As a result, although the following example is considered as valid XML, it is not a valid DID.

The second validation rule makes sure that the `encoding` attribute is only used for `Resources` declared by value.

The last validation rule makes sure that whenever the data is included by value and the `encoding` attribute is present, the `Resource` is base64 encoded. Base64 encoded data has the advantage that it does not contain any illegal characters that could possibly confuse XML parsers. For example, when a piece of binary data would contain the '<' character, an XML parser would not know whether that character is the first character of a starting tag or just data. Base64 encoding makes sure that confusing situations cannot occur by translating the binary format into the plain text base64 format.

Statement
The `Statement` element has the following validation rules:

- If the `ref` attribute is specified, then the `Statement` shall not be defined by value, and vice versa.
- If the `Statement` is not included by value, the `encoding` attribute shall be omitted.
- If the `encoding` attribute is present, the value shall be set to 'base64'.

The validation rules for the `Resource` and the `Statement` are equal and have the same functionality. For an explanation of the validation rules, see `Resource`.

Item
The `Item` element has the following validation rule:

- An `Item` element shall not be conditional on any of its descendant `Selection` elements. In other words, an `Item` shall not contain a `Condition` element specifying a `select_id` value that identifies any descendant `Selection` element within the `Item`.

The reason for having this validation rule can be easily explained by looking at an example. In Example 3.28, which violates this validation rule, there is an `Item` that is conditioned on one of its descendant `Selection` elements. Suppose that at configuration time the value of the *predicate* P1 is still undecided. Now, suppose that a DIDL processor tries to configure the `Choices` in the DID. To do so, it could search through the DID to

```
<?xml version="1.0" encoding="UTF-8"?>
<DIDL xmlns="urn:mpeg:mpeg21:2002:02-DIDL-NS">
  <Item>
    <Condition require="P1"/>
    <Choice>
      <Selection select_id="P1"/>
    </Choice>
  </Item>
</DIDL>
```

Example 3.28   Violation of the first validation rule of the Item element

look for the Choices that are currently available. Since the Item requires the predicate P1 to be true, which is not the case, the DIDL processor will not be allowed to search through that Item (because the Condition is not satisfied). Therefore, the Choice will not be available and it will not be possible to configure that Choice. The result being that the Item will never become available.

Condition
The Condition element has two validation rules:

- Each id value specified in the require and except attributes shall match a select_id attribute value defined in a Selection element located somewhere within an Item element that is an ancestor of the Condition.

- Empty Conditions are not permitted. Therefore, it is not valid for a Condition element to have neither a require attribute nor an except attribute.

The first validation rule of the Condition element requires the Choice mechanism of the DID specification to be a decision tree. In Example 3.29, which violates that rule, the Choice mechanism is no longer a decision tree. There is a circular reference of predicates between both Choices, which makes both the Choices useless. This results

```
<?xml version="1.0" encoding="UTF-8"?>
<DIDL xmlns="urn:mpeg:mpeg21:2002:02-DIDL-NS">
  <Container>
    <Item>
      <Condition require="P1"/>
      <Choice>
        <Selection select_id="P2"/>
      </Choice>
    </Item>
    <Item>
      <Condition require="P2"/>
      <Choice>
        <Selection select_id="P1"/>
      </Choice>
    </Item>
  </Container>
</DIDL>
```

Example 3.29   Violation of the first validation rule of the Condition element

in the fact that both Items will never become available because none of the Choices can be resolved.

The second validation rule makes sure that it is not possible to create Condition elements that do not contain any requirements. Such Condition elements would be meaningless.

Choice

The Choice element has three validation rules:

- The value of the maxSelections attribute shall be no less than the value of the minSelections attribute.
- The value of the minSelections attribute shall be no larger than the number of Selection children.
- The values specified in the default attribute shall each match the select_id value of one of the Selections within this Choice. The number of individual values in the default attribute shall not be less than the value of the minSelections attribute, nor more than the value of the maxSelections attribute.

The first validation rule makes sure that the minSelections and the maxSelections attributes of the Choice element do not require an invalid state. For example, it is not possible to have at least three *predicates* (minSelections = '3') with the value true and at the same time have at most two *predicates* (maxSelections = '2') that have the value true. It would be impossible to have a valid state for such a Choice.

The reason for having the second validation rule is similar to the reason for having the first one. Both validation rules try to avoid the creation of Choices that will never have a valid configuration state. The second validation rule makes sure that it is always possible to set at least the minimal number of *predicates* to the value true. This is realized by requiring that the number of predicates that can be set to true is at least equal to the number of predicates that are required to have the value true.

Because the default attribute contains a list of *predicates* that are set to true 'by default', the third validation rule makes sure that this list does not violate the number of *predicates* that are allowed and/or required to be true. The number of *predicates* that are allowed to be set to true is identified by the maxSelections attribute. Therefore, the number of *predicates* that appears in the list of the default attribute may not be larger than the value of the maxSelections element. The number of *predicates* that must be set to true is identified by the minSelections attribute. Therefore, the number of *predicates* that appears in the list of the default attribute may not be lower than the value of the minSelections attribute. To make sure that the list of *predicates* in the default attribute are *predicates* that are defined within that Choice, the third validation rule also requires that they are defined within the Selection elements contained within that Choice.

Example 3.30 violates this third validation rule because, in the default attribute of the first Choice, there is a *predicate* that is not defined in any of the select_ids of the Selection elements that are a child of this Choice. However, this example will

```
<?xml version="1.0" encoding="UTF-8"?>
<DIDL xmlns="urn:mpeg:mpeg21:2002:02-DIDL-NS">
  <Item>
    <Choice default="P2">
      <Selection select_id="P1"/>
    </Choice>
    <Choice>
      <Selection select_id="P2"/>
    </Choice>
  </Item>
</DIDL>
```

Example 3.30   Violation of the third validation rule of the Choice element

validate against the MPEG-21 DIDL Schema. However, because of the violation of this validation rule, this DID is not a valid DID.

Annotation
The Annotation element has the following validation rules:

- For internal annotation, the values given in the target attribute shall each correspond to some descendant element of the parent element, or that of the parent element itself.
- The contents of an Annotation shall conform to the content model of the targeted element(s). For example, Anchors can be included only if the target attribute values each match the id value of a Component.
- If an Annotation contains an Assertion, then its target attribute values shall each match the id attribute value of an Item.

The first validation rule of the Annotation element makes sure that the Annotation element and its target element have at least a common ancestor, namely, the parent of the Annotation element. This validation rule makes sure that Annotations are grouped together with the elements that are annotated. This validation rule only applies to internal Annotations, allowing the creation of Annotations targeting external documents. If they had required to have a common ancestor, it would have been impossible to put them in different DIDL documents.

This functionality can be useful in the following scenario. Suppose one would like to create a set of personal Annotations (points of interest in a music track and so on.), but one does not have the permission to modify the XML code of the original DID. Without limiting the validation rule to internal annotations, it is not possible to annotate this DID. However, limiting the DID specification allows creating a second DID that contains all of the Annotations to the first DID. This functionality is also useful whenever one wants to store the state of the Choices in a separate DID. With DID, this can be realized by creating a DID with a set of Annotations and Assertions.

The second validation rule makes sure that one does not try to target elements that are not allowed to contain the element that is stored in the Annotation element. In Example 3.31, the Annotation element targets the Choice element. Because a Choice element does not allow the Anchor element as a child element, this is a violation of the second validation rule and therefore this DID is not valid.

```
<?xml version="1.0" encoding="UTF-8"?>
<DIDL xmlns="urn:mpeg:mpeg21:2002:02-DIDL-NS">
  <Item>
    <Choice choice_id="choice_01">
      <Selection select_id="P1"/>
    </Choice>
    <Annotation target="#choice_01">
      <Anchor>
        <Descriptor>
          <Statement mimeType="text/plain">
            John's longest trumpet solo
          </Statement>
        </Descriptor>
        <Fragment fragmentId="mp(/~time('npt','70')"/>
      </Anchor>
    </Annotation>
  </Item>
</DIDL>
```

Example 3.31   Violation of the second validation rule of the `Annotation` element

The last validation rule of the `Annotation` element, combined with the first validation rule of the `Assertion` element, requires `Annotations` that contain `Assertion` elements to target the `Item` that is an ancestor of the `Choice` element that is targeted by the `Assertion` element.

Assertion
The `Assertion` element has the following validation rules:

- The associated `Choice` element shall be a descendant of the `Item` whose id attribute value matches the parent `Annotation`'s `target` attribute value.
- The number of true predicates (i.e., the number of `select_id` values listed in the `true` attribute) shall be less than or equal to the `maxSelections` attribute value in the associated `Choice`.
- The total number of `select_id` values defined in the associated `Choice` minus the number of `select_id` values listed in the `false` attribute, shall be greater than or equal to the `minSelections` attribute value in the associated `Choice`.

The first validation rule of the `Annotation` element makes sure that the `Choice` that is asserted by the `Assertion` element is a descendant of the `Item` that is targeted by the `Annotation` that contains the `Assertion` element.

The second validation rule of the `Assertion` element makes sure that the number of *predicates* that are asserted to the value 'true' is not greater than the value of the `maxSelections` attribute. This is realized by requiring that the number of *predicates* listed in the `true` attribute is not greater than the value of the `maxSelections` attribute. In Example 3.32, this validation rule is violated. The `Assertion` element that targets the `Choice` with `choice_id` 'choice_01' asserts both *predicate* P1 and P2 to 'true'. This is not allowed because the `maxSelections` attribute has the value '1'.

The last validation rule of the `Assertion` element makes sure that after asserting the `Choice` with the `Assertion` element, it is still possible to further configure the

```
<?xml version="1.0" encoding="UTF-8"?>
<DIDL xmlns="urn:mpeg:mpeg21:2002:02-DIDL-NS">
  <Item id="item_01">
    <Choice choice_id="choice_01" maxSelections="1">
      <Selection select_id="P1"/>
      <Selection select_id="P2"/>
    </Choice>
    <Annotation target="#item_01">
      <Assertion target="#choice_01" true="P1 P2"/>
    </Annotation>
  </Item>
</DIDL>
```

Example 3.32 Violation of the second validation rule of the `Assertion` element

`Choice` to a state that can be considered as a valid state according to the value of the `minSelections` attribute. In Example 3.33, the `Assertion` element violates this rule by asserting two *predicates* to false. After the `Assertion`, the `Choice` has the state in which *predicate* P1 and P2 have the value 'false' and *predicate* P3 has the value 'undecided'. This state results in the impossibility to further configure the `Choice` element to a `Choice` with a valid state. After the `Assertion`, only P3 can be asserted to the value 'true' or 'false'. Both the configurations of P3 will not satisfy the requirement that a minimum of two *predicates* need to have the value 'true'.

3.2.3.3 Reference Software

From the sections above, it becomes clear that it is useful to specially develop software that allows validating MPEG-21 DIDs. The MPEG-21 DID reference software [25] implements validation of DIDs using the three-step process that is depicted in Figure 3.2.

```
<?xml version="1.0" encoding="UTF-8"?>
<DIDL xmlns="urn:mpeg:mpeg21:2002:02-DIDL-NS">
  <Item id="item_01">
    <Choice choice_id="choice_01" minSelections="2">
      <Selection select_id="P1"/>
      <Selection select_id="P2"/>
      <Selection select_id="P3"/>
    </Choice>
    <Annotation target="#item_01">
      <Assertion target="#choice_01" false="P1 P2"/>
    </Annotation>
  </Item>
</DIDL>
```

Example 3.33 Violation of the third validation rule of the `Assertion` element

Figure 3.2 Three-step validation process of the MPEG-21 DID reference software

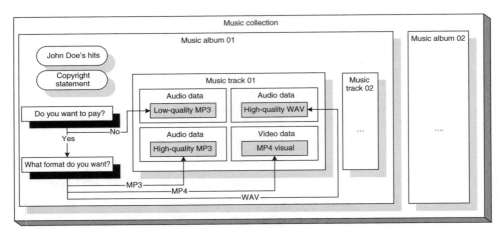

Figure 3.3 A music collection Digital Item Declaration – graphical representation

The input to the validation software is a DID in the XML-based DIDL. During the first step of the validation process, the XML document is processed by an XInclude processor. This processor resolves all XIncludes and prepares the DIDL document for further validation.

During the second step of the validation process, the output of the XInclude processor is checked for being well formed and the XML is validated against the MPEG-21 DIDL Schema. This validation is done using a standard XML Schema validator, Xerces2.

The final step of the validation process is the checking of the validation rules of each of the DIDL elements. In this step, each of the DIDL elements is traversed and checked one by one to see whether they are valid according to their validation rules.

3.2.4 THE USE OF ALL DIDL ELEMENTS IN A REAL-WORLD EXAMPLE

In this section, it is shown how the DIDL elements can be used in an integrated real-world example. The DID below represents an MPEG-21 compliant music collection. In the following paragraphs, this example is discussed in detail. Graphical representation of a music collection Digital Item Declaration is shown in Figure 3.3.

```
<?xml version="1.0" encoding="UTF-8"?>
<DIDL xmlns="urn:mpeg:mpeg21:2002:02-DIDL-NS"
   xmlns:xsi="http://www.w3.org/2001/XMLSchema-instance">
  <Declarations>
    <Descriptor id="copyright">
      <Statement mimeType="text/plain">
        Copyright MyCompany.  All rights reserved.
      </Statement>
    </Descriptor>
  </Declarations>
  <Container id="music_collection">
    <Item id="album_01">
      <Descriptor>
        <xi:include xpointer="copyright/1"/>
```

```
    </Descriptor>
    <Descriptor>
      <Statement mimeType="text/plain">
        All time greatest hits of John Doe
      </Statement>
    </Descriptor>
    <Descriptor>
      <Component>
        <Resource mimeType="image/jpeg"
                  ref="http://server/cover.jpg"/>
      </Component>
    </Descriptor>
    <Choice choice_id="payment" minSelections="1" maxSelections="1">
      <Descriptor>
        <Statement mimeType="text/plain">
          Do you want to pay for listening?
        </Statement>
      </Descriptor>
      <Selection select_id="not_paid">
        <Descriptor>
          <Statement mimeType="text/plain">
            I do not want to pay.
          </Statement>
        </Descriptor>
      </Selection>
      <Selection select_id="paid">
        <Descriptor>
          <Statement mimeType="text/plain">
            I want to pay 10$.
          </Statement>
        </Descriptor>
      </Selection>
    </Choice>
    <Choice choice_id="format" minSelections="1" maxSelections="1"
            default="mp3">
      <Condition require="paid"/>
      <Descriptor>
        <Statement mimeType="text/plain">
          What format do you want?
        </Statement>
      </Descriptor>
      <Selection select_id="mp3">
        <Descriptor>
          <Statement mimeType="text/plain">
            I want MP3 format.
          </Statement>
        </Descriptor>
      </Selection>
      <Selection select_id="wav">
        <Descriptor>
          <Statement mimeType="text/plain">
            I want WAV format.
          </Statement>
        </Descriptor>
```

```
  </Selection>
  <Selection select_id="video">
    <Descriptor>
     <Statement mimeType="text/plain">
        I want to see the video clip in MPEG-4 Visual format.
     </Statement>
    </Descriptor>
  </Selection>
</Choice>
<Item id="track_01">
  <Component>
    <Condition require="not_paid"/>
    <Descriptor>
     <Statement mimeType="text/xml">
        <Mpeg7 xmlns="urn:mpeg:mpeg7:schema:2001">
          <Description xsi:type="MediaDescriptionType">
            <MediaInformation>
              <MediaProfile>
                <MediaFormat>
                  <Content href="http://server/track01_64.mp3"/>
                  <BitRate>64</BitRate>
                </MediaFormat>
              </MediaProfile>
            </MediaInformation>
          </Description>
        </Mpeg7>
     </Statement>
    </Descriptor>
    <Resource mimeType="audio/mpeg"
              ref="http://server/track01_64.mp3"/>
  </Component>
  <Component>
    <Condition require="paid mp3"/>
    <Resource mimeType="audio/mpeg"
              ref="http://server/track01_320.mp3"/>
    <Anchor>
      <Descriptor>
        <Statement mimeType="text/plain">
        John's longest trumpet solo
        </Statement>
      </Descriptor>
      <Fragment fragmentId="mp(/~time('npt','70')"/>
    </Anchor>
  </Component>
  <Component>
    <Condition require="paid wav"/>
    <Resource mimeType="audio/wav"
              ref="http://server/track01.wav"/>
  </Component>
  <Component>
    <Condition require="paid video"/>
    <Descriptor>
     <Statement mimeType="text/xml">
        <Mpeg7 xmlns="urn:mpeg:mpeg7:schema:2001">
          <Description xsi:type="ContentEntityType">
```

```
                    <MultimediaContent xsi:type="VideoType">
                      <Video>
                        <MediaInformation>
                          <MediaProfile>
                            <MediaFormat>
                              <Content href="rtsp://server/track01.mp4"/>
                              <BitRate>257000</BitRate>
                              <VisualCoding>
                                <Frame width="240" height="112"
                                       rate="24"/>
                              </VisualCoding>
                            </MediaFormat>
                          </MediaProfile>
                        </MediaInformation>
                      </Video>
                    </MultimediaContent>
                  </Description>
                </Mpeg7>
              </Statement>
            </Descriptor>
            <Resource mimeType="video/mp4"
                    ref="rtsp://server/track01.mp4"/>
          </Component>
          <Annotation target="#track_01">
            <Descriptor>
              <Statement mimeType="text/plain">
                This is a really cool track.
              </Statement>
            </Descriptor>
          </Annotation>
        </Item>
        <Item id="track_02">
          <!-- ... -->
        </Item>
        <!-- ... -->
        <Annotation target="#album_01">
          <Assertion target="#format" true="wav" false="mp3 video"/>
        </Annotation>
      </Item>
      <Item id="album_02">
        <Descriptor>
          <xi:include xpointer="copyright/1"/>
        </Descriptor>
        <!-- ... -->
      </Item>
    </Container>
</DIDL>
```

Example 3.34 A Music Collection DID – XML representation

The first part of this DID contains a `Declarations` element. This element contains a `Descriptor` with a `Statement`. The `Statement` is a digital representation of the text of a copyright statement. Because the same copyright statement is used for several music albums in the collection, it is declared once in the `Declarations` element and used several times throughout this music collection DID.

The next DIDL element in the music collection is the `Container` element. The `Container` element is the digital representation of a CD rack. It is a grouping of different music albums into one collection.

Within this music collection, there are currently two different music albums. The different music albums are represented by `Item` elements. In this example, the `Items` with id 'album_01' and id 'album_02' are representations of music albums.

The first music album contains three different `Descriptors`. The first occurrence of the `Descriptor` element is a `Descriptor` that is declared by reference. This `Descriptor` contains an `include` element that points to the `Statement` with the copyright statement, which is located in the `Declarations` element. It is an example of a plain text description describing an `Item`.

Another example of a plain text `Descriptor` in an `Item` is the second `Descriptor`. This `Descriptor` contains a description of the content of the `Item`. It describes that the music album has the title 'All time greatest hits of John Doe'. The third `Descriptor` that is contained in the `Item` is an example of a `Descriptor` that contains a `Component` with a `Resource`. This `Descriptor` contains a graphical description of the music album. It contains the cover of the music album in JPEG format.

The `Choice` elements that allow run-time configuration of the digital music album are located immediately after the `Descriptors`. The first `Choice` element allows a User to choose whether to pay for listening to the music album or not. If the User responds, 'I do not want to pay' to the question 'Do you want to pay for listening'? the *predicate* 'not_paid' located in the `Selection` with `select_id` 'not_paid' will become true. On the other hand, if the User responds 'I want to pay 10$' the predicate 'paid' will become true after the User has paid for listening. This `Choice` contains both a `minSelections` and a `maxSelections` element, both with the value '1'. To configure this `Choice` in a valid state, it is necessary that there is exactly one *predicate* that has the value 'true'. All of the other *predicates* can have the value 'false' or 'undecided'.

If the User has chosen not to pay for listening to the content, only a low-quality preview of the tracks will become available. In that case, the bitrate of the available `Component` in the `Item` with id 'track_01' will be only 64 kbps. If the User chooses to pay for listening, the second question in the music album becomes available.

This `Choice` element allows the User to choose between two different formats for the high-resolution content. If the User responds, 'I want the MP3 version' to the question 'What format do you want'? the *predicate* 'mp3' will be assigned the value 'true'. On the other hand, if the response is 'I want the WAV version', the *predicate* 'wav' will be assigned the value 'true'. Or if the response is 'I want to see a video clip in MPEG-4 Visual format' the 'video' predicate will be assigned the value true. On the basis of the outcome of the second question, the high-resolution `Components` in WAV, MP3 or MPEG-4 format will become available for listening/viewing. If the User does not respond to the question in due time, the value of the *predicate* 'mp3' will be automatically set to true. This is the case because the value of the `default` attribute of the `Choice` element is equal to 'mp3'.

After the definition of the different `Choices`, which allow the configuration of the music album, there is a collection of sub-`Items` containing the different tracks of the music album. For example, both the `Items` with id 'track_01' and 'track_02' represent a track in the music album. The music track Item is constructed out of four different Components.

The first one is a Component with a low-quality preview of the actual track. This Component is available when the User does not pay for the audio content. It allows a content creator to give a free preview of the actual music album. This Component contains a Descriptor that describes the Resource with the low-quality audio. The Descriptor in the Component contains a Statement with MPEG-7 data, which states that the bitrate of the track equals to 64 kbps. The actual Resource is defined by reference to a web server.

The second Component becomes available whenever the User decides to pay and selects the MP3 format. This Component contains an Anchor that indicates a point of interest in the audio resource. This point of interest, 'John's longest trumpet solo', is located at 1 min and 10 s in the audio track.

The third Component becomes available whenever the User has paid and selected the WAV format. The Component contains a Resource that points to the audio data, which is located somewhere on a web server.

The fourth Component becomes available whenever the User has paid and selected the MP4 format. The Component contains a Resource that points to the video data, which is located somewhere on a Real-Time Streaming Protocol (RTSP) [26] enabled server.

At the end of the Item that represents the music track, there is an Annotation element. This Annotation element contains a description that was added by the User. This description is represented by the Descriptor element and contains a Statement with the text 'This is a really cool track'. Annotations allow Users to add custom information to the music album without modifying the content of the original album.

At the end of the Item that contains the music album, there is an Annotation element with an Assertion element. This Annotation element contains the state of the Choice with choice_id 'format'. The Annotation element stores the Selections that were made by the User. Using this Annotation element, the Choice can be reconfigured whenever the User wants to play the music album in the future. This makes sure that it is not required to ask the same question every time the User listens to the music. It should be noted that this Annotation element does not allow the evasion of payment of the royalties for listening to the music. In this example, the Annotation element is just a tool to allow the storage of the configuration state of a Choice. For example, whenever the Choice with choice_id 'payment' is configured, the Intellectual Property Management and Protection (IPMP) [27] system of the MPEG-21 terminal will make sure that the necessary payments are done. If this is not the case, asserting the Choice with the Assertion element will fail.

3.2.5 FIRST AND SECOND EDITION OF THE DIGITAL ITEM DECLARATION

Since the publication of the DID version 1 in 2003, there has been an evolution in the technologies used in the DID. Therefore, a second edition of the DID was created. Because, there were quite some improvements to the specification, only the second edition has been discussed in this chapter. An extensive overview of the differences between the first and second edition can be found in Annex B of the DID [2]. This annex also contains a pseudo algorithm for converting a first edition DIDL document to a second edition DIDL document.

3.3 DIGITAL ITEM IDENTIFICATION

In the second part of this chapter, ISO/IEC 21000–3, the Digital Item Identification (DII) is discussed. The scope of the MPEG-21 DII specification includes the following:

- how to uniquely identify DIs and parts thereof (including resources);
- how to uniquely identify intellectual property (IP) related to the DIs (and parts thereof), for example, abstractions;
- how to use identifiers to link DIs with related information, such as, descriptive *meta-data*;
- how to identify different types of DIs;

To realize this unique identification, the DII specification defines three different XML elements: the `Identifier`, the `RelatedIdentifier` and the `Type` element. These elements are discussed in the following sections.

The purpose of the DII specification is not to reinvent existing identification schemes, but to provide the hooks for the integration of existing identification schemes into the DID specification. This allows the reuse of existing technologies. For example, the DII specification allows for the integration of the International Standard Book Number (ISBN) [28] within DIDs.

The DII specification contains an informative annex with a long (nonexhaustive) list of existing identification and description schemes that are good candidates to be used within the DII elements. While some of the identification systems that are used to uniquely identify 'content', have the capabilities to resolve an identifier on-line to appropriate *metadata* (for example, cIDf [29], DOI [30]), others do not have the capability to be resolved on-line (for example, ISBN, ISRC [31]). The latter identification systems are still usable in the DII specification. The actual resolving of the identifier to the content, that is, retrieving the content based on the identifier, can be done off-line or by developing a new on-line resolving mechanism. The DII specification does not require or specify such resolving mechanisms. It specifies how identifiers can be used for DIs. However, the DII specification also provides examples of how on-line resolving *could* be realized.

3.3.1 LINK BETWEEN DII AND DID

Before starting with the actual discussion of the different elements that are defined in the DII specification, the relationship between the DII specification and the DID specification is given.

In the DID model, a *statement* was defined as follows: 'A *statement* is a literal textual value that contains information, but not an asset'. Because DII information is a literal textual value (an identifier is not supposed to change), the *statement* DID model element is suited for including DII information.

Therefore MPEG specified that, within DIDs, DII information will be contained in `Statement` elements. Such `Statement` elements appear in the DIDL as children of the `Descriptor` element. This also means that it is possible to include DII information

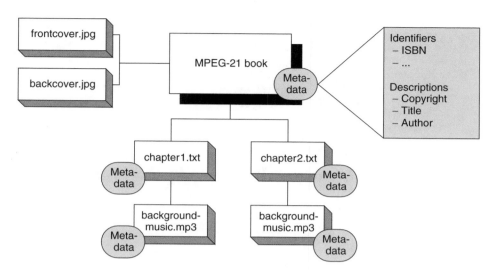

Figure 3.4 Digital Item Identification – the relationship between DID & DII

wherever a `Descriptor` element can appear in the DIDL. Each of these `Descriptors` may contain a `Statement` element with one DII identifier relating to the parent element of the `Descriptor`.

Figure 3.4 depicts the link between the DII specification and the DID specification by using an example. In this example, an MPEG-21 book is presented. With the DIDL, this book will typically be constructed using an `Item` element. This `Item` element contains several `Descriptor` elements, the ones on the left-hand side containing `Components` with the covers of the book, and the ones on the right-hand side containing the identifiers and additional descriptive information. The DII information resides in the `Descriptor` on the right-hand side. In this example, the DII identifier is the ISBN number of the book.

DII information can also appear in the different sub-`Items` of the MPEG-21 book (for example, in the different chapters of the book). DII information could be used to identify the type (for example, a chapter type) of the sub-`Items` or even to uniquely identify each different chapter.

3.3.2 DII ELEMENTS

In this section, the elements that can be used to include identifiers in DIs are discussed. Some examples of how they can be used together with the DIDL are given.

3.3.2.1 Identifier

DIs and their parts within the MPEG-21 Multimedia Framework are identified by encapsulating Uniform Resource Identifiers into the `Identifier` element. This `Identifier` element contains a single URI which uniquely identifies the parent element of the `Descriptor` element that contains the `Identifier` element. Note that it is possible for a single DI to have multiple `Identifiers`, for example, coming from different identification schemes.

```
<?xml version="1.0" encoding="UTF-8"?>
<DIDL xmlns="urn:mpeg:mpeg21:2002:02-DIDL-NS"
      xmlns:dii="urn:mpeg:mpeg21:2002:01-DII-NS">
  <Item>
    <Descriptor>
      <Statement mimeType="text/plain">
        All time greatest hits of John Doe
      </Statement>
    </Descriptor>
    <Descriptor>
      <Statement mimeType="text/xml">
        <dii:Identifier>
          myRegistrationAuthority:AlbumID:A1-888999-0029733-22-F
        </dii:Identifier>
      </Statement>
    </Descriptor>
    <!-- ... -->
  </Item>
</DIDL>
```

Example 3.35   Adding unique identifiers to a Digital Item

In Example 3.35, an `Identifier` element is used in a DID. The `Identifier` element is contained within a `Statement` element, which is a child of a `Descriptor` element. This `Descriptor` element associates the DII information with its parent element, namely, the `Item`. In this example, the music album, which is declared in this Item, is uniquely identified by the DII identifier 'myRegistrationAuthority:AlbumID:A1-888999-0029733-22-F'. This URI, which contains several parts, could possibly be resolved as described here. Consider a registration authority called *myRegistrationAuthority*. In this example, the registration authority keeps track of several different types of content, including music albums. Resolving the identifier would be the same as asking the registration authority the following: Give me the `Item` containing a music album with the id 'A1-888999-0029733-22-F'.

From the content creators' point of view, whenever they create a music album, they request an 'AlbumID' from the registration authority. The Album ID is then included by the content creators in the DID by means of a `Descriptor/Statement` element containing the DII `Identifier` element.

3.3.2.2 RelatedIdentifier

While the DII `Identifier` element is intended to enable the unique identification of DIs (or parts thereof), the `RelatedIdentifier` element allows for the identification of information that is related to the DI (or parts thereof). Note that the same comment as for the `Identifier` applies here, it is possible for a single DI to have multiple `RelatedIdentifier`.

For example, the DII `RelatedIdentifier` allows for the identification of the abstraction of a work. The abstraction of the work is related to the `Resource` that corresponds to the recording of the musical work. The recording would typically be identified with a DII `Identifier`, identifying that the work itself is done, from within the same `Component`, with the DII `RelatedIdentifier` element.

The `RelatedIdentifier` element allows associating identifiers that are related to the DI *(item), container, component* and/or *fragments* thereof, but do not identify the DI (or part thereof) directly. Similar to the `Identifier` element, the `Related-Identifier` contains a URI. Both elements can therefore be resolved in the same way. However, the `RelatedIdentifier` element may not be used for identifying the DI (or part thereof) itself. This should be done using the DII `Identifier` element as specified above.

The `RelatedIdentifier` element has a `relationshipType` attribute that allows expressing the relationship of the `Resource` identified by a `RelatedIdentifier` to the DI (or part thereof) bearing the `RelatedIdentifier`. The value shall be in the form of a URI. An attribute x shall be identified with the URI urn:mpeg:mpeg 21:2002:01-RDD-NS:x, where x is the headword or unique identifier for a `Relator` defined in the ISO/IEC 21000−6 Rights Data Dictionary [32] or as registered by the MPEG-21 Registration Authority [33].

The difference between the `RelatedIdentifier` and the `Identifier` is clarified with the following example. In the `Item` of Example 3.36, there is a `Component` element that contains an audio track of a music album. This `Component` contains two `Descriptors`. The first `Descriptor` contains a DII `Identifier`. This `Identifier` is a Digital Object Identifier (DOI) [30], which identifies the actual recording of the resource, namely, the MP3 sound recording. The second `Descriptor` contains a DII `RelatedIdentifier`. This `RelatedIdentifier` is an International Standard Musical Work Code (ISWC) [34] identifier, which identifies the underlying music work.

```xml
<?xml version="1.0"?>
<DIDL xmlns="urn:mpeg:mpeg21:2002:02-DIDL-NS"
      xmlns:dii="urn:mpeg:mpeg21:2002:01-DII-NS">
  <Item>
    <Component>
      <Descriptor>
        <Statement mimeType="text/xml">
          <!-- ISRC identifying the sound recording -->
          <dii:Identifier>
            urn:mpegRA:mpeg21:dii:doi:10.1000/123456789
          </dii:Identifier>
        </Statement>
      </Descriptor>
      <Descriptor>
        <Statement mimeType="text/xml">
          <!-- ISWC identifying the underlying musical work -->
          <dii:RelatedIdentifier
    relationshipType="urn:mpeg:mpeg21:2002:01-RDD-NS:IsAbstractionOf">
            urn:mpegRA:mpeg21:dii:iswc:T-034.524.680-1
          </dii:RelatedIdentifier>
        </Statement>
      </Descriptor>
      <Resource ref="track01.mp3" mimeType="audio/mpeg"/>
    </Component>
  </Item>
</DIDL>
```

Example 3.36   Use of RelatedIdentifier versus Identifier

This underlying music work could have multiple recordings, for example, MP3, WMA, each with its own DOI.

3.3.2.3 Type

The DII Type element allows for the identification of special types of DIs. For example, suppose the music industry decides to restrict the generic format of the DIDL to create a simpler format for MPEG-21 music albums. Example 3.37 illustrates what such a format could look like. The top-level Item contains a Descriptor with a Component/Resource containing the cover of the music album in jpeg format. Finally, it contains sub-Items for each track containing a Component/Resource in mp3 format.

The DII Type identifier can be used to identify that the Item is created according to the format described above. This can be done by adding the DII Type element as a Descriptor to the music album Item. Whenever an MPEG-21 terminal receives an Item with the DII Type identifier 'urn:MusicIndusty:MusicAlbum', it can be sure that the Item has this predefined format.

Note that the DII specification does not require a predefined format for the Item that contains the DII Type element. The only requirements for the DII Type element are that the parent element of the Descriptor containing the DII Type element must be an Item element, and that the data contained within the DII Type element needs to be a URI.

```xml
<?xml version="1.0"?>
<DIDL xmlns="urn:mpeg:mpeg21:2002:02-DIDL-NS"
   xmlns:dii="urn:mpeg:mpeg21:2002:01-DII-NS">
  <!-- Music album -->
  <Item>
    <!-- DII Type identifier -->
    <Descriptor>
      <Statement mimeType="text/xml">
        <dii:Type>urn:MusicIndusty:MusicAlbum</dii:Type>
      </Statement>
    </Descriptor>
    <!-- Cover -->
    <Descriptor>
      <Component>
        <Resource mimeType="image/jpeg" ref="cover.jpg"/>
      </Component>
    </Descriptor>
    <!-- Music Track -->
    <Item>
      <Component>
        <Resource mimeType="audio/mpeg" ref="track01.mp3"/>
      </Component>
    </Item>
    <!-- ... -->
  </Item>
</DIDL>
```

Example 3.37   Typing of Digital Items

At the time of writing, no DII Types have been defined by MPEG. DII Types will typically be defined from within the industry to identify a certain type of DI, for example, a music album type or a movie collection type.

3.4 SUMMARY

In this chapter, the role of so-called 'Digital Items' within the MPEG-21 Multimedia Framework has been discussed. It was explained that the DID specification includes three parts, the Abstract Model, its representation in XML, and the corresponding MPEG-21 DID Schemas. These parts have been described in detail. In addition, the strength of the MPEG-21 DID specification was illustrated through a concrete example of building and configuring a digital music album.

It was also discussed how the DII specification can be used to uniquely identify, or type DIs and/or parts thereof. These concepts were illustrated with several examples.

REFERENCES

[1] ISO/IEC, "ISO/IEC 21000-2:2003 Information technology – Multimedia framework (MPEG-21) – Part 2: Digital Item Declaration," March 2003.

[2] ISO/IEC, "ISO/IEC FDIS 21000-2 Information technology – Multimedia framework (MPEG-21) – Part 2: Digital Item Declaration second edition," 2005.

[3] ISO/IEC, "ISO/IEC 21000-3:2003 Information technology – Multimedia framework (MPEG-21) – Part 3: Digital Item Identification," March 2003.

[4] I. Burnett, R. Van de Walle, K. Hill, J. Bormans, and F. Pereira "MPEG-21: Goals and Achievements", *IEEE Multimedia, IEEE Computer Society*, vol. **10** no. 4, pp. 60–70, 2003.

[5] ISO/IEC, "ISO/IEC TR 21000-1:2004 Information technology – Multimedia framework (MPEG-21) – Part 1: Vision, Technologies and Strategy," November 2004.

[6] World Wide Web Consortium, "Extensible Markup Language (XML) 1.0 (2nd Edition)," W3C Recommendation, 6 October 2000.

[7] World Wide Web Consortium, "XML Schema Part 0: Primer", W3C Recommendation, 2 May 2001.

[8] ISO/IEC, "ISO/IEC 11172-3:1993 Information technology – Coding of moving pictures and associated audio for digital storage media at up to about 1,5 Mbit/s – Part 3: Audio," 1993.

[9] ISO/IEC, "ISO/IEC 14977:1996 Information technology – Syntactic metalanguage – Extended BNF," 1996.

[10] B.S. Manjunath, P. Salembier, T. Sikora, "Introduction to MPEG-7 – Multimedia Content Description Interface," Wiley, NJ, 2003.

[11] World Wide Web Consortium, "XML-Signature Syntax and Processing," W3C Recommendation, 12 February 2002.

[12] Internet Engineering Task Force, "Multipurpose Internet Mail Extensions (MIME) Part One: Format of Internet Message Bodies," November 1996.

[13] Internet Engineering Task Force, "The audio/mpeg Media Type," November 2000.

[14] Internet Engineering Task Force, "GZIP file format specification version 4.3," May 1996.

[15] Internet Engineering Task Force, "Uniform Resource Identifiers (URI): Generic Syntax," August 1998.

[16] ISO/IEC, "ISO/IEC 15938-4:2002 Information technology – Multimedia content description interface – Part 4: Audio," July 2002.

[17] ISO/IEC, "ISO/IEC 10918-1:1994 Information technology – Digital compression and coding of continuous-tone still images: Requirements and guidelines," 1994.

[18] Adobe Systems Incorporated, "Tag(ged) Image File Format Revision 6.0," June 1992.

[19] World Wide Web Consortium, "XPointer Framework," W3C Recommendation, 25 March 2003.

[20] World Wide Web Consortium, "XML Path Language (XPath) Version 1.0," W3C Recommendation, 16 November 1999.

[21] MPEG, "Introducing MPEG-21 Part 17 – an Overview," ISO/JTC1/SC29/WG11/N7221, Busan, April 2005.

[22] World Wide Web Consortium, "XML Inclusions (XInclude) Version 1.0," W3C Recommendation, 20 December 2004.

[23] The Apache XML Project, "Xerces2 Java Parser," (http://xercesz.apache.org/xerces2-j/index.html), accessed 2005.

[24] Microsoft, "Microsoft XML Core Services," (http://msdn.microsoft.com/library/en-us/xmlsdk/html/xmmscXML.asp), accessed 2005.

[25] ISO/IEC, "ISO/IEC 21000-8:2005 Information technology – Multimedia framework (MPEG-21) – Part 8: Reference Software," 2005.

[26] Internet Engineering Task Force, "Real Time Streaming Protocol (RTSP)," April 1998.

[27] MPEG, "ISO/IEC 21000-4 IPMP Components FCD," ISO/JTC1/SC29/WG11/N7196, Busan, April 2005.

[28] ISO, "ISO 2108:1992 Information and documentation – International standard book numbering (ISBN)," 1992.

[29] Content ID Forum, http://www.cidf.org, accessed 2004.

[30] ANSI/NISO, "ANSI/NISO Z39.84–2000 Syntax for the Digital Object Identifier," December 2000.

[31] ISO, "ISO 3901:2001 Information and documentation – International Standard Recording Code (ISRC)," September 2001.

[32] ISO/IEC, "ISO/IEC 21000-6:2004 Information technology – Multimedia framework (MPEG-21) – Part 6: Rights Data Dictionary," May 2004.

[33] MPEG-21 Registration Authority, http://www.mpegra.org/, accessed 2005.

[34] ISO, "ISO 15707:2001 Information and documentation – International Standard Musical Work Code (ISWC)," November 2001.

4

IPMP Components

Shane Lauf and Eva Rodriguez

4.1 BACKGROUND AND OBJECTIVES

The IPMP Components part of MPEG-21 is a response to a demand from the content industry for a part of the MPEG-21 framework that can deliver a Digital Item (DI) securely but which does not bring yet another DRM to an already crowded marketplace. Thus, after much discussion during the requirements setting stage of the standardization, the IPMP (Intellectual Property Management and Protection) Components standard provides a mechanism for Users to protect a DI and its declaration using a specified scheme [1]. This allows DIs to be used in conjunction with many DRM schemes and also to offer a degree of interoperability between such schemes by acting as 'rich' containers, which can describe content and its availability in one of several DRM formats.

MPEG-21 IPMP Components shares the idea of using Tools with MPEG-4 IPMP [2], but adds the ability to leverage Rights expressions using MPEG-21 Rights Expression Language (REL) [3] and MPEG-21 Rights Data Dictionary (RDD) [4] and the ability to leverage external DRMs. A further difference is that MPEG-21 IPMP Components is designed for use in DIs, a container of all types of content, whereas MPEG-4 IPMP was designed to work specifically with MPEG-4 content.

The goal of IPMP Components is simple: to provide a means of expressing governance (Tools, Rights, signatures) over a specific part of a DI hierarchy to be governed, while maintaining the transactability and schematic validity as a DI [5]. This goal implies two key areas. The first is a means to attach protection metadata to a specific part of the DI hierarchy, and this is made possible by IPMP DIDL (Section 4.2), a mapping of the Digital Item Declaration Language (DIDL) [6] explained in Chapter 3, to a language which allows the protection of DIs at arbitrary points in the hierarchy. The second is the metadata itself, expressing governance in a flexible and extensible manner, and this role is fulfilled by the IPMP Info section (Section 4.3).

The MPEG-21 Book   Ian S Burnett, Fernando Pereira, Rik Van de Walle, Rob Koenen
© 2006 John Wiley & Sons, Ltd

4.2 IPMP DIDL

As seen in Chapter 3, the DI hierarchy is represented in DIDL [6], which is defined by an XML schema [7]. One issue with DIDL when dealing with DI content that may be sensitive or controlled is that the Digital Item Declaration (DID) (Chapter 3) is a clear XML document. This means that the structure, contents and metadata (possibly including Rights expressions) included in the declared DI are vulnerable to unauthorized perusal. While a clear DID document can be signed, allowing its authenticity to be verified, DIDL itself does not offer a simple solution for ensuring that parts of the Declaration can be 'protected'. One possibility would be to use the DIDL element `didl:Statement` but this would give very limited flexibility, as `Statements` are restricted in their positioning in a DI hierarchy.

It is in response to this need that IPMP DIDL was developed. The IPMP DIDL is a schema that facilitates the representation of a protected DI structure within a DID document by encapsulating protected DIDL elements and linking appropriate IPMP Info (Section 4.3) to them, allowing for encryption and other forms of protection over DIDL hierarchy but maintaining the DI as a transactable container.

The IPMP DIDL rests on a simple concept: replacing sensitive DIDL elements in a DID document with alternative representations of those same entities, for example, an `ipmpdidl:Container` for a `didl:Container`. The new elements encapsulate the original DIDL element alongside additional information (IPMP Info), and remain schematically valid when they are used in place of the original DIDL elements. This mechanism is illustrated in Figure 4.1.

Thus, a `didl:Item` containing sensitive information might be replaced with an `ipmp-didl:Item`, which includes an encrypted version of the original `didl:Item` as well as decryption keys (or access mechanisms for them) protected by appropriate Rights expressions. To a User or Peer not authorized to view the contents of this governed *Item*, it simply appears in the DIDL as an *Item* of unknown content; when authorization exists, however, the `ipmpdidl:Item` may be browsed and consumed in the same way as a normal `didl:Item`.

The IPMP DIDL is the schema that defines these alternative elements. An equivalent IPMP DIDL element exists for each DIDL element that corresponds to an entity in the DID model, for example, *Resource* or *Item*, (see Chapter 3 for details on the DID model). To protect an existing section of a DI hierarchy expressed in DIDL, the XML section is packaged underneath the equivalent IPMP DIDL element, along with information about the protection (such as Rights expressions and protection Tools), and then this IPMP DIDL

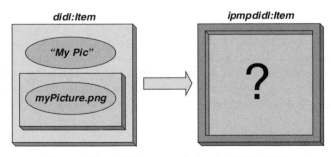

Figure 4.1   DIDL encapsulation in IPMP DIDL

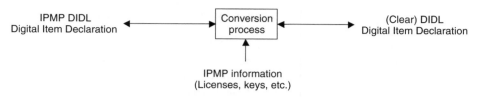

Figure 4.2   IPMP DIDL processing [5]

element is placed back into the DI hierarchy in place of the original DIDL element. The resulting structure remains valid as a transactable DI, since `ipmpdidl:Item` is still a valid *Item* even to a Peer that is unable to consume its contents. To a Peer with appropriate authorization, a DI using IPMP DIDL for part or all of its hierarchy should be convertible to a full, clear DIDL DI by use of licenses and keys, as shown in Figure 4.2. It is expected that these licenses would generally be expressed using the MPEG-21 REL [3].

4.2.1 ELEMENTS IN THE IPMP DIDL

Each of the IPMP DIDL elements below represents the corresponding entity in the DID model, and corresponds to and has the same semantics as an element defined in the DIDL representation.

- `ipmpdidl:Container`
- `ipmpdidl:Item`
- `ipmpdidl:Descriptor`
- `ipmpdidl:Statement`
- `ipmpdidl:Component`
- `ipmpdidl:Anchor`
- `ipmpdidl:Fragment`
- `ipmpdidl:Condition`
- `ipmpdidl:Choice`
- `ipmpdidl:Selection`
- `ipmpdidl:Resource`
- `ipmpdidl:Annotation`
- `ipmpdidl:Assertion`

Chapter 3 includes a detailed description of the semantics of each DIDL element. It is worth recalling the key elements that make up much of DI structure, which are as follows:

- *Resources* that link to individually identifiable assets, such as audio or video files
- *Components* encapsulating and linking metadata with *Resources*
- *Items* encapsulating Components and sub-*Items*
- *Descriptors* which attach descriptive information to *Components* and *Items*
- *Containers* encapsulating a collection of *Items*

All of these, as well as the other elements listed above, are available as IPMP DIDL protected versions.

One further element is defined specific to IPMP DIDL, and which does not have an equivalent in DIDL. This is the `ipmpdidl:ProtectedAsset` element. The `ipmpdidl:ProtectedAsset` element is designed for use when an author wishes to protect an asset – that is, the binary data of an image, video or other multimedia file, referenced by a `didl:Resource` – by encapsulating it in IPMP DIDL. Note that this is not the same as protecting the `didl:Resource` element itself (which only references the asset that itself may reside elsewhere). Whereas `ipmpdidl:Resource` expresses governance on that part of the DI hierarchy expressed in a `didl:Resource`, `ipmpdidl:ProtectedAsset` expresses governance information that applies only to the asset referred to by the `didl:Resource`. For example, a DI author making use of a proprietary DRM on associated asset files may not have to govern the `didl:Resources` to protect the location of an asset but could use `ipmpdidl:ProtectedAsset` to express governance information indicating the fact that each asset is protected with that specific DRM. Note that when `ipmpdidl:ProtectedAsset` is placed beneath a `didl:Resource` element, the mimeType attribute of the `didl:Resource` tag changes to 'application/ipmp' to reflect the inclusion of IPMP instead of a direct link to the digital asset. These elements are part of the namespace URI defined as 'urn:mpeg:mpeg21:2004:01-IPMPDIDL-NS'.

4.2.2 USING THE IPMP DIDL

As an example of how IPMP DIDL elements may be used to express governance at different levels of a DI hierarchy, the DI expressed in Figure 4.3, which contains four audio tracks, will be considered throughout this discussion. By first applying IPMP DIDL to governed parts, and then attaching IPMP Info information about this governance to the IPMP DIDL structure, a complete IPMP Components DI will be constructed.

For the purposes of this example, it will be assumed that the author has the following requirements for managing access to each track using IPMP Components.

- *Track 1*: both descriptive information and the track resource location are governed and only visible to possessors of a specific Right
- *Track 2*: track description and low-quality version are available to all; high-quality version is governed and available only to possessors of a specific Right
- *Track 3*: visible to all, but the audio data is managed by a DRM
- *Track 4*: free unrestricted access to all

To achieve this, the DI is reexpressed in the following way:

- The *Item* for the whole of Track 1 is placed in an `ipmpdidl:Item`
- The *Component* for the high-quality version of Track 2 is placed in an `ipmpdidl:Component`
- The digital asset for Track 3 is placed in an `ipmpdidl:ProtectedAsset`
- Track 4 is left unchanged

This is shown in Figure 4.4.

```
<DIDL xmlns="urn:mpeg:mpeg21:2002:02-DIDL-NS"
      xmlns:xsi="http://www.w3.org/2001/XMLSchema-instance">
  <Container>
    <Item>
      <Descriptor id="Track1_TITLE">
        <Statement mimeType="text/plain">Save It</Statement>
      </Descriptor>
      <Component>
        <Resource ref="http://www.dmu.com/always_red/01_Save_It.mp3"
          mimeType="audio/mp3"/>
      </Component>
    </Item>
    <Item>
      <Descriptor id="Track2_TITLE">
        <Statement mimeType="text/plain">I Haven't been Anywhere</Statement>
      </Descriptor>
      <Component>
        <Descriptor>
          <Statement mimeType="text/plain"> low-quality version </Statement>
        </Descriptor>
        <Resource ref=
          "http://www.dmu.com/always_red/02_L_I_Haven't_been_Anywhere.mp3"
          mimeType="audio/mp3"/>
      </Component>
      <Component>
        <Descriptor>
          <Statement mimeType="text/plain"> high-quality version </Statement>
        </Descriptor>
        <Resource ref=
          "http://www.dmu.com/always_red/02_H_I_Haven't_been_Anywhere.mp3"
          mimeType="audio/mp3"/>
      </Component>
    </Item>
    <Item>
      <Descriptor id="Track3_TITLE">
        <Statement mimeType="text/plain">Sawdust and Sticks</Statement>
      </Descriptor>
      <Component>
        <Resource ref=
          "http://www.dmu.com/always_red/03_Sawdust_and_Sticks.mp3"
          mimeType="audio/mp3"/>
      </Component>
    </Item>
    <Item>
      <Descriptor id="Track4_TITLE">
        <Statement mimeType="text/plain">When the Thistle Blooms</Statement>
      </Descriptor>
      <Component>
        <Resource
          ref="http://www.dmu.com/always_red/04_When_the_Thistle_Blooms.mp3"
          mimeType="audio/mp3"/>
      </Component>
    </Item>
  </Container>
</DIDL>
```

Figure 4.3 Example Digital Item including four audio tracks

```
<DIDL xmlns="urn:mpeg:mpeg21:2002:02-DIDL-NS"
      xmlns:ipmpdidl="urn:mpeg:mpeg21:2004:01-IPMPDIDL-NS"
      xmlns:xsi="http://www.w3.org/2001/XMLSchema-instance">
  <Container>
    <ipmpdidl:Item> ... </ipmpdidl:Item>
    <Item>
      <Descriptor id="Track2_TITLE">
        <Statement mimeType="text/plain">I Haven't been Anywhere</Statement>
      </Descriptor>
      <Component>
        <Descriptor>
          <Statement mimeType="text/plain"> low-quality version </Statement>
        </Descriptor>
        <Resource ref=
          "http://www.dmu.com/always_red/02_L_I_Haven't_been_Anywhere.mp3"
          mimeType="audio/mp3"/>
      </Component>
      <ipmpdidl:Component> ... </ipmpdidl:Component>
    </Item>
    <Item>
      <Descriptor id="Track3_TITLE">
        <Statement mimeType="text/plain">Sawdust and Sticks</Statement>
      </Descriptor>
      <Component>
        <Resource mimeType="application/ipmp">
          <ipmpdidl:ProtectedAsset mimeType="audio/mp3"> ...
          </ipmpdidl:ProtectedAsset>
        </Resource>
      </Component>
    </Item>
    <Item>
      <Descriptor id="Track4_TITLE">
        <Statement mimeType="text/plain">When the Thistle Blooms</Statement>
      </Descriptor>
      <Component>
        <Resource
          ref="http://www.dmu.com/always_red/04_When_the_Thistle_Blooms.mp3"
          mimeType="audio/mp3"/>
      </Component>
    </Item>
  </Container>
</DIDL>
```

Figure 4.4   Example Digital Item using top-level IPMP DIDL elements

4.2.3 STRUCTURE OF IPMP DIDL ELEMENTS

The structure of IPMP DIDL elements reflects the mechanism by which the governed content is encapsulated within them. Each includes both the governed content itself, in the ipmpdidl:Contents element, and a placeholder for information about the protection in the ipmpdidl:Info element (envisioned to hold IPMP Info structure, as detailed in Section 4.3). Though licenses protecting the content will generally be carried under ipmp-didl:Info, this may not always be the case, and so to allow licenses located externally to refer to the content they govern, the optional child element ipmpdidl:Identifier is

Figure 4.5   Structure of IPMP DIDL elements

also included. This is designed to hold an identifier from a suitable scheme such as Digital Item Identification (DII) [8].

The structure of IPMP DIDL elements is reflected in XML as shown in Figure 4.5.

Each of the child elements constituting this structure is considered in more detail below.

- *Identifier*: Contains an identifier from an appropriate identification scheme (such as DII [8]) that may be used, for example, when a REL Grant [3] located outside the IPMP DIDL element needs to refer to the protected content about which it expresses Rights

- *Info*: Contains expression of mechanisms and licenses involved in the protection of and access to content. This element may link directly to an existing DRM, or contain an `IPMPDescriptor` from the IPMP Info schema (see Section 4.3)

- *Contents*: Contains the governed DIDL structure or asset itself, possibly with encryption, watermarking or other protection Tools applied

Conceptually, this three-element structure facilitates the encapsulation and protection of DIDL structure in IPMP DIDL elements as shown in Figure 4.6.

Thus, the example audio-track DI in Section 4.2.2 may be expressed in more detail as shown in Figure 4.7. Note that, for particularly sensitive DI contents, the `ipmp-didl:Contents` may be encrypted or otherwise protected, with a Tool described in `ipmpdidl:Info`. In this case, the structure and contents of `ipmpdidl:Contents` will not be visible, as is the case for the first two tracks in the example in Figure 4.7. It will be seen later how IPMP Info can be used to fill in the `ipmpdidl:Info` elements to provide information on the application of Tools to the contents.

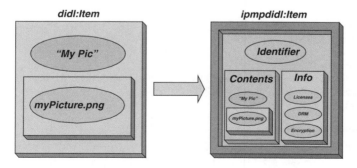

Figure 4.6   DIDL encapsulation in IPMP DIDL

```
<DIDL xmlns="urn:mpeg:mpeg21:2002:02-DIDL-NS"
      xmlns:ipmpdidl="urn:mpeg:mpeg21:2004:01-IPMPDIDL-NS"
      xmlns:xsi="http://www.w3.org/2001/XMLSchema-instance">
  <Container>
    <ipmpdidl:Item>
      <ipmpdidl:Identifier> ... </ipmpdidl:Identifier>
      <ipmpdidl:Info> ... </ipmpdidl:Info>
      <ipmpdidl:Contents>3E674F632A56BD56...</ipmpdidl:Contents>
    </ipmpdidl:Item>
    <Item>
      <Descriptor id="Track2_TITLE">
        <Statement mimeType="text/plain">I Haven't been Anywhere</Statement>
      </Descriptor>
      <Component>
        <Descriptor>
          <Statement mimeType="text/plain"> low-quality version </Statement>
        </Descriptor>
        <Resource ref=
          "http://www.dmu.com/always_red/02_L_I_Haven't_been_Anywhere.mp3"
          mimeType="audio/mp3"/>
      </Component>
      <ipmpdidl:Component>
        <ipmpdidl:Identifier> ... </ipmpdidl:Identifier>
        <ipmpdidl:Info> ... </ipmpdidl:Info>
        <ipmpdidl:Contents>4A5F6326BD563E67...</ipmpdidl:Contents>
      </ipmpdidl:Component>
    </Item>
    <Item>
      <Descriptor id="Track3_TITLE">
        <Statement mimeType="text/plain">Sawdust and Sticks</Statement>
      </Descriptor>
      <Component>
        <Resource mimeType="application/ipmp">
          <ipmpdidl:ProtectedAsset mimeType="audio/mp3">
            <ipmpdidl:Info>...</ipmpdidl:Info>
            <ipmpdidl:Contents
              ref="http://www.dmu.com/always_red/03_Sawdust_and_Sticks"/>
          </ipmpdidl:ProtectedAsset>
        </Resource>
      </Component>
    </Item>
    <Item>
      <Descriptor id="Track4_TITLE">
        <Statement mimeType="text/plain">When the Thistle Blooms</Statement>
      </Descriptor>
      <Component>
        <Resource
          ref="http://www.dmu.com/always_red/04_When_the_Thistle_Blooms.mp3"
          mimeType="audio/mp3"/>
      </Component>
    </Item>
  </Container>
</DIDL>
```

Figure 4.7   Example Digital Item using IPMP DIDL

4.3 IPMP INFO

The information relating to the protection of a DI falls into two categories. The first is information that pertains to the whole DI, including collections of general licenses and a list of Tools used in the DI. The second category is information about the specific protection applied to a certain part of DI hierarchy protected with IPMP DIDL – that is, the specific Tool applied, keys, a license specific to that content, and so on. In the IPMP Info schema, these two categories of information are expressed with two top-level elements: `IPMPGeneralInfoDescriptor` and `IPMPInfoDescriptor`, respectively.

The `IPMPGeneralInfoDescriptor` element contains general information about protection and governance related to a complete DI. IPMP Info Tool lists and licenses packaged in a DI can be included under this element. The `IPMPInfoDescriptor`, on the other hand, is designed to contain information about governance of a specific section of a DI hierarchy, and is designed to be attached to that section through the use of IPMP DIDL (Section 4.2 above). Both these elements are part of the namespace URI defined as '`urn:mpeg:mpeg21:2004:01-IPMPINFO-NS`'.

4.3.1 USING THE IPMPGENERALINFODESCRIPTOR

The `IPMPGeneralInfoDescriptor` has two purposes: to allow a DI author to provide a list of IPMP Tools used in governance that can then be referred to from `IPMP InfoDescriptors`; and to provide a container for licenses carried in the DI. Strictly speaking, a protected DI could be constructed without the use of an `IPMPGeneral-InfoDescriptor` by including licenses and Tool definitions locally (i.e., within each `IPMPInfoDescriptor` contained within IPMP DIDL protected parts of the DI hierarchy), but when Tools are used in multiple places in the DI, or licenses include Rights for more than one piece of the protected hierarchy, it is more practical to include these under an `IPMPGeneralInfoDescriptor`.

4.3.1.1 Defining a List of IPMP Tools

An IPMP Tool is a module that performs one or more IPMP functions such as authentication, decryption, watermarking, and so on. An IPMP Tool has the granularity that it can be a single protection module, for example, a single decryption Tool, and can also be a collection of Tools, that is, a complete IPMP system. It also may coordinate the other IPMP Tools. By defining these Tools in a list upfront and referring to them from each `IPMPInfoDescriptor`, the need to repeat definitions is avoided. The `ToolList` element and its child elements are defined as shown in Figure 4.8.

The list of IPMP Tools can be expressed by attaching one `ToolDescription` element under `ToolList` for each Tool to be included. An example `ToolList` is shown in Figure 4.9.

Each Tool defined is given a LocalID that is unique for each of the IPMP Tools in a DI, and this is used to reference the Tool from `IPMPInfoDescriptors` elsewhere in the DI. IPMP Tools also have a universal identifier and this is specified by the `IPMPToolID`.

In some situations, the author might wish to refer to a number of Tools, any of which can be used in processing governance. An example might be Rivest–Shamir–Adelman (RSA) decryption engines from different companies. In this case, the author can define a

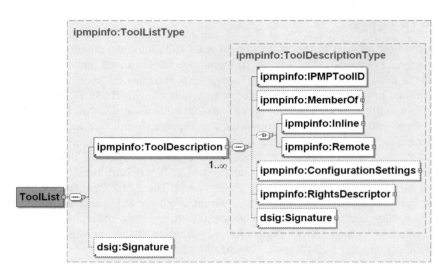

Figure 4.8 Structure of the ToolList element

```
<ipmpinfo:IPMPGeneralInfoDescriptor>
  <ipmpinfo:ToolList>
    <ipmpinfo:ToolDescription localID="Tool1">
      <ipmpinfo:IPMPToolID>
        urn:mpegRA:mpeg21:IPMP:ABC003:77:29
      </ipmpinfo:IPMPToolID>
      <ipmpinfo:MemberOf>
        <ipmpinfo:AlternateGroup groupID="50"/>
      </ipmpinfo:MemberOf>
      <ipmpinfo:Remote ref="urn:IPMPToolsServer:ToolEnc003-3484"/>
    </ipmpinfo:ToolDescription>
    <ipmpinfo:ToolDescription localID="Tool2">
      <ipmpinfo:IPMPToolID>
        urn:mpegRA:mpeg21:IPMP:ABC005:77:29
      </ipmpinfo:IPMPToolID>
      <ipmpinfo:MemberOf>
        <ipmpinfo:AlternateGroup groupID="50"/>
      </ipmpinfo:MemberOf>
      <ipmpinfo:Remote ref="urn:IPMPToolsServer:ToolEnc005-3484"/>
    </ipmpinfo:ToolDescription>
    <ipmpinfo:ToolDescription localID="Tool3">
      <ipmpinfo:IPMPToolID>
        urn:mpegRA:mpeg21:IPMP:ABC064:55:86
      </ipmpinfo:IPMPToolID>
      <ipmpinfo:Inline>
        <ipmpinfo:Binary>74F632X0j9q99y...</ipmpinfo:Binary>
      </ipmpinfo:Inline>
    </ipmpinfo:ToolDescription>
  </ipmpinfo:ToolList>
</ipmpinfo:IPMPGeneralInfoDescriptor>
```

Figure 4.9 Example ToolList declaration

group of interchangeable Tools by declaring each as a `MemberOf` a group with a certain GroupID; thus, in Figure 4.9, the first and the second Tools exist in the same group and may be substituted for each other.

Of course, the actual code for the IPMP Tool must also be specified, and this may be done either by reference to a remote location or inline. If the DI does not carry the IPMP Tool, the remote location from where it can be retrieved can be specified in the `Remote` element, as in the first two `ToolDescriptions` in Figure 4.9. On the other hand, for inline definitions, the `Inline` element is used as a container for the binary of the Tool, and its `Binary` child element is designed to contain a Base64-encoded IPMP Tool. The third `ToolDescription` in Figure 4.9 demonstrates an inline Tool.

Three other optional child elements exist for each `ToolDescriptor`.

- *ConfigurationSettings*: If specific configuration settings are necessary, these can be specified using the `ConfigurationSettings` element
- *RightsDescriptor*: If the Tool itself is governed, the `RightsDescriptor` element can include information about the License that governs the IPMP Tool (Section 4.3.1.2)
- *Signature*: To ensure integrity and authenticity for the description information provided for the list of IPMP Tools, the `Signature` element can be used to include a digital signature for the complete list of Tools or for each one of the Tools, using the signature element under the `ToolList` or `ToolDescription` elements

The second function of the `IPMPGeneralInfoDescriptor` is to provide a container for licenses carried in the DI. When licenses identify their own targets, it is not necessary to carry them in immediate proximity to the target content; in this case, they can be packaged in the `LicenseCollection` element under the `IPMPGeneralInfoDescriptor`. The `LicenseCollection` contains one or more `RightsDescriptor` elements, as shown in Figure 4.10. Note that the `LicenseCollection` element also can contain licenses associated with the whole protected DID.

4.3.1.2 Expressing Rights with the RightsDescriptor

The `RightsDescriptor` element can declare licenses in three ways: by inline inclusion in the `License` element; by reference to the `LicenseReference` element; or by reference via a license service using the `LicenseService` element. Note that licenses

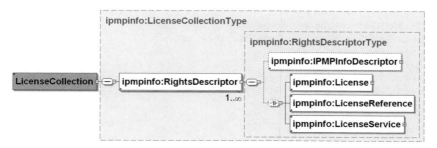

Figure 4.10   LicenseCollection and RightsDescriptor elements

themselves may be protected, and in this case an `IPMPInfoDescriptor` element may be included under the `RightsDescriptor`.

Figures 4.11–4.13 demonstrate these three means of declaring MPEG-21 REL Licenses, under a `RightsDescriptor`. Figure 4.11 shows how such a license can be included directly under the `ipmpinfo:License` element.

Figure 4.12 shows how a license may be referenced by using the `ipmpinfo:LicenseReference` element.

Figure 4.13 shows how license services may be referenced by using the `ipmpinfo:LicenseService` element and including a REL `serviceReference` element as a child element.

4.3.2 POSITIONING THE IPMPGENERALINFODESCRIPTOR IN A DIGITAL ITEM

To associate general protection and governance information with the entirety of a DI containing parts that are governed, it should be carried at the outmost place of the protected DIDL. For example, if one wants to associate this information with a protected DI with

```
<ipmpinfo:IPMPGeneralInfoDescriptor>
  <ipmpinfo:ToolList>. . . </ipmpinfo:ToolList>
  <ipmpinfo:LicenseCollection>
    <ipmpinfo:RightsDescriptor>
      <ipmpinfo:License>
        <r:license>
          <r:encryptedLicense
            Type="http://www.w3.org/2001/04/xmlenc#Content">
            <enc:EncryptionMethod
              Algorithm="http://www.w3.org/2001/04/xmlenc#3des-cbc"/>
            <dsig:KeyInfo>
              <dsig:KeyName>SymmetricKey</dsig:KeyName>
            </dsig:KeyInfo>
            <enc:CipherData>
              <enc:CipherValue>
              Ktd63SDfkDWEjeSdkj39872A5ToQ...</enc:CipherValue>
            </enc:CipherData>
          </r:encryptedLicense>
        </r:license>
      </ipmpinfo:License>
    </ipmpinfo:RightsDescriptor>
  </ipmpinfo:LicenseCollection>
</ipmpinfo:IPMPGeneralInfoDescriptor>
```

Figure 4.11   Inline inclusion of a license using the License element

```
<ipmpinfo:IPMPGeneralInfoDescriptor>
  <ipmpinfo:ToolList>. . . </ipmpinfo:ToolList>
  <ipmpinfo:LicenseCollection>
    <ipmpinfo:RightsDescriptor>
      <ipmpinfo:LicenseReference>
        urn:RELLicsServer:LicEnc002-7484
      </ipmpinfo:LicenseReference>
    </ipmpinfo:RightsDescriptor>
  </ipmpinfo:LicenseCollection>
</ipmpinfo:IPMPGeneralInfoDescriptor>
```

Figure 4.12   Reference to a license via the LicenseReference element

```
<ipmpinfo:IPMPGeneralInfoDescriptor>
  <ipmpinfo:ToolList>. . . </ipmpinfo:ToolList>
  <ipmpinfo:LicenseCollection>
    <ipmpinfo:RightsDescriptor>
      <ipmpinfo:LicenseService>
        <r:serviceReference>
          <sx:uddi>
            <sx:serviceKey>
              <sx:uuid>ee1398c0-8abe-11d7-a735-b8a03c50a862</sx:uuid>
            </sx:serviceKey>
          </sx:uddi>
          <r:serviceParameters>
            <r:datum> ... </r:datum>
          </r:serviceParameters>
        </r:serviceReference>
      </ipmpinfo:LicenseService>
    </ipmpinfo:RightsDescriptor>
  </ipmpinfo:LicenseCollection>
</ipmpinfo:IPMPGeneralInfoDescriptor>
```

Figure 4.13   Reference to a License service using the LicenseService element

a `didl:Item` contained directly below the `didl:DIDL` element, the `IPMPGeneral-InfoDescriptor` may either be carried under the hierarchy `DIDL/Declarations Descriptor/Statement`, as shown in Figure 4.14, or placed as a child element of a `DIDLInfo` element, as shown in Figure 4.15.

```
<DIDL xmlns="urn:mpeg:mpeg21:2002:02-DIDL-NS"
      xmlns:ipmpdidl="urn:mpeg:mpeg21:2004:01-IPMPDIDL-NS"
      xmlns:ipmpinfo="urn:mpeg:mpeg21:2004:01-IPMPINFO-NS"
      xmlns:xsi="http://www.w3.org/2001/XMLSchema-instance">
  <Declarations>
    <Descriptor>
      <Statement mimeType="text/xml">
        <ipmpinfo:IPMPGeneralInfoDescriptor>
          //General protection and governance information
        </ipmpinfo:IPMPGeneralInfoDescriptor>
      </Statement>
    </Descriptor>
  </Declarations>
  <ipmpdidl:Item>
    //Protected Content
  </ipmpdidl:Item>
</DIDL>
```

Figure 4.14   Association of general protection information with a Digital Item in Declarations

On the other hand, if the protected DI is composed of a `Container` of a number of `Items`, the `IPMPGeneralInfoDescriptor` should be carried under the hierarchy `DIDL/Container/Descriptor/Statement`. This is shown in the example of Figure 4.16.

4.3.3 USING THE INFODESCRIPTOR

The `IPMPInfoDescriptor` contains information about governance that applies to a specific piece of content – generally, a part of the DI hierarchy. The `Tool` child element

```
<DIDL xmlns="urn:mpeg:mpeg~21:2002:02-DIDL-NS"
      xmlns:ipmpdidl="urn:mpeg:mpeg~21:2004:01-IPMPDIDL-NS"
      xmlns:ipmpinfo="urn:mpeg:mpeg~21:2004:01-IPMPINFO-NS"
      xmlns:xsi="http://www.w3.org/2001/XMLSchema-instance">
  <DIDLInfo>
    <ipmpinfo:IPMPGeneralInfoDescriptor>
      //General protection and governance information
    </ipmpinfo:IPMPGeneralInfoDescriptor>
  </DIDLInfo>
  <ipmpdidl:Container>
    //Protected contents
  </ipmpdidl:Container>
</DIDL>
```

Figure 4.15 Association of general protection information to a Digital Item in DIDLInfo

```
<DIDL xmlns="urn:mpeg:mpeg21:2002:02-DIDL-NS"
      xmlns:ipmpdidl="urn:mpeg:mpeg21:2004:01-IPMPDIDL-NS"
      xmlns:ipmpinfo="urn:mpeg:mpeg21:2004:01-IPMPINFO-NS"
      xmlns:xsi="http://www.w3.org/2001/XMLSchema-instance">
  <Container>
    <Descriptor>
      <Statement mimeType="text/xml">
        <ipmpinfo:IPMPGeneralInfoDescriptor>
          //General protection and governance information
        </ipmpinfo:IPMPGeneralInfoDescriptor>
      </Statement>
    </Descriptor>
    <ipmpdidl:Item>
      //Protected Content
    </ipmpdidl:Item>
    <ipmpdidl:Item>
      //Protected Content
    </ipmpdidl:Item>
  </Container>
</DIDL>
```

Figure 4.16 Association of general protection information with a Digital Item in a Descriptor under a Container

specifies a Tool that can be used to unprotect the content; the RightsDescriptor child element contains licenses governing the content. A digital signature for the entire element may also be attached. The IPMPInfoDescriptor element structure is shown in Figure 4.17.

4.3.3.1 Expressing Protection Mechanisms

There are two ways to express the Tools needed in unprotecting a governed piece of content. If the author has defined the Tool in a ToolList (see above), the ToolRef element is used to reference this definition, as shown in Figure 4.18.

It is also possible to make a local definition of a Tool explicitly within the IPMP InfoDescriptor, using the ToolBaseDescription element instead. This is similar in structure to the ToolDescription used in the ToolList of IPMPGeneralInfo

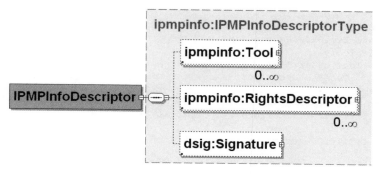

Figure 4.17   Structure of the IPMPInfoDescriptor element

```
<ipmpinfo:Tool>
  <ipmpinfo:ToolRef localidref="Tool1"/>
  ...
</ipmpinfo:Tool>
```

Figure 4.18   Tool reference example using the ToolRef element

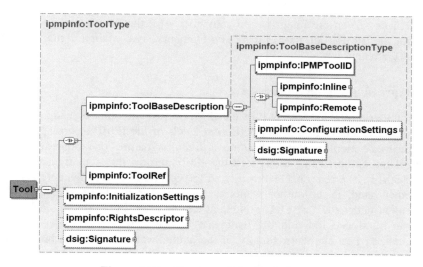

Figure 4.19   Structure of the Tool element

Descriptor, but some child elements that are not relevant for a local definition (e.g.,
MemberOf) are removed.

Figure 4.19 shows how either ToolBaseDescription or ToolRef can be used to
declare Tools.

One piece of protected content may require multiple Tools to access it (for example,
content that has been signed and then encrypted) and the Tool element provides the
order attribute to specify the order in which the Tools are to be applied. This is shown
in Figure 4.20. Note that the IPMP Tools with lower values of the 'order' attribute are
executed first.

```
<ipmpinfo:IPMPInfoDescriptor>
  <ipmpinfo:Tool order="2">
    <ipmpinfo:ToolBaseDescription> ... </ipmpinfo:ToolBaseDescription>
  </ipmpinfo:Tool>
  <ipmpinfo:Tool order="3">
    <ipmpinfo:ToolBaseDescription> ... </ipmpinfo:ToolBaseDescription>
  </ipmpinfo:Tool>
  <ipmpinfo:Tool order="1">
    <ipmpinfo:ToolRef localidref="ToolEnc-01"/>
  </ipmpinfo:Tool>
</ipmpinfo:IPMPInfoDescriptor>
```

Figure 4.20   Tool order definition example

Three other optional elements can be included under the Tool element:

- *InitializationSettings*: InitializationSettings is designed to hold information required to initialize the Tool, and the format of this data will depend on the nature of the Tool itself
- *RightsDescriptor*: When the Tool itself is governed by a License, the Rights-Descriptor is used to include this License when available (Section 4.3.1.2)
- *Signature*: To ensure integrity and authenticity for the description information provided for the IPMP Tool, the Signature element may contain a digital Signature for the Tool element

4.3.3.2 Expressing Digital Signatures

To ensure integrity and authenticity for the IPMP information that is being described, the Signature element can be used at different levels of the IPMP information described. The Signature element always contains a digital signature for its parent element. For example, if the author wants to sign all the IPMP information, the signature under the IPMPInfoDescriptor or IPMPGeneralInfoDescriptor elements can be used. On the other hand, if the author wants to sign the list of IPMP Tools, the Signature element under the ToolList element can be used. Furthermore, each one of the Tools can be signed individually, using a Signature element under the Tool or ToolDescription elements. Finally, if the author wants to sign the binary Tool, the Signature element under the Inline element can be used.

4.3.3.3 Configuration Settings

IPMP Components provides the ConfigurationSettings element to specify detailed configuration settings for a specific IPMP Tool. Any child element is legal, to allow the inclusion of Tool-specific configuration settings. As one possible child element, the IPMP Components standard provides the Update element, which specifies a location from which a Tool can be updated, and conditions under which this update may be performed. This is shown in Figure 4.21.

Figure 4.22 shows how the updating of an IPMP Tool may be specified. Note that both the timing conditions and the platform conditions (hardware, software, and so on) may be

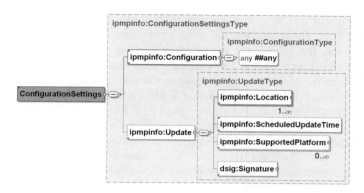

Figure 4.21 ConfigurationSettings element

```
<ipmpinfo:Tool>
  <ipmpinfo:ToolBaseDescription>
    ...
    <ipmpinfo:ConfigurationSettings>
      ...
      <ipmpinfo:Update>
        <ipmpinfo:Location
          ref="urn:IPMPToolsUpdServer1:ToolPartEnc002-9090-NewVers"/>
        <ipmpinfo:ScheduledUpdateTime periodic="P1D">2005-03-07T00:00:00
        </ipmpinfo:ScheduledUpdateTime>
        <ipmpinfo:SupportedPlatform
          xmlns:mpeg4ipmp="urn:mpeg:mpeg4:IPMPSchema:2002">
          <mpeg4ipmp:TerminalID>
            <mpeg4ipmp:TerminalType> ... </mpeg4ipmp:TerminalType>
            <mpeg4ipmp:OperatingSystem> ... </mpeg4ipmp:OperatingSystem>
            <mpeg4ipmp:CPU> ... </mpeg4ipmp:CPU>
            <mpeg4ipmp:Memory> ... </mpeg4ipmp:Memory>
          </mpeg4ipmp:TerminalID>
        </ipmpinfo:SupportedPlatform>
      </ipmpinfo:Update>
    </ipmpinfo:ConfigurationSettings>
  </ipmpinfo:ToolBaseDescription>
</ipmpinfo:Tool>
```

Figure 4.22 Updating of an IPMP Tool example

specified and the Tool will only be updated if all the conditions are met; in this example, the platform characteristics have been specified as in [2].

4.4 USING IPMP COMPONENTS IN A DIGITAL ITEM

As described above, IPMP DIDL elements can be used to express governance over sections of a DI hierarchy, and IPMP Info provides metadata describing that governance. The following section returns to the audio-track DI example introduced in Section 4.2.2 to see how these two parts can be integrated.

Recall that the aim was to govern the descriptive information and track resource location of Track 1 (i.e., governance at the *Item* level); make available the description and low-quality version, but govern the high-quality version of Track 2 (i.e., governance at the *Component* level); express the fact that the audio asset for Track 3 is managed by a DRM (i.e., governance at the `ProtectedAsset` level); and leave Track 4 ungoverned. Figure 4.23 shows how this is realized using IPMP DIDL and IPMP Info.

In this example, a `ToolList` and an encrypted REL License are included in the `IPMP-GeneralInfoDescriptor`, allowing them to be referenced throughout the DI. The hierarchy for Track 1 is packaged in its entirety into an `ipmpdidl:Item`, and the Tool defined locally as ToolEnc001 (the Tool defined in the `IPMPGeneral-InfoDescriptor`) is required to decrypt and access the original hierarchy. The empty `RightsDescriptor` indicates that the User's Right to access Track 1 must be validated, but that the license is not carried or referenced (it may instead reside in the `LicenseCollection` or outside this DI). The hierarchy for the high-quality version of Track 2 is packaged into an `ipmpdidl:Component`, also requiring ToolEnc001 for access; in addition a `LicenseReference` is provided to a REL License, which can be checked to see if it contains Rights for the current User to access this governed hierarchy. The reference to the digital asset for Track 3 is packaged into a `ProtectedAsset`, and the external DRM is specified as a Tool. Note that since the `RightsDescriptor` is absent, it is understood that the IPMP system does not need to perform a Rights check (it may be expected that the external DRM would handle this).

```
<DIDL xmlns="urn:mpeg:mpeg21:2002:02-DIDL-NS"
      xmlns:dii="urn:mpeg:mpeg21:2002:01-DII-NS"
      xmlns:ipmpdidl="urn:mpeg:mpeg21:2004:01-IPMPDIDL-NS"
      xmlns:ipmpinfo="urn:mpeg:mpeg21:2004:01-IPMPINFO-NS"
      xmlns:r="urn:mpeg:mpeg21:2003:01-REL-R-NS"
      xmlns:enc="http://www.w3.org/2001/04/xmlenc#"
      xmlns:dsig="http://www.w3.org/2000/09/xmldsig#"
      xmlns:xsi="http://www.w3.org/2001/XMLSchema-instance">
  <Container>
    <Descriptor>
      <Statement mimeType="text/xml">
        <ipmpinfo:IPMPGeneralInfoDescriptor>
          <ipmpinfo:ToolList>
            <ipmpinfo:ToolDescription localID="ToolEnc001">
              <ipmpinfo:IPMPToolID>
                urn:mpegRA:mpeg21:IPMP:FGA423:44:77
              </ipmpinfo:IPMPToolID>
              <ipmpinfo:Inline>
                <ipmpinfo:Binary>74F632X0j9q99y...</ipmpinfo:Binary>
              </ipmpinfo:Inline>
            </ipmpinfo:ToolDescription>
          </ipmpinfo:ToolList>
          <ipmpinfo:LicenseCollection>
            <ipmpinfo:RightsDescriptor>
              <ipmpinfo:License>
                <r:license>
                  <r:encryptedLicense
                    Type="http://www.w3.org/2001/04/xmlenc#Content">
                    <enc:EncryptionMethod
```

```
                        Algorithm=
                          "http://www.w3.org/2001/04/xmlenc#3descbc"/>
                      <dsig:KeyInfo>
                        <dsig:KeyName>SymmetricKey</dsig:KeyName>
                      </dsig:KeyInfo>
                      <enc:CipherData>
                        <enc:CipherValue>
                        Ktd63SDfkDWEjeSdkj39872A5ToQ...</enc:CipherValue>
                      </enc:CipherData>
                    </r:encryptedLicense>
                  </r:license>
                </ipmpinfo:License>
              </ipmpinfo:RightsDescriptor>
            </ipmpinfo:LicenseCollection>
          </ipmpinfo:IPMPGeneralInfoDescriptor>
        </Statement>
    </Descriptor>
    <ipmpdidl:Item>
      <ipmpdidl:Identifier>
        <dii:Identifier>urn:mpegRA:mpeg21:dii:IPMPDI0001:AC</dii:Identifier>
      </ipmpdidl:Identifier>
      <ipmpdidl:Info>
        <ipmpinfo:IPMPInfoDescriptor>
          <ipmpinfo:Tool>
            <ipmpinfo:ToolRef localidref="ToolEnc001"/>
          </ipmpinfo:Tool>
          <ipmpinfo:Tool>
            <ipmpinfo:ToolBaseDescription>
              <ipmpinfo:IPMPToolID>
                urn:mpegRA:mpeg21:IPMP:ABC005:77:29
              </ipmpinfo:IPMPToolID>
              <ipmpinfo:Remote ref="urn:IPMPToolsServer:ToolWat005-3484"/>
            </ipmpinfo:ToolBaseDescription>
          </ipmpinfo:Tool>
          <ipmpinfo:RightsDescriptor/>
        </ipmpinfo:IPMPInfoDescriptor>
      </ipmpdidl:Info>
      <ipmpdidl:Contents>3E674F632A56BD56...</ipmpdidl:Contents>
    </ipmpdidl:Item>
    <Item>
      <Descriptor id="Track2_TITLE">
        <Statement mimeType="text/plain">I Haven't been Anywhere</Statement>
      </Descriptor>
      <Component>
        <Descriptor>
          <Statement mimeType="text/plain"> low-quality version </Statement>
        </Descriptor>
        <Resource ref=
          "http://www.dmu.com/always_red/02_L_I_Haven't_been_Anywhere.mp3"
          mimeType="audio/mp3"/>
      </Component>
      <ipmpdidl:Component>
        <ipmpdidl:Identifier>
          <dii:Identifier>
            urn:mpegRA:mpeg21:dii:IPMPDI0406:33
```

```
            </dii:Identifier>
          </ipmpdidl:Identifier>
          <ipmpdidl:Info>
            <ipmpinfo:IPMPInfoDescriptor>
              <ipmpinfo:Tool>
                <ipmpinfo:ToolRef localidref="ToolEnc001"/>
              </ipmpinfo:Tool>
              <ipmpinfo:RightsDescriptor>
                <ipmpinfo:LicenseReference>
                 urn:RELLicsServer:LicEnc002-7484
                </ipmpinfo:LicenseReference>
              </ipmpinfo:RightsDescriptor>
            </ipmpinfo:IPMPInfoDescriptor>
          </ipmpdidl:Info>
          <ipmpdidl:Contents>4A5F6326BD563E67...</ipmpdidl:Contents>
        </ipmpdidl:Component>
      </Item>
      <Item>
        <Descriptor id="Track3_TITLE">
          <Statement mimeType="text/plain">Sawdust and Sticks</Statement>
        </Descriptor>
        <Component>
          <Resource mimeType="application/ipmp">
            <ipmpdidl:ProtectedAsset mimeType="audio/mp3">
              <ipmpdidl:Info>
                <ipmpinfo:IPMPInfoDescriptor>
                  <ipmpinfo:Tool>
                    <ipmpinfo:ToolBaseDescription>
                     <ipmpinfo:IPMPToolID>
                       urn:mpegRA:mpeg21:IPMP:GFTR977
                     </ipmpinfo:IPMPToolID>
                     <ipmpinfo:Remote ref="urn:IPMPToolsServer:DRMS06565_FGR"/>
                    </ipmpinfo:ToolBaseDescription>
                  </ipmpinfo:Tool>
                </ipmpinfo:IPMPInfoDescriptor>
              </ipmpdidl:Info>
              <ipmpdidl:Contents
                ref="http://www.dmu.com/always_red/03_Sawdust_and_Sticks"/>
            </ipmpdidl:ProtectedAsset>
          </Resource>
        </Component>
      </Item>
      <Item>
        <Descriptor id="Track4_TITLE">
          <Statement mimeType="text/plain">When the Thistle Blooms</Statement>
        </Descriptor>
        <Component>
          <Resource
            ref="http://www.dmu.com/always_red/04_When_the_Thistle_Blooms.mp3"
            mimeType="audio/mp3"/>
        </Component>
      </Item>
    </Container>
  </DIDL>
```

Figure 4.23   Example Digital Item using IPMP DIDL and IPMP Info

By using IPMP Info and IPMP DIDL to express and attach appropriate governance information, the above DI achieves each of the author's governance requirements expressed in Section 4.2.2, while remaining a structurally valid and transactable DI. Note that even a Peer without IPMP DIDL capabilities, which skips over all IPMP governed sections, will still be able to access the low-quality version of Track 2 and all of Track 4 as per the author's requirements, as these sections remain pure DIDL.

4.5 RELATIONSHIP BETWEEN IPMP COMPONENTS AND THE OTHER PARTS OF MPEG-21

The MPEG-21 Multimedia Framework is composed of various parts linked by the common unit of transaction, the DI. The relationship between IPMP Components and four of the most relevant parts, Part 2: Digital Item Declaration, Part 3: Digital Item Identification, Part 5: Rights Expression Language and Part 7: Digital Item Adaptation, will be examined here.

4.5.1 RELATIONSHIP BETWEEN IPMP COMPONENTS AND MPEG-21 PART 2: DIGITAL ITEM DECLARATION

As has been detailed above, Part 2 of MPEG-21, the Digital Item Declaration (DID) specification [6], defines a set of entities (*Item, Component, Resource,* etc.) that make up the DID Model, and the DIDL defines clear XML elements and attributes that correspond to these. The IPMP DIDL part of IPMP Components provides an alternative set of elements corresponding to these elements (`ipmpdidl:Item`, `ipmpdidl:Component`, `ipmp-mdidl:Resource` etc.) that links a section of governed DI hierarchy with information about that governance.

4.5.2 RELATIONSHIP BETWEEN IPMP COMPONENTS AND MPEG-21 PART 3: DIGITAL ITEM IDENTIFICATION

Part 3 of MPEG-21, Digital Item Identification (DII) [8], defines identifiers that may be associated with parts of DIs (including the DI as a whole) by including them in the DI structure. These identifiers may be used when, for example, a REL License refers to its target content. However, the use of IPMP DIDL to govern parts of DI hierarchy means that identifiers located within the governed hierarchy may be hidden – particularly when the governed hierarchy is encrypted. For this reason, the IPMP DIDL specifies the `ipmpdidl:Identifier` element as a container for identifiers to be placed.

4.5.3 RELATIONSHIP BETWEEN IPMP COMPONENTS AND ISO/IEC 21000-5 RIGHTS EXPRESSION LANGUAGE

MPEG-21 Part 5, REL [3], specifies metadata for expressing the rights of a User to interact with and consume assets, including DIs and the contents contained within them. REL defines Licenses, which define the rights a user has with respect to a specific resource (asset). IPMP Components allows these Licenses to be associated with their targets by including them (directly, by reference, or via a license service) in IPMP Info structure.

4.5.4 RELATIONSHIP BETWEEN IPMP COMPONENTS AND ISO/IEC 21000-7 DIGITAL ITEM ADAPTATION

MPEG-21 Part 7, Digital Item Adaptation (DIA) [9], allows for the adaptation of DIs and their contents to specific Users and consumption environments. The metadata used to personalize DIs in this way may be sensitive. IPMP Components allows this metadata to be protected and governed as necessary, by packaging the DIDL parent of any DIA hierarchy into IPMP DIDL.

4.6 FUTURE OUTLOOK

Work on IPMP Components was concluded at the end of 2005, and the International Standard is due for publication in 2006. One additional area of functionality was added during the late stages of standardization. This was the addition of an `ipmpdidl:ContentInfo` element as an optional child for IPMP DIDL elements such as `didl:Item`, `didl:Resource`, and so on. This element would be neighboring `ipmpdidl:Content`, `ipmpdidl:Info` and `ipmpdidl:Identifier`, and would allow authors to specify additional information about the content being governed.

4.7 SUMMARY

This chapter has examined the aims and functionality of Part 4 of MPEG-21, IPMP Components. It has been shown how IPMP Components addresses the need for a mechanism allowing governance to be expressed over a specific part of the DI hierarchy. This is achieved by expressing this governance information in IPMP Info structure and attaching this to the governed hierarchy using IPMP DIDL. Using the example of a DI containing four audio tracks, each with differing governance requirements, it has been demonstrated how IPMP Info and IPMP DIDL can be used while maintaining a transactable DI. Additional functionality included in the final standard was also considered.

REFERENCES

[1] MPEG, "Requirements for MPEG-21 IPMP", ISO/IEC JTC1/SC29/WG11/N6389, Munich, March 2004.
[2] ISO/IEC, "ISO/IEC 14496-13:2004 Intellectual Property Management and Protection (IPMP) extensions", September 2004.
[3] ISO/IEC, "ISO/IEC 21000-5:2004 Information technology – Multimedia framework (MPEG-21) – Part 5: Rights Expression Language", April 2004.
[4] ISO/IEC, "ISO/IEC 21000-6:2004 Information technology – Multimedia framework (MPEG-21) – Part 6: Rights Data Dictionary", May 2004.
[5] MPEG, "ISO/IEC 21000-4 IPMP Components FCD", ISO/JTC1/SC29/WG11/N7196, Busan, April 2005.
[6] ISO/IEC, "ISO/IEC FDIS 21000-2 Information technology – Multimedia framework (MPEG-21) – Part 2: Digital Item Declaration second edition", 2005.
[7] World Wide Web Consortium, "Extensible Markup Language (XML) 1.0 (2nd Edition)", W3C Recommendation, 6 October 2000.
[8] ISO/IEC, "ISO/IEC 21000-3:2003 Information technology – Multimedia framework (MPEG-21) – Part 3: Digital Item Identification", March 2003.
[9] ISO/IEC, "ISO/IEC 21000-7: 2004 Information technology – Multimedia framework (MPEG-21) – Part 7: Digital Item Adaptation", 2004.

5

Rights Expression Language

Thomas DeMartini, Jill Kalter, Mai Nguyen,
Edgar Valenzuela and Xin Wang

5.1 ABOUT THIS CHAPTER

The purpose of this chapter is to provide a general understanding of the Rights Expression Language specification, ISO/IEC 21000-5 [1]. This section introduces the organization and conventions of this chapter. Section 5.2 introduces Rights Expression Languages. To demonstrate ways that the Rights Expression Language can be used, this chapter provides a background in Section 5.2 and throughout the chapter about systems that use Rights Expressions. This background information is intended only for illustrative purposes; the Rights Expression Language can be used in many different domains and applications.

Sections 5.3–5.5 give a conceptual introduction to the important concepts and features so that readers will obtain a basic understanding that can be used as a foundation for future exploration. Section 5.3 introduces Licenses and Section 5.4 introduces authorization. The features of ISO/IEC 21000-5 that are covered in Section 5.5 were selected to foster an overall understanding of the Rights Expression Language and the capabilities it offers. The reader is encouraged to review the materials listed in Section 5.6.2 to obtain more information. Section 5.6.2.1 describes the content and organization of ISO/IEC 21000-5, and Section 5.6.2.2 lists online resources providing additional examples, use cases and technical papers.

To aid the comprehension of important Rights Expression Language concepts, elements and types, Sections 5.3–5.5 present a number of examples. Each example is meant as a teaching tool and is designed to clearly and succinctly illustrate concepts presented in the section in which it appears. As such, the example may not represent the most effective use of the language to address a particular use case in a deployed system.

Section 5.6 concludes by describing ongoing standards work, reference materials available for further reading and a short summary. Section 5.7 provides an index of the

The MPEG-21 Book   Ian S Burnett, Fernando Pereira, Rik Van de Walle, Rob Koenen
© 2006 John Wiley & Sons, Ltd

elements described in this chapter and gives the number of the section in which they
are introduced.

The sections in this chapter build upon each other, with later sections using and
elaborating on concepts explained in earlier sections. For optimal understanding, it is
recommended that the sections be read in the order in which they are presented.

5.1.1 NAMESPACE CONVENTIONS

This chapter uses prefixes to identify the namespaces in which elements are defined.
Table 5.1 lists the prefixes and corresponding XML namespaces.

5.1.2 EXAMPLE CONVENTIONS

The examples in this chapter are adjusted for readability.

- The namespace declarations in all examples are omitted and replaced with ellipses (. . .).
- Ellipses (. . .) are used in many of the examples in place of some of the details within
  the elements. This abbreviated form of the elements is used to improve clarity and
  focus.
- New, changed or otherwise important parts of the examples appear in bold face to draw
  attention to these parts.
- Extra white space is inserted in examples 5.7, 5.11, 5.12, 5.14, 5.15, and 5.28; this
  white space must be removed to correctly process the examples.

Valid versions of all of the examples in this chapter are available as part of this book's
supplementary materials [2].

5.1.3 RIGHTS EXPRESSION LANGUAGE ELEMENT AND TYPE CONVENTIONS

This chapter uses the following typographical conventions:

- Rights Expression Language elements appear in mixed case with an initial lowercase letter. Example elements are `r:grant`, `r:keyHolder`, and `r:validityInterval`.

Table 5.1 Prefixes and namespaces

Prefix	Namespace
r	`urn:mpeg:mpeg 21:2003:01-REL-R-NS`
sx	`urn:mpeg:mpeg 21:2003:01-REL-SX-NS`
mx	`urn:mpeg:mpeg 21:2003:01-REL-MX-NS`
c	`urn:mpeg:mpeg 21:2003:01-REL-SX-NS:2003:currency`
didl	`urn:mpeg:mpeg 21:2002:01-DIDL-NS`
dii	`urn:mpeg:mpeg 21:2002:01-DII-NS`
dsig	`http://www.w3.org/2000/09/xmldsig#`
xsd	`http://www.w3.org/2001/XMLSchema`

- Rights Expression Language types appear in mixed case with an initial uppercase letter. Example types are `r:Grant`, `r:KeyHolder`, and `r:ValidityInterval`.
- A related group of Rights Expression Language elements appears with an article, such as 'a' or 'the' and the name of a conceptually abstract element, beginning with an upper-case letter. For example, 'a Principal' refers to the group of elements that includes the `r:principal` element and all elements that can substitute for `r:principal`. Note that this convention differs slightly from the Rights Expression Language specification, which would use `r:Principal` to refer to a similar group of elements.
- Variables referenced within the text appear in bold italic. For example, 'the variable *x* is defined as follows: `<r:forAll varName="x"/>`'.

5.1.4 ELEMENT INTRODUCTION CONVENTIONS

Whenever a new element is first introduced, it is highlighted in grey. For example, the following paragraph introduces a fictitious element named `my:element`. These new elements also appear in the element index in Section 5.7.

The `my:element` element represents one of four cases: Earth, Air, Fire or Water.

5.2 INTRODUCTION TO RIGHTS EXPRESSION LANGUAGES

The unprecedented popularity of the Internet has created an environment in which digital content distribution has proliferated. Text, images, audio and video are distributed and shared, often indiscriminately, among Internet users. Indeed, many owners of valuable multimedia assets (such as publishers, record companies, video distributors) have contended that this indiscriminate sharing of copyrighted materials has resulted in a loss of revenue that is very damaging to their interests, and ultimately to those of the consumers.

The solution to this situation is to harness the public's desire to obtain and share digital content in such a way that all parties in the information exchange benefit. Digital content distribution offers enormous opportunities to the owners of multimedia assets. Using the Internet as a distribution medium, they can reach many more consumers, both faster and at a much lower cost than is possible with traditional content distribution techniques. In addition, they can leverage the consumers' file-sharing proclivities to create superdistribution channels, in effect enlisting the consumers as a sales force/distribution channel. To support these new and exciting business models, mechanisms are needed that enable content owners to specify the terms and conditions under which content that they own may be obtained, used and distributed.

The Internet also acts as a major channel of distribution for high-value digital content distributed other than 'for profit'. Enterprises and individual owners of confidential information need to enable information flow to the people who need it while restricting it from public access. For example, health care professionals (such as doctors) need to share confidential information about their patients to other health care providers (such as hospitals). To cite another example, legal and financial documents need to be exchanged between companies and their lawyers. Content distribution scenarios such as these could be greatly enhanced if a trustworthy mechanism existed to protect this confidential content from unauthorized use throughout the life cycle of the digital material.

The desire to take advantage of the opportunities offered by digital content distribution has fueled an ever-increasing demand for enabling technologies. Since the mid-1990s, companies such as Xerox®, IBM®, Microsoft®, Adobe® and InterTrust®[1] have been working on various systems for addressing this need.

One of these required technologies is a mechanism for expressing the rights, terms and conditions associated with each piece of content. In the last five years, standards groups such as MPEG, the Open eBook Forum (OeBF) [3], and the Open Mobile Alliance (OMA) [4] have taken on the task of standardizing rights expression technologies in the areas of multimedia and mobile data services, respectively.

For a rights expression technology to address the needs of all participants in the value chain, MPEG determined that the rights expression technology must provide the following features:

- It must express the rights, terms and conditions that apply throughout the content's life cycle.
- It must provide a common, unambiguous, machine-interpretable mechanism for expressing this information that is shared among all parties.
- It must address security issues, such as the integrity of rights expressions, trust and authentication.
- It must be both flexible and extensible to address as many business models as possible.

ISO/IEC 21000-5 meets these requirements in its specification of a Rights Expression Language, which provides the grammar used to express the rights that a particular entity has or is given regarding a certain resource, under a set of conditions. Each such expression is termed a *Rights Expression*. The language addresses trust and authentication issues by providing mechanisms for identifying the Rights Expression's issuer and verifying that the Rights Expression has not been tampered with.

Using a Rights Expression Language, content owners can associate a Rights Expression with their content that unambiguously describes all the rights and conditions that downstream parties (such as distributors and end-users) may exercise in relation to that content over the entire course of its life cycle.

For example, if Alice holds copyright to a video file, she can associate a Rights Expression with it that expresses the following rights:

- A distributor, Bob, may distribute the video file to consumers if he pays Alice a fee each time he does so.
- A consumer may play that video file beginning from 2006.

In this way, Alice can originate the rights that all downstream participants have to her video file, not just those parties with whom Alice interacts directly. In addition, Alice can sign the Rights Expression, which enables downstream participants to verify its integrity.

---

[1] Xerox is a registered trademark of Xerox Corporation.
IBM is a registered trademark of IBM in the United States.
Microsoft is a registered trademark of Microsoft Corporation.
Adobe is a registered trademark of Adobe Systems Incorporated in the United States and/or other countries.
InterTrust is a registered trademark of InterTrust Technologies Corp.

A Rights Expression Language should address a broad range of business models, including the following:

- rights to use resources, such as multimedia content;
- rights to distribute resources; and
- rights to own characteristics or properties, such as memberships or email addresses.

In addition to governing the use of digital content, a Rights Expression Language can be used in many other domains and applications, including as a generic authorization language. For example, a Rights Expression Language can be used to express rights over a web service or to express a digital certificate independent of its use in conjunction with specifying the usage rights of digital content.

A Rights Expression Language should also support new business models through extensions and optimization through profiles. Extensions add elements to the language that are needed to address the requirements of a particular business segment. Profiles, on the other hand, identify a subset of the elements in the language that is sufficient to address the needs of a particular business segment.

The Rights Expression Language is typically one part of a system that enables digital content distribution while managing the rights of its Users. Such systems are usually called *Digital Rights Management (DRM) systems.*

To better understand the scope of a Rights Expression Language, it is instructive to contrast it with the scope of systems that use Rights Expressions. In addition to Rights Expression generation and interpretation, systems that use Rights Expressions typically also provide several means, both technical and nontechnical (laws, incentives and the like), to address each of the following additional functionalities:

- ensuring the availability of Rights Expressions and resources;
- logically associating Rights Expressions to resources;
- authenticating Rights Expressions, resources, Users, systems and devices;
- protecting resources from being used without rights; and
- enabling resources to be used consistent with rights.

However, it is not uncommon to find systems that use Rights Expressions yet provide only very rudimentary means, if any at all, to address some of these functionalities. For example, one system might not provide any way to ensure the availability of resources. A different system might have some way to ensure the availability of resources but might rely only on laws (and their enforcement) to protect the resources from being used without the rights to do so.

The scope of ISO/IEC 21000-5 does not include standardizing any of the means listed above; it is limited strictly to the syntax and semantics of the language. For example, although the language defined by ISO/IEC 21000-5 provides the syntax and semantics for each License to identify the resources to which it applies and although that information is useful in finding those resources, verifying them and associating them with the License, ISO/IEC 21000-5 does not dictate how such finding, verification and association take place. Similarly, ISO/IEC 21000-5 specifies an element `r:validityInterval` used to express a limit on the exercise of a right to within a specified interval of time. This element simply specifies the two boundaries of the validity interval (when it begins and ends). It does not stipulate anything about the use of the `r:validityInterval` condition in a deployed system. For instance, it does *not* stipulate how the system determines the current time or what security precautions it takes while doing so.

5.3 UNDERSTANDING LICENSES

Rights Expressions are sometimes referred to as *Licenses*, for short. It is understood from the technical context that the term *Licenses* refers to machine-interpretable expressions written in the Rights Expression Language and not to legal licenses.

ISO/IEC 21000-5 defines an `r:license` element to represent a Rights Expression. The two most important parts of a License are the following:

- an `r:grant`, which describes the terms of the License and
- an `r:issuer`, which identifies the party issuing the License.

For example, suppose Bob issues a License to Cathy that allows her to play a particular video file. At the most basic level, the `r:license` Bob issues would contain an `r:grant` that says 'Cathy may play the video file' and an `r:issuer` that identifies Bob as the party issuing the License.

Sections 5.3.1 and 5.3.2 describe the `r:grant` and `r:issuer` elements and demonstrate how Bob can use them to construct a License.

5.3.1 GRANTS

The `r:grant` element is a fundamental construct of the Rights Expression Language and encapsulates much of the information necessary to construct Rights Expressions. Sections 5.3.1.1 through 5.3.1.4 describe the basics of an `r:grant` element's structure and provide some examples.

5.3.1.1 Understanding the Basic Grant Structure

When issuing a License, an issuer includes those `r:grant` elements necessary to convey the intent of the License.

Figure 5.1 illustrates the basic structure of an `r:grant`.

Each `r:grant` may contain a Principal, Right, Resource and Condition (as well as other elements that will be introduced later in this chapter).

- A Principal is the 'subject' of an `r:grant`. It identifies an entity capable of acting in a system. Such an entity is typically identified using information unique to that entity, which often includes an associated authentication mechanism by which the entity can prove its identity. Example principals include people, members of a club, devices and repositories.

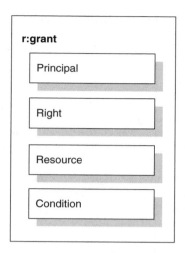

Figure 5.1   Grant structure

- A Right is the 'verb' of an `r:grant`. It identifies an act. Example rights include play, print, issue and obtain.

- A Resource is the 'object' in an `r:grant`. It can identify anything that can be referred to by a noun, such as an entity, quality, event, state, concept or substance. Example resources include multimedia files, web services, names and email addresses.

- A Condition specifies a permission limitation. It encapsulates the information necessary to determine whether the Condition is satisfied. Example conditions include fees, validity time intervals, and limits on the number of times rights may be exercised.

ISO/IEC 21000-5 defines four conceptually abstract elements that represent the four `r:grant` components listed above in a general way. These conceptually abstract elements are `r:principal`, `r:right`, `r:resource` and `r:condition`. For instance, the `r:right` element represents the abstract concept of an act.

Extensions to these elements are defined that represent particular, concrete instances of these abstractions. These extension elements can substitute for the conceptually abstract elements from which they are derived. For this reason, each conceptually abstract element is also known as a '*substitution head*'.

Table 5.2 illustrates this concept. Each element listed in the column 'Example extension element' can appear in place of the corresponding conceptually abstract element in the column 'Substitution head'.

Although this table lists only one example element that can substitute for each conceptually abstract element, the Rights Expression Language has several elements that can substitute for each of these conceptually abstract elements. For example, the Rights Expression Language has several Rights, each of which represents a particular, concrete act.

Table 5.2   Substitution heads and example extension elements

Substitution head	Example extension element
r:principal	r:keyHolder Used to refer to the holder of a secret key, such as the private key of a public/private key pair. The secret key is typically not provided inside the r:keyHolder element. Instead, either the name of the secret key or the public key of a public/private key pair is provided.
r:right	mx:play Used to refer to the act of deriving a transient and directly perceivable representation of a resource, such as playing an audio file or viewing a text file.
r:resource	mx:diReference Used to refer to a Digital Item or a component, such as a media file, using its Digital Item identifier.
r:condition	r:validityInterval Used to refer to a time interval (with the specified starting and/or ending dates and times) for which this Condition is satisfied.

Sections 5.3.1.2 through 5.3.1.4 provide some simple examples that use the elements r:keyHolder, mx:play, mx:diReference and r:validityInterval to demonstrate how r:grant elements can be constructed.

5.3.1.2 Constructing a Simple Grant

To begin with a simple example, the following r:grant contains only a Right and a Resource. It omits the Principal and Condition components of an r:grant. Without a Principal or Condition, this r:grant does not impose constraints as to whom or under which conditions it applies. That is, this r:grant would be usable by any entity in any situation.

```
<r:grant...>
   <mx:play/>
   <mx:diReference>
      <mx:identifier>urn:example:001</mx:identifier>
   </mx:diReference>
</r:grant>
```

Example 5.1

This example r:grant contains two elements:

- a concrete Right, mx:play;
- a concrete Resource, an mx:diReference element that refers to a Digital Item identified as urn:example:001.

Although a Resource is optional within an r:grant, specifying a Resource is appropriate in this case. Since this r:grant contains the mx:play Right, specifying a Resource answers the question 'what may be played?' For example, if urn:example:001 refers to a video file, a License containing this `r:grant` would express that the License issuer allows that video file to be played.

Sections 5.3.1.3 and 5.3.1.4 expand this simple example to include Principal and Condition elements, respectively.

5.3.1.3 Constraining a Simple Grant to a Specific Party

This example adds an `r:keyHolder` Principal to the simple `r:grant` in Example 5.1. In this example, the `r:keyholder` element represents a Principal named Cathy. Unlike Example 5.1, this `r:grant` can be used to authorize only Cathy to play the video file.

```
<r:grant...>
   <r:keyHolder>
      <r:info>
         <dsig:KeyValue>
            <dsig:RSAKeyValue>
               <dsig:Modulus>Cat7QzxAprs=</dsig:Modulus>
               <dsig:Exponent>AQABAA==</dsig:Exponent>
            </dsig:RSAKeyValue>
         </dsig:KeyValue>
      </r:info>
   </r:keyHolder>
   <mx:play/>
   <mx:diReference>
      <mx:identifier>urn:example:001</mx:identifier>
   </mx:diReference>
</r:grant>
```

Example 5.2

The `r:keyHolder` element uses W3C XML digital signature [5] technology to represent a public key. However, it is not necessary to understand the details of W3C digital signatures to understand the examples presented in this chapter. To use this example `r:grant`, Cathy must establish that she holds the corresponding private key to play the specified video file.

5.3.1.4 Constructing a Conditional Simple Grant

This example adds an `r:validityInterval` Condition to the simple `r:grant` in Example 5.2. Unlike Example 5.2, this `r:grant` authorizes Cathy to play the video file only during the specified validity interval. It does not authorize Cathy to play the video file at any other time.

```
<r:grant...>
   <r:keyHolder>
      <r:info>
         <dsig:KeyValue>
            <dsig:RSAKeyValue>
               <dsig:Modulus>Cat7QzxAprs=</dsig:Modulus>
               <dsig:Exponent>AQABAA==</dsig:Exponent>
            </dsig:RSAKeyValue>
         </dsig:KeyValue>
      </r:info>
   </r:keyHolder>
```

```
    </r:keyHolder>
      <mx:play/>
      <mx:diReference>
         <mx:identifier>urn:example:001</mx:identifier>
      </mx:diReference>
      <r:validityInterval>
         <r:notBefore>2003-01-01T00:00:00</r:notBefore>
         <r:notAfter>2004-01-01T00:00:00</r:notAfter>
      </r:validityInterval>
    </r:grant>
```

Example 5.3

The `r:validityInterval` element specifies a time period to which the `r:grant` applies. It contains two elements:

- `r:notBefore`: identifies the start of the time period. The `r:grant` may not be used to authorize actions occurring before this date and time.
- `r:notAfter`: identifies the end of the time period. The `r:grant` may not be used to authorize actions occurring after this date and time.

An `r:validityInterval` may omit either of these elements. For example, omitting the `r:notAfter` element results in an open-ended validity interval – that is, the `r:grant` would be usable to authorize actions occurring any time after the start time stipulated by `r:notBefore`.

To use this example `r:grant`, the time that Cathy plays the specified video file must fall between midnight of January 1, 2003, and midnight of January 1, 2004.

5.3.2 ISSUER

When issuing a License, the issuer (Bob, for example) includes an `r:issuer` element that identifies him and usually contains his signature over the License.

The `r:issuer` element in the example below contains a `dsig:Signature` [5] with three children:

- `dsig:SignedInfo`: specifies the information that is being signed. The transform algorithm `urn:mpeg:mpeg21:2003:01-REL-R-NS:licenseTransform` defined in ISO/IEC 21000-5 indicates that the containing License is being selected for inclusion in the signature. The transform algorithm `urn:uddi-org:schemaCentricC14N:2002-07-10` [6] indicates that Schema Centric Canonicalization is used to canonicalize the containing License before it is digested. The `dsig:DigestValue` gives the value of the digested canonicalized License.
- `dsig:SignatureValue`: specifies the value of the signature of the `dsig:SignedInfo` by the key given in the `dsig:KeyInfo`.
- `dsig:KeyInfo`: indicates the License issuer's public key. For example, if Bob issued a License, the `dsig:KeyInfo` element contains Bob's public key.

In the example below and in other example Licenses in this chapter, the values of `dsig:DigestValue` and `dsig:SignatureValue` elements are for illustration purposes only. They are not cryptographically correct.

```
<r:issuer...>
   <dsig:Signature>
      <dsig:SignedInfo>
         <dsig:CanonicalizationMethod
            Algorithm="http://www.w3.org/TR/2001/REC-xml-c14n-20010315"/>
         <dsig:SignatureMethod
            Algorithm="http://www.w3.org/2000/09/xmldsig#rsa-sha1"/>
         <dsig:Reference>
            <dsig:Transforms>
               <dsig:Transform Algorithm=
                  "urn:mpeg:mpeg21:2003:01-REL-R-NS:licenseTransform"/>
               <dsig:Transform Algorithm=
                  "urn:uddi-org:schemaCentricC14N:2002-07-10"/>
            </dsig:Transforms>
            <dsig:DigestMethod
               Algorithm="http://www.w3.org/2000/09/xmldsig#sha1"/>
            <dsig:DigestValue>L05dJk9QdTExKO4ltbj1/Q==</dsig:DigestValue>
         </dsig:Reference>
      </dsig:SignedInfo>
      <dsig:SignatureValue>L05qOQhAprs=</dsig:SignatureValue>
      <dsig:KeyInfo>
         <dsig:KeyValue>
            <dsig:RSAKeyValue>
               <dsig:Modulus>Bob7QzxAprs=</dsig:Modulus>
               <dsig:Exponent>AQABAA==</dsig:Exponent>
            </dsig:RSAKeyValue>
         </dsig:KeyValue>
      </dsig:KeyInfo>
   </dsig:Signature>
</r:issuer>
```

Example 5.4

5.3.3 EXAMPLE LICENSE

The following example License combines the `r:grant` element from Example 5.3 with
the `r:issuer` element from Example 5.4. In this example License, Bob (the issuer)
authorizes Cathy to play a video file within a specified validity interval.

```
<r:license...>
   <r:grant>
      <r:keyHolder>
         <r:info>
            <dsig:KeyValue>
               <dsig:RSAKeyValue>
                  <dsig:Modulus>Cat7QzxAprs=</dsig:Modulus>
                  <dsig:Exponent>AQABAA==</dsig:Exponent>
               </dsig:RSAKeyValue>
            </dsig:KeyValue>
         </r:info>
      </r:keyHolder>
      <mx:play/>
      <mx:diReference>
         <mx:identifier>urn:example:001</mx:identifier>
      </mx:diReference>
```

```
        <r:validityInterval>
            <r:notBefore>2003-01-01T00:00:00</r:notBefore>
            <r:notAfter>2004-01-01T00:00:00</r:notAfter>
        </r:validityInterval>
    </r:grant>
    <r:issuer>
        <dsig:Signature>
            <dsig:SignedInfo>
                <dsig:CanonicalizationMethod
                    Algorithm="http://www.w3.org/TR/2001/REC-xml-c14n-20010315"/>
                <dsig:SignatureMethod
                    Algorithm="http://www.w3.org/2000/09/xmldsig#rsa-sha1"/>
                <dsig:Reference>
                    <dsig:Transforms>
                        <dsig:Transform Algorithm=
                            "urn:mpeg:mpeg21:2003:01-REL-R-NS:licenseTransform"/>
                        <dsig:Transform Algorithm=
                            "urn:uddi-org:schemaCentricC14N:2002-07-10"/>
                    </dsig:Transforms>
                    <dsig:DigestMethod
                        Algorithm="http://www.w3.org/2000/09/xmldsig#sha1"/>
                    <dsig:DigestValue>L05dJk9QdTExKO4ltbj1/Q==</dsig:DigestValue>
                </dsig:Reference>
            </dsig:SignedInfo>
            <dsig:SignatureValue>L05qOQhAprs=</dsig:SignatureValue>
            <dsig:KeyInfo>
                <dsig:KeyValue>
                    <dsig:RSAKeyValue>
                        <dsig:Modulus>Bob7QzxAprs=</dsig:Modulus>
                        <dsig:Exponent>AQABAA==</dsig:Exponent>
                    </dsig:RSAKeyValue>
                </dsig:KeyValue>
            </dsig:KeyInfo>
        </dsig:Signature>
    </r:issuer>
</r:license>
```

Example 5.5

5.4 UNDERSTANDING AUTHORIZATION

Section 5.3.1 explained the r:grant element. The r:grant element plays an important role in the authorization model, which is described in more detail later in this section. r:grant elements reside in two main places:

- in Licenses, as described above;
- in a set of root grants, as described in 5.4.1.

The r:grant elements in Licenses and root grants are used in accordance with the authorization model to convey authorization.

Beginning with this section, many of the examples use a notation that omits some of the detail within the elements to improve clarity and focus of the examples. For instance, the following abbreviated XML shows how Example 5.3 can be collapsed to focus attention on the new element, r:validityInterval:

```
<r:grant...>
   <r:keyHolder>...Cat...</r:keyHolder>
   <mx:play/>
   <mx:diReference>...</mx:diReference>
   <r:validityInterval>
       <r:notBefore>2003-01-01T00:00:00</r:notBefore>
       <r:notAfter>2004-01-01T00:00:00</r:notAfter>
   </r:validityInterval>
</r:grant>
```

In this abbreviated XML, 'Cat' within the `r:keyHolder` element is used to differentiate this `r:keyHolder` (which represents Cathy) from other `r:keyHolder` elements that may be included in the abbreviated XML.

5.4.1 ROOT GRANTS

A set of root grants is just a set of `r:grant` elements that is determined by the application as containing all those `r:grant` elements that do not need to appear in Licenses so as to be in effect. For example, a root grant may specify that a company that holds copyright for a particular book may issue Licenses to anyone, granting any rights to that book under any conditions. In this case, the copyright holder does not need to present a License proving the holder has the right to issue the Licenses.

There are a number of business, social and legal considerations that go into designing the set of root grants that an application will use.

ISO/IEC 21000-5 does not standardize any particular set of root grants. Application designers might be required by law to include or exclude certain `r:grant` elements from their set of root grants. Additionally, application designers might be required by contract (for instance, by the contract they must sign to get access to secret keys to access content) to include or exclude certain `r:grant` elements from their set of root grants. Other `r:grant` elements may be included or excluded from the set of root grants simply because of customer demand or other business considerations to make an application more attractive in the market.

Regardless of how the application's set of root grants is determined, the authorization model accounts for them as part of an authorization request, as described in the following text.

5.4.2 UNDERSTANDING THE AUTHORIZATION MODEL

ISO/IEC 21000-5 defines an authorization model to specify the semantics of Licenses. The two most prominent concepts in the authorization model are the authorization request and the authorization proof. The authorization request is an important introductory concept and is explained in more detail in Section 5.4.2.1.

The concept of an authorization proof is more involved and is beyond the scope of this chapter. Conceptually, if the answer to an authorization request is 'yes', there must be an authorization proof for that authorization request to prove that the answer is 'yes'. Conversely, if there is no authorization proof for an authorization request, the answer to that authorization request is 'no'.

In most of the common cases, it is possible to think about an authorization request and quickly determine by human intuition whether the answer to the authorization request is 'yes' or 'no', so it is not necessary to resort to creating an authorization proof. However, human intuition is sometimes faulty or biased, so ISO/IEC 21000-5 fully specifies the details of an authorization proof in a mathematical way. If there is any doubt or ambiguity as to the answer to an authorization request, one should refer to the mathematical description of authorization proof given in sub-clause 5.7 of ISO/IEC 21000-5. The goal of this chapter on authorization model is to develop the reader's intuition as to whether the answer to an authorization request is 'yes' or 'no'.

5.4.2.1 Authorization Request

An authorization request represents a question that can be answered with an 'yes' or 'no', with 'yes' being the reply if an authorization proof for the authorization request can be found and 'no' being the reply otherwise. An example authorization request would be 'Is Cathy permitted to play a video file (the DI identified as `urn:example:001`) between midnight of 2003-09-01 and midnight of 2003-09-02 according to her Licenses and the root grants that are in force?'

An authorization request consists of seven components. In order of their appearance in an authorization request, they are a Principal, a Right, a Resource, an interval of time, an authorization context, a set of Licenses and a set of root grants. Licenses and root grants were explained earlier. The other five components (a Principal, a Right, a Resource, an interval of time and an authorization context) are used to represent an activity that has occurred or may potentially occur. The Principal identifies the doer of the activity (Cathy). The Right identifies the act taking place (play). The Resource identifies the object of the activity (the DI identified as `urn:example:001`). The interval of time stretches from the start of the activity to the end of the activity (between midnight of 2003-09-01 and midnight of 2003-09-02). The authorization context contains other information about the activity that might be useful in evaluating the authorization request. ISO/IEC 21000-5 provides a rigorous definition of the authorization context. Conceptually, the authorization context includes information about what devices are used in the activity, where the activity takes place, how many similar activities take place before this activity, what fees has been paid for the activity, and so on.

The question represented by an authorization request is 'is the activity represented by the first five components (a Principal, a Right, a Resource, an interval of time, and an authorization context) permitted according to the sixth and the seventh components (the set of Licenses and the set of root grants)?'

5.4.2.2 Evaluating an Authorization Request by Intuition

This section describes a two-step process by which a simple authorization request can be intuitively evaluated.

1. Check whether the activity in an authorization request corresponds to one of the root grants given in that authorization request.
   a. If it does, the answer to the authorization request is 'yes'.
   b. Otherwise, go to Step 2.
2. Check whether the activity in an authorization request corresponds to one of the `r:grant` elements in one of the Licenses in the authorization request.
   a. If it does, the next step is to verify that the corresponding `r:grant` was properly issued. Using a Principal, a Right, a Resource, an interval of time, and an

authorization context, describe the activity of that corresponding `r:grant` being included in that License by the issuer of that License. Go to Step 1 using the new activity.

b. Otherwise, the answer to the authorization request is 'no'.

These steps are depicted as diamonds in the flow chart in Figure 5.2.

Both the steps in this process involve checking for correspondence with an activity. As will be seen in examples throughout this chapter, checking for correspondence often involves checking for equality between two elements (for example, the Principal from the activity and the Principal in a root grant). A simple string comparison of two elements may not be the best way to determine equality, because it is possible for two elements that are not syntactically identical to convey the same meaning at the XML level. At the same time, it is possible for two elements that are unequal at the XML level to represent or identify the same object or entity at the application level. To strike an appropriate balance to allow some flexibility in representation without requiring full application semantic knowledge in all instances, equality is defined in sub-clause 6.1 of ISO/IEC 21000-5 to occur at the XML infoset level.

The following two sections describe two examples of using the process shown in Figure 5.2 to evaluate an authorization request.

Evaluating a Simple Authorization Request by Intuition

This section describes the evaluation of a simple authorization request by intuition. In this example, Cathy wants to play a video file (the DI identified as `urn:example:001`) on September 1, 2003. In this simple example, Cathy does not have any Licenses, but a root grant is in force.

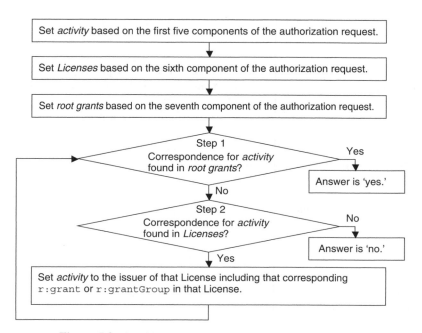

Figure 5.2   Intuitive evaluation of an authorization request

Table 5.3 Simple authorization request

Component	Value
Principal	`<r:keyHolder>...Cat...</r:keyHolder>`
Right	`<mx:play/>`

Resource	`<mx:diReference>`
	`<mx:identifier>urn:example:001</mx:identifier>`
	`</mx:diReference>`

Interval of time	Midnight of 2003-09-01 through midnight of 2003-09-02
Authorization context	...
Set of Licenses	The empty set

Set of root grants	`<r:grant...>`
	`<r:keyHolder>...Cat...</r:keyHolder>`
	`<mx:play/>`
	`<mx:diReference>`
	`<mx:identifier>urn:example:001</mx:identifier>`
	`</mx:diReference>`
	`<r:validityInterval>`
	`<r:notBefore>2003-01-01T 00:00:00</r:notBefore>`
	`<r:notAfter> 2004-01-01T 00:00:00</r:notAfter>`
	`</r:validityInterval>`
	`</r:grant>`

Authorization Request
The simple authorization request outlined in Table 5.3 is used for this demonstration.

Step 1

The first step is to check whether the activity given by the Principal, Right, Resource, interval of time and authorization context components of the authorization request corresponds to one of the root grants given in the authorization request.

In the above authorization request, the Principal equals the Principal in the root grant, the Right equals the Right in the root grant, the Resource equals the Resource in the root grant, and the `r:validityInterval` in the root grant is satisfied because the interval of time (midnight of September 1, 2003, through midnight of September 2, 2003) falls within the validity interval. Therefore, the activity in the authorization request corresponds to the root grant in the authorization request, so the authorization request should be answered 'yes', and there is no need to proceed to Step 2.

Step 2

Step 2 does not apply to this simple authorization request, since a correspondence was found in Step 1.

Evaluating a More Typical Authorization Request by Intuition
This section describes the evaluation of a more typical authorization request by intuition. As in the example in the previous section, Cathy wants to play a video file (the DI identified as `urn:example:001`) on September 1, 2003. However, in this

example, Cathy has a License that must be evaluated. In addition, a different root grant is in force.

Authorization Request

The more typical authorization request outlined in Table 5.4 is used for this demonstration.

At this point, it is not important to understand 'trustedRootIssuers Bob' beyond understanding that it is shorthand for a set of root grants that say 'Bob may put whatever he wants in the Licenses he issues'.

Step 1

As before, the first step in evaluating an authorization request by intuition is to check whether the activity given by the Principal, Right, Resource, interval of time and authorization context components of the authorization request corresponds to one of the root grants given in the authorization request.

Table 5.4 More typical authorization request

Component	Value
Principal	`<r:keyHolder>...Cat...</r:keyHolder>`
Right	`<mx:play/>`

Resource	`<mx:diReference>`
	`  <mx:identifier>urn:example:001</mx:identifier>`
	`</mx:diReference>`

Interval of time	Midnight of 2003-09-01 through midnight of 2003-09-02
Authorization context	...

Set of Licenses	`<r:license...>`
	`  <r:grant>`
	`  <r:keyHolder>...Cat...</r:keyHolder>`
	`  <mx:play/>`
	`  <mx:diReference>`
	`   <mx:identifier>urn:example:001</mx:identifier>`
	`  </mx:diReference>`
	`  <r:validityInterval>`
	`   <r:notBefore>2003-01-01T00:00:00</r:notBefore>`
	`   <r:notAfter>2004-01-01T00:00:00</r:notAfter>`
	`  </r:validityInterval>`
	`  </r:grant>`
	`  <r:issuer>`
	`   <dsig:Signature>`
	`   <dsig:SignedInfo>...</dsig:SignedInfo>`
	`   <dsig:SignatureValue>L05qOQhAprs=</dsig:SignatureValue>`
	`   <dsig:KeyInfo>...Bob...</dsig:KeyInfo>`
	`   </dsig:Signature>`
	`  </r:issuer>`
	`  </r:license>`

Set of root grants	trustedRootIssuers Bob

In this case, no correspondence is found in this step because the activity, 'Cathy playing the DI identified as `urn:example:001` on September 1, 2003', does not correspond to the root grants, which state that 'Bob may put whatever he wants in the Licenses he issues'.

Evaluation proceeds to Step 2.

Step 2

The second step in intuitively evaluating an authorization request (if a correspondence for the activity is not found in one of the root grants) is to check for a correspondence in one of the `r:grant` elements in one of the Licenses in the authorization request.

In this example, a correspondence is found in Step 2 because the activity, which is 'Cathy playing the DI identified as `urn:example:001` on September 1, 2003', corresponds with the `r:grant` element in the License, which states that 'Cathy may play the DI identified as `urn:example:001` within 2003'.

However, Step 2 is not complete because Bob might not be permitted to issue such Licenses. Since a correspondence was found in a License issued by Bob, a new activity comes into question: the activity of Bob putting that `r:grant` in the License he issued. A correspondence must now be found for this activity, starting at Step 1.

Step 1 (Issuance chaining link 1)

When Step 1 (checking if a correspondence for the activity is found in one of the root grants) is performed with the new activity, a correspondence is found because the activity is 'Bob putting Cathy's `r:grant` in the License he issued' and the root grants state that 'Bob may put whatever he wants in the Licenses he issues'.

Step 2 (Issuance chaining link 1)

Step 2 does not apply for the first link of issuance chaining because a correspondence was found in Step 1.

Comments on Issuance Chaining

This process of checking issuance by going back to Step 1 is sometimes called '*issuance chaining*', and it continues from activity to activity as long as the correspondence for the activity is found as part of Step 2 (in a License). Once a correspondence is found as part of Step 1 (in the root grants), as in the case of 'Bob may put whatever he wants in the Licenses he issues' above, the chain is said to be complete, and the authorization request should be answered 'yes'.

If a correspondence is not found in Step 2, the authorization request should be answered 'no'. If a correspondence is found in Step 2 but the 'issuance chaining' recurs infinitely, the authorization request should also be answered 'no'.

Most of the examples in this chapter involve only one link of 'issuance chaining', as is the case with this example. The exceptions are in the Section '*May Cathy Play?*' (see 5.5.4.7) and Section 5.5.11. The former includes an example of two links of 'issuance chaining' using the `r:issue` element. Section 5.5.11 includes an example of two links of 'issuance chaining' using the `r:delegationControl` element.

5.5 SELECTED FEATURES

Sections 5.3 and 5.4 describe the most fundamental features of the Rights Expression Language defined by ISO/IEC 21000-5. An understanding of these concepts provides a foundation for understanding and using all other features that the language offers.

The following sections cover additional features that were selected because they foster a deeper understanding of the language and because they are broadly applicable to many usage scenarios. For information on language features not covered in these sections, the reader is encouraged to consult the materials listed in Section 5.6.2.

5.5.1 USING A CONJUNCTIVE PRINCIPAL

Section 5.3.1.3 describes how to specify a single Principal in an `r:grant`. In some situations, however, it may be necessary to represent an entity that must prove it has multiple identities. For instance, Cathy may hold two private keys:

- a private key corresponding to a public key with modulus 'Cat7QzxAprs=' and exponent 'AQABAA=='
- a private key corresponding to a public key with modulus 'Cat8dTzx4Q==' and exponent 'AQABAA=='

Bob may issue Cathy a License that permits her to play a secure slide show presentation (a DI) during a certain time period. For added security, Bob may require Cathy to present both her keys, in case one of them is compromised. The following example License demonstrates this case.

```
<r:license...>
   <r:grant>
      <r:allPrincipals>
         <r:keyHolder>
            <r:info>
               <dsig:KeyValue>
                  <dsig:RSAKeyValue>
                     <dsig:Modulus>Cat7QzxAprs=</dsig:Modulus>
                     <dsig:Exponent>AQABAA==</dsig:Exponent>
                  </dsig:RSAKeyValue>
               </dsig:KeyValue>
            </r:info>
         </r:keyHolder>
         <r:keyHolder>
            <r:info>
               <dsig:KeyValue>
                  <dsig:RSAKeyValue>
                     <dsig:Modulus>Cat8dTzx4Q==</dsig:Modulus>
                     <dsig:Exponent>AQABAA==</dsig:Exponent>
                  </dsig:RSAKeyValue>
               </dsig:KeyValue>
            </r:info>
         </r:keyHolder>
      </r:allPrincipals>
      <mx:play/>
```

```
        <mx:diReference>...</mx:diReference>
        <r:validityInterval>...</r:validityInterval>
     </r:grant>
     <r:issuer>
        <dsig:Signature>
           <dsig:SignedInfo>...</dsig:SignedInfo>
           <dsig:SignatureValue>L06qOQhAprs=</dsig:SignatureValue>
           <dsig:KeyInfo>...Bob...</dsig:KeyInfo>
        </dsig:Signature>
     </r:issuer>
  </r:license>
```

Example 5.6

The `r:allPrincipals` element represents an entity that holds several identifications. If an entity does not prove its identity for all of the identifications given within an r:allPrincipals, then an entity does not prove its identity for the r:allPrincipals as a whole. An r:allPrincipals can contain any number of Principals, and the order of the Principals is not significant. An empty r:allPrincipals element has the same effect as having no Principal at all. An r:allPrincipals element can substitute for an r:principal element, so it may be used anywhere that an r:principal can be used. In Example 5.6, an r:allPrincipals is used to substitute the r:principal inside an r:grant.

If Cathy uses this License to play the specified DI, she must establish that she holds both the private keys corresponding to a public key with modulus 'Cat7QzxAprs=' and exponent 'AQABAA==' and the private key corresponding to a public key with modulus 'Cat8dTzx4Q==' and exponent 'AQABAA==.'

5.5.2 USING A CONJUNCTIVE CONDITION

Section 5.3.1.4 describes how to add a single Condition to an r:grant. Many use cases require that more than one Condition be satisfied for a permission to be granted. For instance, Cathy might be permitted to play a video file (a DI) between midnight of January 1, 2003, and midnight of January 1, 2004, and with a maximum of five times on the basis of one License. The following example License demonstrates this case.

```
  <r:license...>
     <r:grant>
        <r:keyHolder>...Cat...</r:keyHolder>
        <mx:play/>
        <mx:diReference>...</mx:diReference>
        <r:allConditions>
           <r:validityInterval>
              <r:notBefore>2003-01-01T00:00:00</r:notBefore>
              <r:notAfter>2004-01-01T00:00:00</r:notAfter>
           </r:validityInterval>
           <sx:exerciseLimit>
             <sx:count>5</sx:count>
           </sx:exerciseLimit>
        </r:allConditions>
     </r:grant>
     <r:issuer>
```

```
<dsig:Signature>
   <dsig:SignedInfo>...</dsig:SignedInfo>
   <dsig:SignatureValue>L07qOQhAprs=</dsig:SignatureValue>
   <dsig:KeyInfo>...Bob...</dsig:KeyInfo>
</dsig:Signature>
   </r:issuer>
</r:license>
```

Example 5.7

This example uses two new Conditions: `r:allConditions` and `sx:exerciseLimit`.

The `r:allConditions` element is a container for Conditions. It is satisfied only if all of the contained Conditions are satisfied. An `r:allConditions` can contain any number of Conditions, and the order of the Conditions is not significant. An empty `r:allConditions` element has the same effect as having no Condition at all. An `r:allConditions` element can substitute for an `r:condition` element, so it may be used anywhere that an `r:condition` can be used. In this example, an `r:allConditions` substitutes for the `r:condition` in an `r:grant`.

If Cathy uses this License to play the specified DI, both the `r:validityInterval` and the `sx:exerciseLimit` Conditions must be satisfied. This means that the time of exercise must be between midnight of January 1, 2003 and midnight of January 1, 2004 and the number of exercises based on this License must not be more than the five exercises specified by the `sx:exerciseLimit`.

As Section 5.4.2.1 explains, the interval of time when an activity takes place is carried within an authorization request as the latter's fourth component. This interval of time is used to determine if an `r:validityInterval` Condition is satisfied. An `sx:exerciseLimit` Condition works in much the same way. The fifth component of an authorization request is called the *authorization context*, and it carries miscellaneous information. In particular, it would carry the information as to how many exercises are based on a particular License. This information can then be used to determine if an `sx:exerciseLimit` Condition is satisfied by comparing the value in the authorization request to the value inside the `sx:exerciseLimit` Condition.

This example illustrates a particular usage of `sx:exerciseLimit` that stipulates the number of exercises per License. This is a convenient usage for situations in which the flow of these Licenses is restricted to a limited number of implementations that have some communication mechanism by which they can determine how many exercises are performed for each License. In situations where Licenses are more widely distributed and accessed, a more open approach is needed. An alternate usage of `sx:exerciseLimit` stipulates the number of exercises per `r:serviceReference`. Any implementation needing to know how many exercises there have been for that `r:serviceReference` can contact the referenced service to find out the information. This alternate usage is discussed in more detail in Section 5.5.3.

5.5.3 KEEPING STATE USING A SERVICE

Section 5.5.2 describes how state can be kept on a per-License basis and refers to this section for details about keeping state on a per-`r:serviceReference` basis. Keeping state on a per-r:serviceReference basis is useful if, among other cases, state is to be shared between two Users with two different Licenses or state is to be kept in common for one User using one License on two different systems.

For instance, Cathy might be permitted to play a specified DI a maximum of five times on the basis of a r:serviceReference. If she plays twice on one system, those two exercises should be recorded with the service described in the r:serviceReference, so that when she moves to another system she has only three exercises remaining. The following example License demonstrates this case.

```
<r:license...>
   <r:grant>
       <r:keyHolder>...Cat...</r:keyHolder>
       <mx:play/>
       <mx:diReference>...</mx:diReference>
       <sx:exerciseLimit>
           <r:serviceReference>
               <sx:uddi>
                   <sx:serviceKey>
                       <sx:uuid>B7DB5111-D1C2-3964-D049-D3D11332CE14</sx:uuid>
                   </sx:serviceKey>
               </sx:uddi>
               <r:serviceParameters>
                   <r:datum>
                       <sx:stateDistinguisher>
                           B70150EE-8F0B-3f47-827C-06959EE527E9
                       </sx:stateDistinguisher>
                   </r:datum>
               </r:serviceParameters>
           </r:serviceReference>
           <sx:count>5</sx:count>
       </sx:exerciseLimit>
   </r:grant>
   <r:issuer>
     <dsig:Signature>
         <dsig:SignedInfo>...</dsig:SignedInfo>
         <dsig:SignatureValue>L08qOQhAprs=</dsig:SignatureValue>
         <dsig:KeyInfo>...Bob...</dsig:KeyInfo>
     </dsig:Signature>
   </r:issuer>
</r:license>
```

Example 5.8

As mentioned in Section 5.5.2, the sx:exerciseLimit Condition indicates the maximum number of exercises. In this example License, the sx:count element within the sx:exerciseLimit sets this maximum to five. The r:serviceReference element within the sx:exerciseLimit indicates that the state for this count is to be kept on a per-r:serviceReference basis rather than a per-License basis.

The r:serviceReference encapsulates the information necessary to interact with a service. Within an sx:exerciseLimit, this information is used to retrieve and update state information. Any software or hardware, local or remote, with which an implementation interacts can be described using the r:serviceReference element.

The `r:serviceReference` element has two children:

- *A service description*: Services can be described using a variety of technologies. ISO/IEC 21000-5 explicitly provides for some of these technologies, while others are likely to be handled in extensions. In this example, the `sx:uddi` element describes the service using Universal Description, Discovery and Integration of Web Services (UDDI) [7]. The `sx:uddi` element contains an `sx:serviceKey` element that specifies a service key, which is a unique identifier for the service and can be bound to service metadata such as information about how to interact with the service.

- *An `r:serviceParameters` element*: This element specifies the parameters that should be passed to the service to enable the service to distinguish the context in which it is being called. In this example, a simple (but effective) approach is used. The service is passed a globally unique identifier (as part of `sx:stateDistinguisher`) that identifies this `r:serviceReference`. Since a service can maintain any number of states, passing the `sx:stateDistinguisher` enables the service to determine which state applies in each case.

By using the information in the `r:serviceReference`, an implementation can contact a service to determine how many exercises are based on a particular `r:serviceReference`. This information is then placed into an authorization context inside an authorization request. When the authorization request is evaluated, the information can be used to determine if an `sx:exerciseLimit` Condition is satisfied by comparing the number of exercises for a particular `r:serviceReference` in the authorization request with the number of exercises allowed for the `r:serviceReference` in the `sx:exerciseLimit` Condition.

Although state interaction is the focus of this section, it may be noted that `r:service Reference` has a wide variety of uses. Another use is mentioned in Section 5.5.6.1.

5.5.4 USING CERTIFICATES

A certificate is a License that binds a property to a Principal. The property is a descriptive data item that the issuer of the certificate bestows upon the Principal, such as an email address, an organizational role, a group membership, or a device characteristic.

Multiple certificates (possibly from different issuers) may be used to assign a set of properties to a Principal. For instance, Alice might have a certificate identifying her as a member of a video club. She might have another certificate that characterizes her as a university student. A third certificate might characterize Alice as a customer of a bank. Alice might use these certificates to download videos, check out books, or access her account.

This section describes the following concepts relating to how certificates are issued and used:

- The unique features of certificate Licenses are described in Section 5.5.4.1. The section describes how a certificate License binds a property to a Principal, for example, how Dave can say that Cathy is a gold club member in November.
- The mechanisms for predicating a grant on a certificate License are described in Section 5.5.4.2. The section describes how to require possession of a certificate License

to use a particular `r:grant`, for example, how can Bob say that Cathy can play a video file as long as Dave says she is a gold club member.

- An example authorization request using a certificate License is presented in Section 5.5.4.3. The section describes how to determine whether a principal is authorized to perform an activity when a certificate License is involved, for example, how to determine whether Cathy can play the video file in November given the above two Licenses.

5.5.4.1 Certificate Licenses

A certificate License is structurally no different from any other License. The most prominent distinguishing feature is the presence of the `r:possessProperty` Right, as shown in the following example certificate License. This example certificate License allows Cathy to claim ownership of a gold club membership (identified as `urn:example:clubMember: gold`) for the month of November 2003.

```
<r:license...>
   <r:grant>
      <r:keyHolder>...Cat...</r:keyHolder>
      <r:possessProperty/>
      <sx:propertyUri definition="urn:example:clubMember:gold"/>
      <r:validityInterval>
         <r:notBefore>2003-11-01T00:00:00</r:notBefore>
         <r:notAfter>2003-12-01T00:00:00</r:notAfter>
      </r:validityInterval>
   </r:grant>
   <r:issuer>
      <dsig:Signature>
         <dsig:SignedInfo>...</dsig:SignedInfo>
         <dsig:SignatureValue>L09qOQhAprs=</dsig:SignatureValue>
         <dsig:KeyInfo>...Dav...</dsig:KeyInfo>
      </dsig:Signature>
   </r:issuer>
</r:license>
```

Example 5.9

A certificate uses the `r:possessProperty` Right, which allows the associated principal to claim ownership of the property specified by the Resource. In this example certificate License, Cathy is the principal. The `r:possessProperty` element represents her right to possess a property. The property she possesses, a gold club membership, is represented by an `sx:propertyUri` Resource. The `sx:propertyUri` element simply represents a property identified using a URI. In this example, `urn:example:clubMember:gold` represents a gold club membership.

ISO/IEC 21000-5 also defines Resources for specifying other properties, such as an email name or a Domain Name System (DNS) name, and additional Resources for specifying more properties can be defined by extension.

A certificate License by itself is of little use. For instance, in the example above, Cathy receives little benefit from simply being able to state that she is a gold club member during the month of November. Her membership is only of value to her if there are some additional rights she can access as a member. Section 5.5.4.2 demonstrates how these additional rights (that are only available to Cathy if she is a member) can be expressed in a License.

5.5.4.2 Licenses Requiring a Certificate License

The following License allows Cathy to play a video file (a DI) if she is a gold member of a club.

```
<r:license...>
   <r:grant>
      <r:keyHolder>...Cat...</r:keyHolder>
      <mx:play/>
      <mx:diReference>...</mx:diReference>
      <r:prerequisiteRight>
         <r:keyHolder>...Cat...</r:keyHolder>
         <r:possessProperty/>
         <sx:propertyUri definition="urn:example:clubMember:gold"/>
         <r:trustedRootIssuers>
            <r:keyHolder>...Dav...</r:keyHolder>
         </r:trustedRootIssuers>
      </r:prerequisiteRight>
   </r:grant>
   <r:issuer>
      <dsig:Signature>
         <dsig:SignedInfo>...</dsig:SignedInfo>
         <dsig:SignatureValue>L10qOQhAprs=</dsig:SignatureValue>
         <dsig:KeyInfo>...Bob...</dsig:KeyInfo>
      </dsig:Signature>
   </r:issuer>
</r:license>
```

Example 5.10

To use the contained `r:grant` to play the specified video file, the `r:prerequisite Right` Condition must be satisfied. In this case, the `r:prerequisiteRight` requires that Cathy have a certificate rooted from Dave certifying she is a gold club member. As long as that certificate is valid, she can continue to play the movie.

When evaluating an authorization request, an `r:prerequisiteRight` Condition is satisfied only if the answer to a new authorization request constructed according to Table 5.5 is 'yes'. Because this new authorization request needs to be evaluated to determine the answer to the first authorization request, this new authorization request is called a *recursive authorization request*.

Section 5.5.4.3 illustrates in detail how an authorization request that uses a certificate is evaluated.

5.5.4.3 Evaluating an Authorization Request That Uses a Certificate by Intuition

When Cathy wishes to play a video file pursuant to her gold club membership, her authorization request contains the following:

- The components of the authorization request that describe the activity (Principal, Right, Resource, interval of time and authorization context).

Table 5.5 Recursive authorization request

Recursive authorization request component	Value determined by
Principal	the Principal inside the `r:prerequisiteRight`
Right	the Right inside the `r:prerequisiteRight`
Resource	the Resource inside the `r:prerequisiteRight`
Interval of time	the interval of time of the authorization request being evaluated
Authorization context	the authorization context of the authorization request being evaluated
Set of Licenses	the set of Licenses of the authorization request being evaluated
Set of root grants	the set of `r:grant` elements determined by the TrustRoot inside the `r:prerequisiteRight` Note: the set of `r:grant` elements determined by an `r:trustedRootIssuers` element containing one Principal element is a set of `r:grant` elements that allows the entity identified by that Principal to put whatever it wants in the Licenses it issues.

- Cathy's licenses (the usage License described in 5.5.4.2 and the certificate License described in 5.5.4.1).
- Any root grants that are in force.

Because the play right granted in Cathy's usage License is predicated on her possession of the certificate License, evaluating this authorization request is a bit more complicated than the authorization requests presented earlier in this chapter. This section describes how to perform this evaluation by intuition.

The Original Authorization Request
Initially, this authorization request appears similar to the authorization requests described earlier in this chapter. It seeks to determine whether Cathy can play the video file on November 7, 2003.

Authorization Request
The authorization request outlined in Table 5.6 asks whether Cathy can play the DI identified as `urn:example:001` on November 7, 2003, according to the specified Licenses and root grants.

As before, 'trustedRootIssuers Bob' is shorthand for a set of root grants that say 'Bob may put whatever he wants in the Licenses he issues'.

Step 1

The first step in evaluating an authorization request by intuition is to check whether the activity given by the Principal, Right, Resource, interval of time and authorization context components of the authorization request corresponds to one of the root grants given in the authorization request.

Table 5.6 Can Cathy play a specific digital item?

Component	Value
Principal	`<r:keyHolder>...Cat...</r:keyHolder>`
Right	`<mx:play/>`

Resource	`<mx:diReference>`
	`<mx:identifier>urn:example:001</mx:identifier>`
	`</mx:diReference>`

Interval of time	Midnight of 2003-11-07 through midnight of 2003-11-08
Authorization context	...

Set of Licenses	`<r:license...>`
	`<r:grant>`
	`<r:keyHolder>...Cat...</r:keyHolder>`
	`<mx:play/>`
	`<mx:diReference>`
	`<mx:identifier>urn:example:001</mx:identifier>`
	`</mx:diReference>`
	`<r:prerequisiteRight>`
	`<r:keyHolder>...Cat...</r:keyHolder>`
	`<r:possessProperty/>`
	`<sx:propertyUri`
	`definition="urn:example:clubMember:gold"/>`
	`<r:trustedRootIssuers>`
	`<r:keyHolder>...Dav...</r:keyHolder>`
	`</r:trustedRootIssuers>`
	`</r:prerequisiteRight>`
	`</r:grant>`
	`<r:issuer>`
	`<dsig:Signature>`
	`<dsig:SignedInfo>...</dsig:SignedInfo>`
	`<dsig:SignatureValue>L10qOQhAprs=</dsig:SignatureValue>`
	`<dsig:KeyInfo>...Bob...</dsig:KeyInfo>`
	`</dsig:Signature>`
	`</r:issuer>`
	`</r:license>`
	
	`<r:license...>`
	`<r:grant>`
	`<r:keyHolder>...Cat...</r:keyHolder>`
	`<r:possessProperty/>`
	`<sx:propertyUri`
	`definition="urn:example:clubMember:gold"/>`
	`<r:validityInterval>`
	`<r:notBefore>2003-11-01T00:00:00</r:notBefore>`

(continued overleaf)

Table 5.6 (*continued*)

Component	Value
	`<r:notAfter>2003-12-01T00:00:00</r:notAfter>`
	`</r:validityInterval>`
	`</r:grant>`
	`<r:issuer>`
	`<dsig:Signature>`
	`<dsig:SignedInfo>...</dsig:SignedInfo>`
	`<dsig:SignatureValue>L09qOQhAprs=</dsig:SignatureValue>`
	`<dsig:KeyInfo>...Dav...</dsig:KeyInfo>`
	`</dsig:Signature>`
	`</r:issuer>`
	`</r:license>`
Set of root grants	trustedRootIssuers Bob

In this case, no correspondence is found in this step because the activity, 'Cathy playing the DI identified as `urn:example:001` on November 7, 2003', does not correspond to the root grants, which state that 'Bob may put whatever he wants in the Licenses he issues'.

Evaluation proceeds to Step 2.

Step 2

The second step in intuitively evaluating an authorization request (if a correspondence for the activity is not found in one of the root grants) is to check for a correspondence in one of the `r:grant` elements in one of the Licenses in the authorization request.

In this example, a potential correspondence is found in Step 2 because the activity, which is 'Cathy playing the DI identified as `urn:example:001` on November 7, 2003', potentially corresponds with the `r:grant` element in the first License, which states that 'Cathy may play the DI identified as `urn:example:001` subject to some `r:prerequisiteRight` Condition'.

To determine whether this potential correspondence is an actual correspondence, a recursive authorization request is constructed using the information from the `r:prerequisite Right` Condition and the current authorization request. The recursive authorization request is shown in Table 5.7. According to the semantics of the `r:prerequisite Right` Condition, the first three components of the recursive authorization request are taken from the first three children of the `r:prerequisiteRight` Condition, the next three components of the recursive authorization request are copied from the current authorization request, and the last component of the recursive authorization request is determined from the last child of the `r:prerequisiteRight` Condition.

For the sake of clarity, the evaluation of the recursive authorization request is presented in the following section rather than inline here. The curious reader is encouraged to jump now to that section to read how the recursive authorization request is evaluated and to return here after learning the answer to the recursive authorization request.

Because the answer to the recursive authorization request is 'yes', the r:prerequisite Right Condition is satisfied, so there is indeed a correspondence between the activity of 'Cathy playing the DI identified as urn:example:001 on November 7, 2003' and the r:grant element in the first License.

However, Step 2 is not complete because Bob might not be permitted to issue such Licenses. Since the correspondence was found in a License issued by Bob, a new activity comes into question: the activity of Bob putting that r:grant in the License he issued. A correspondence must now be found for this activity, starting at Step 1.

Step 1 (Issuance chaining link 1)

When Step 1 (checking whether a correspondence for the activity is found in one of the root grants) is performed with the new activity, a correspondence is found because the activity is 'Bob putting Cathy's r:grant in the License he issued' and the root grants state that 'Bob may put whatever he wants in the Licenses he issues'. The authorization request is answered 'yes'.

Step 2 (Issuance chaining link 1)

Step 2 does not apply for the first link of issuance chaining because a correspondence was found in Step 1.

The Recursive Authorization Request

This recursive authorization request determines whether Cathy is authorized to possess a gold club membership. The result of this recursive authorization request determines whether the r:prerequisiteRight Condition in Cathy's usage License (shown in the previous section) is satisfied.

Authorization Request

The recursive authorization request is constructed using the information from the r:prerequisiteRight Condition and the original authorization request (shown in the previous section). According to the semantics of the r:prerequisiteRight Condition, the recursive authorization request is composed of the following:

- Its first three components (Principal, Right and Resource) are taken from the first three children of the r:prerequisiteRight Condition (stated informally: Cathy, possess property, gold club membership).
- Its next three components (interval of time, authorization context and Licenses) are copied from the current authorization request (November 11, 2003, . . ., Cathy's certificate License and usage License).
- Its last component (the set of root grants) is determined from the last child of the r:prerequisiteRight Condition (trustedRootIssuers Dave).

The recursive authorization request is shown in Table 5.7.

As before, 'trustedRootIssuers Dave' is shorthand for a set of root grants that say 'Dave may put whatever he wants in the Licenses he issues'.

Table 5.7 Recursive authorization request

Component	Value
Principal	`<r:keyHolder>...Cat...</r:keyHolder>`
Right	`<r:possessProperty/>`
Resource	`<sx:propertyUri definition="urn:example:clubMember:gold"/>`
Interval of time	Midnight of 2003-11-07 through midnight of 2003-11-08
Authorization context	...
Set of Licenses	

```
<r:license...>
 <r:grant>
 <r:keyHolder>...Cat...</r:keyHolder>
 <mx:play/>
 <mx:diReference>
  <mx:identifier>urn:example:001</mx:identifier>
 </mx:diReference>
 <r:prerequisiteRight>
  <r:keyHolder>...Cat...</r:keyHolder>
  <r:possessProperty/>
  <sx:propertyUri
   definition="urn:example:clubMember:gold"/>
  <r:trustedRootIssuers>
   <r:keyHolder>...Dav...</r:keyHolder>
  </r:trustedRootIssuers>
 </r:prerequisiteRight>
 </r:grant>
 <r:issuer>
  <dsig:Signature>
   <dsig:SignedInfo>...</dsig:SignedInfo>
   <dsig:SignatureValue>L10qOQhAprs=</dsig:SignatureValue>
   <dsig:KeyInfo>...Bob...</dsig:KeyInfo>
  </dsig:Signature>
 </r:issuer>
</r:license>
<r:license...>
 <r:grant>
 <r:keyHolder>...Cat...</r:keyHolder>
 <r:possessProperty/>
 <sx:propertyUri
  definition="urn:example:clubMember:gold"/>
 <r:validityInterval>
  <r:notBefore>2003-11-01T00:00:00</r:notBefore>
  <r:notAfter>2003-12-01T00:00:00</r:notAfter>
 </r:validityInterval>
 </r:grant>
```

Table 5.7   (*continued*)

Component	Value
	`<r:issuer>`
	`<dsig:Signature>`
	`<dsig:SignedInfo>...</dsig:SignedInfo>`
	`<dsig:SignatureValue>LO9qOQhAprs=</dsig:SignatureValue>`
	`<dsig:KeyInfo>...Dav...</dsig:KeyInfo>`
	`</dsig:Signature>`
	`</r:issuer>`
	`</r:license>`
Set of root grants	trustedRootIssuers Dave

Step 1

The first step in evaluating an authorization request by intuition is to check whether the activity given by the Principal, Right, Resource, interval of time and authorization context components of the authorization request corresponds to one of the root grants given in the authorization request.

In this case, no correspondence is found in this step because the activity, 'Cathy possessing the property defined by `urn:example:clubMember:gold` on November 7, 2003, does not correspond to the root grants, which state that 'Dave may put whatever he wants in the Licenses he issues'.

Evaluation proceeds to Step 2.

Step 2

The second step in intuitively evaluating an authorization request (whether a correspondence for the activity is not found in one of the root grants) is to check for a correspondence in one of the `r:grant` elements in one of the Licenses in the authorization request.

In this example, a correspondence is found in Step 2 because the activity, which is 'Cathy possessing the property defined by `urn:example:clubMember:gold` on November 7, 2003', corresponds with the `r:grant` element in the second License, which states that 'Cathy may possess the property defined by `urn:example:clubMember:gold` within November 2003'.

However, Step 2 is not complete because Dave might not be permitted to issue such Licenses. Since the correspondence was found in a License issued by Dave, a new activity comes into question: the activity of Dave putting that `r:grant` in the License he issued. A correspondence must now be found for this activity, starting at Step 1.

Step 1 (Issuance chaining link 1)

When Step 1 (checking whether a correspondence for the activity is found in one of the root grants) is performed with the new activity, a correspondence is found because the activity is 'Dave putting Cathy's `r:grant` in the License he issued' and the root grants state that 'Dave may put whatever he wants in the Licenses he issues'. The recursive authorization request is answered 'yes'.

Step 2 (Issuance chaining link 1)

Step 2 does not apply for the first link of issuance chaining because a correspondence was found in Step 1.

5.5.5 USING VARIABLES AND SPECIFYING CRITERIA

Variables add flexibility to Rights Expressions and increase their expressiveness. This section describes several uses for variables, including using a variable as a Principal and using a variable as a Resource. This section also describes some Conditions that can be used to specify criteria on variable Principals and Resources.

In more complex business models (for example, those that involve redistribution or repackaging of content), variables can be used as placeholders to be replaced by the downstream participants with choices appropriate for their business model. For more information on these types of scenarios, see Section 5.5.7.

5.5.5.1 Issuing a Club Member Special for a Single Digital Item

In the example in Section 5.5.4.2, Bob issued a License specifically to Cathy that she could use while she had a valid gold club membership. In the following example, Bob issues a License that is not specific to any User. However, he still wants to make sure that the User has a valid gold club membership.

```
<r:license...>
   <r:grant>
      <r:forAll varName="x"/>
      <r:keyHolder varRef="x"/>
      <mx:play/>
      <mx:diReference>...</mx:diReference>
      <r:prerequisiteRight>
         <r:keyHolder varRef="x"/>
         <r:possessProperty/>
         <sx:propertyUri definition="urn:example:clubMember:gold"/>
         <r:trustedRootIssuers>
            <r:keyHolder>...Dav...</r:keyHolder>
         </r:trustedRootIssuers>
      </r:prerequisiteRight>
   </r:grant>
   <r:issuer>
      <dsig:Signature>
         <dsig:SignedInfo>...</dsig:SignedInfo>
         <dsig:SignatureValue>L11qOQhAprs=</dsig:SignatureValue>
         <dsig:KeyInfo>...Bob...</dsig:KeyInfo>
      </dsig:Signature>
   </r:issuer>
</r:license>
```

Example 5.11

To construct an `r:grant` that is not specific to any User, Bob uses a variable to represent the Principal. The variable is defined as follows:

```
<r:forAll varName="x"/>
```

Syntactically, a variable is defined using `r:forAll`. Semantically, a variable has a name, a scope, a set of bindings, and a set of eligible bindings.

- The name of a variable is specified by its `r:varName` attribute. The example above defines a variable called '*x*'.
- The scope of a variable starts at the beginning of the variable declaration and ends at the end of the parent element (`r:grant` in this example). Within this scope, a variable may be overridden by a local definition (or a re-declaration). That is, if a variable's scope contains another variable with the same name, the scope of the outside variable is its normal scope minus the scope of the contained variable.
- The set of bindings and set of eligible bindings of a variable are used to determine what values a variable can take on. Additional information about the set of bindings and set of eligible bindings will be introduced gradually during the remainder of this chapter.

Once a variable is defined, it can be referenced by any element within the scope of the variable. A variable is referenced using the `r:varRef` attribute, and the referencing elements impact the set of eligible bindings. In the example above, the variable is referenced as the Principal of the `r:grant` and as the Principal for the `r:prerequisiteRight` as follows:

```
<r:keyHolder varRef="x"/>
```

Since both of these `r:keyHolder` elements reference the same variable, they must both take on the same value whenever the variable is resolved.

The referencing of the variable *x* by `r:keyHolder` elements limits the set of eligible bindings to `r:keyHolder` elements or elements that can substitute for an `r:keyHolder` element. An example of an eligible binding for variable *x* would be the `r:keyHolder` identifying Cathy. If the variable *x* were resolved to the `r:keyHolder` identifying Cathy, the `r:grant` would look like the one in Example 5.10. By using variables, Bob can avoid ever mentioning Cathy explicitly and can allow all gold club members to play the DI.

5.5.5.2 Issuing a Club Member Special Based on Digital Item Style

In the example in Section 5.5.5.1, Bob issued a License to all gold club members allowing them to play a specific DI. In the following example, Bob issues a License that is not specific to any DI. However, he wants the License to be limited to DIs that are animations.

```
<r:license...>
   <r:grant>
      <r:forAll varName="x"/>
      <r:forAll varName="y"/>
      <r:keyHolder varRef="x"/>
      <mx:play/>
```

```
        <mx:diReference varRef="y"/>
        <r:allConditions>
            <r:prerequisiteRight>
                <r:keyHolder varRef="x"/>
                <r:possessProperty/>
                <sx:propertyUri definition="urn:example:clubMember:gold"/>
                <r:trustedRootIssuers>
                    <r:keyHolder>...Dav...</r:keyHolder>
                </r:trustedRootIssuers>
            </r:prerequisiteRight>
            <mx:diCriteria>
                <mx:diReference varRef="y"/>
                <r:anXmlExpression>
                    didl:Descriptor[didl:Descriptor/didl:Statement="Style"]
                    /didl:Statement="Animated"
                </r:anXmlExpression>
            </mx:diCriteria>
        </r:allConditions>
    </r:grant>
    <r:issuer>
        <dsig:Signature>
            <dsig:SignedInfo>...</dsig:SignedInfo>
            <dsig:SignatureValue>L12qOQhAprs=</dsig:SignatureValue>
            <dsig:KeyInfo>...Bob...</dsig:KeyInfo>
        </dsig:Signature>
    </r:issuer>
</r:license>
```

Example 5.12

In much the same way that the r:prerequisiteRight element is used to require the r:keyHolder to identify a club member, the mx:diCriteria element is used to require the mx:diReference to identify an animation. The mx:diCriteria Condition contains an mx:diReference and one or more elements that can substitute for r:anXmlPatternAbstract, which is the abstract element that represents a pattern. A pattern is an XML element that defines a set of other XML elements, which are said to *match* the pattern. Each pattern defines exactly what its matching function is. For instance, an r:anXmlExpression (used in the example above) is a pattern that matches anything over which the contained XPath expression evaluates to true. The mx:diCriteria Condition is satisfied if the didl:Item element declaring the DI identified by the contained mx:diReference matches all of the contained patterns.

To determine whether the mx:diCriteria Condition is satisfied, the variables must be resolved and the declaration for the identified DI must be available. If x is resolved to identify Cathy and y is resolved to identify the DI identified by urn:example:001, the r:grant would be as in Example 5.13.

```
<r:grant...>
    <r:keyHolder>...Cat...</r:keyHolder>
    <mx:play/>
    <mx:diReference>
        <mx:identifier>urn:example:001</mx:identifier>
```

```
    </mx:diReference>
    <r:allConditions>
      <r:prerequisiteRight>
        <r:keyHolder>...Cat...</r:keyHolder>
        <r:possessProperty/>
        <sx:propertyUri definition="urn:example:clubMember:gold"/>
        <r:trustedRootIssuers>
          <r:keyHolder>...Dav...</r:keyHolder>
        </r:trustedRootIssuers>
      </r:prerequisiteRight>
      <mx:diCriteria>
        <mx:diReference>
          <mx:identifier>urn:example:001</mx:identifier>
        </mx:diReference>
        <r:anXmlExpression>
          didl:Descriptor[didl:Descriptor/didl:Statement="Style"]
          /didl:Statement="Animated"
        </r:anXmlExpression>
      </mx:diCriteria>
    </r:allConditions>
  </r:grant>
```

Example 5.13

The following is an example DI declaration declaring six DIs (the part of the DI declaration that declares the DI identified by `urn:example:001` is in bold):

```
<didl:DIDL...>
  <didl:Container>
    <didl:Item>
      <didl:Descriptor>
        <didl:Statement mimeType="text/xml">
          <dii:Identifier>urn:example:000</dii:Identifier>
        </didl:Statement>
      </didl:Descriptor>
      <didl:Item>
        <didl:Descriptor>
          <didl:Statement mimeType="text/xml">
            <dii:Identifier>urn:example:001</dii:Identifier>
          </didl:Statement>
        </didl:Descriptor>
        <didl:Descriptor>
          <didl:Descriptor>
            <didl:Statement mimeType="text/plain">Style</didl:Statement>
          </didl:Descriptor>
          <didl:Statement mimeType="text/plain">Animated</didl:Statement>
        </didl:Descriptor>
        <didl:Component>
          <didl:Resource mimeType="video/mpeg"
            ref="http://www.example.com/001.mpeg"/>
        </didl:Component>
      </didl:Item>
      <didl:Item>
        <didl:Descriptor>
```

```
          <didl:Statement mimeType="text/xml">
            <dii:Identifier>urn:example:002</dii:Identifier>
          </didl:Statement>
        </didl:Descriptor>
        <didl:Descriptor>
          <didl:Descriptor>
            <didl:Statement mimeType="text/plain">Style</didl:Statement>
          </didl:Descriptor>
          <didl:Statement mimeType="text/plain">Filmed</didl:Statement>
        </didl:Descriptor>
        <didl:Component>
          <didl:Resource mimeType="video/mpeg"
            ref="http://www.example.com/002.mpeg"/>
        </didl:Component>
      </didl:Item>
      <didl:Item>
        <didl:Descriptor>
          <didl:Statement mimeType="text/xml">
            <dii:Identifier>urn:example:003</dii:Identifier>
          </didl:Statement>
        </didl:Descriptor>
        <didl:Descriptor>
          <didl:Descriptor>
            <didl:Statement mimeType="text/plain">Style</didl:Statement>
          </didl:Descriptor>
          <didl:Statement mimeType="text/plain">Animated</didl:Statement>
        </didl:Descriptor>
        <didl:Component>
          <didl:Resource mimeType="video/quicktime"
            ref="http://www.example.com/003.mov"/>
        </didl:Component>
      </didl:Item>
      <didl:Component>
        <didl:Resource mimeType="text/html"
          ref="http://www.example.com/000.html"/>
      </didl:Component>
    </didl:Item>
    <didl:Item>
      <didl:Descriptor>
        <didl:Statement mimeType="text/xml">
          <dii:Identifier>urn:example:100</dii:Identifier>
        </didl:Statement>
      </didl:Descriptor>
      <didl:Item>
        <didl:Descriptor>
          <didl:Statement mimeType="text/xml">
            <dii:Identifier>urn:example:101</dii:Identifier>
          </didl:Statement>
        </didl:Descriptor>
        <didl:Descriptor>
          <didl:Descriptor>
            <didl:Statement mimeType="text/plain">Style</didl:Statement>
          </didl:Descriptor>
          <didl:Statement mimeType="text/plain">Animated</didl:Statement>
        </didl:Descriptor>
```

```
            <didl:Component>
              <didl:Resource mimeType="video/mpeg"
                ref="http://www.example.com/101.mpeg"/>
            </didl:Component>
          </didl:Item>
          <didl:Component>
            <didl:Resource mimeType="text/html"
              ref="http://www.example.com/100.html"/>
          </didl:Component>
        </didl:Item>
      </didl:Container>
  </didl:DIDL>
```

Example 5.14

When the XPath expression specified in the mx:diCriteria is evaluated over the declaration for the DI identified by urn:example:001, the result is true. Had *y* been resolved to urn:example:003 or urn:example:101, the result would be true also. However, had *y* been resolved to urn:example:000, urn:example:002 or urn:example:100, the result would be false. Thus, the License in Example 5.12 would allow Cathy to view urn:example:001, urn:example:003 and urn:example:101. It would also allow her to view other animations, and it would also allow other gold club members to view urn:example:001, urn:example:003, urn:example:101 and other animations.

5.5.5.3 Issuing a Club Member Special for Digital Items in a Magazine Based on Digital Item Style

It is also possible for Bob to scope a License to only those animations inside a particular online magazine (rather than all animations in the world). The following example is the same as Example 5.12 with the added restriction that the DIs must be part of the magazine DI identified by urn:example:000.

```
<r:license...>
   <r:grant>
      <r:forAll varName="x"/>
      <r:forAll varName="y"/>
      <r:keyHolder varRef="x"/>
      <mx:play/>
      <mx:diReference varRef="y"/>
      <r:allConditions>
         <r:prerequisiteRight>
            <r:keyHolder varRef="x"/>
            <r:possessProperty/>
            <sx:propertyUri definition="urn:example:clubMember:gold"/>
            <r:trustedRootIssuers>
                <r:keyHolder>...Dav...</r:keyHolder>
            </r:trustedRootIssuers>
         </r:prerequisiteRight>
         <mx:diCriteria>
            <mx:diReference varRef="y"/>
```

```
        <r:anXmlExpression>
            didl:Descriptor[didl:Descriptor/didl:Statement="Style"]
            /didl:Statement="Animated"
        </r:anXmlExpression>
      </mx:diCriteria>
      <mx:diPartOf>
          <mx:diReference varRef="y"/>
          <mx:diReference>
              <mx:identifier>urn:example:000</mx:identifier>
          </mx:diReference>
      </mx:diPartOf>
    </r:allConditions>
  </r:grant>
  <r:issuer>
    <dsig:Signature>
        <dsig:SignedInfo>...</dsig:SignedInfo>
        <dsig:SignatureValue>L15qOQhAprs=</dsig:SignatureValue>
        <dsig:KeyInfo>...Bob...</dsig:KeyInfo>
    </dsig:Signature>
  </r:issuer>
</r:license>
```

Example 5.15

In this example, the `mx:diPartOf` Condition contains two `mx:diReference` elements: the first identifies the animation and the second identifies the magazine (`urn:example:000`). This Condition is satisfied if the first DI (the animation) is part of the second DI (the magazine).

Using the DI declaration given in Example 5.14, this `mx:diPartOf` Condition would be satisfied if *y* were resolved to `urn:example:001`, `urn:example:002`, or `urn:example:003`. However, it would not be satisfied if, for instance, *y* were resolved to `urn:example:101`. The `mx:diCriteria` Condition and the `mx:diPartOf` Condition would *both* be satisfied if *y* were resolved to `urn:example:001` or `urn:example:003`. As expected, this corresponds to the two animations inside the magazine identified by `urn:example:000`. All gold club members would be allowed to play these two animations.

5.5.6 REVOCATION

Sometimes the issuer of a License wishes to reserve the right to revoke that License at a later time. The Rights Expression Language provides two elements that are particularly useful in such cases: `r:revocationMechanism` and `r:revocationFreshness`.

5.5.6.1 Specifying a Revocation Mechanism

There are several options for specifying a revocation mechanism, all of which are represented by the `r:revocationMechanism` element. The `r:revocationMechanism` element includes a child element that describes the particular mechanism being represented. Conceptually, every revocation mechanism provides a basic functionality: that of mapping a License and issuer to a guaranteed fresh-through time (a time through which it is guaranteed that the issuer's issuance of the License has not been revoked). The specific method for achieving that basic functionality changes from mechanism to mechanism, but the basic functionality remains the same.

Example revocation mechanisms include the following:

- A newsletter published on the first of each month listing all the Licenses that have been revoked and the issuer that revoked each one. In this case, the guaranteed fresh-through time of any License and issuer pair not on this month's list can be mapped to the first of the following month. If a License and issuer pair does not appear on the following month's newsletter, its guaranteed fresh-through time can be updated for another month, and so on.

- A scrolling ticker that lists a License, an issuer, and a guaranteed fresh-through time, another License, another issuer, and another guaranteed fresh-through time, and so on. In this case, any License and issuer can be mapped to the guaranteed fresh-through time immediately following it in the ticker.

- A phone number that, when called, prompts for the License and issuer and then states whether that License has been revoked by that issuer. If this mechanism states that the License has not been revoked by the issuer, the guaranteed fresh-through time would be the time of the call.

Without extension, the Rights Expression Language supports one kind of revocation mechanism: a revocation service. A revocation service is specified using an `r:service` `Reference`, which was introduced in Section 5.5.3. The specific interactions with the service are defined by the service description and can resemble any of the example revocation mechanisms described earlier. In addition, the Rights Expression Language can be extended to accommodate other revocation mechanisms.

5.5.6.2 Specifying Revocation Freshness

The `r:revocationMechanism` element describes the mechanism used to determine the guaranteed fresh-through time for a License and issuer pair. The `r:revocation` `Freshness` Condition prohibits activities associated with the `r:grant` elements in a License within a specified time in relation to the guaranteed fresh-through time. The following example demonstrates `r:revocationMechanism` and `r:revocation` `Freshness`.

```
<r:license...>
   <r:grant>
      <r:keyHolder>...Cat...</r:keyHolder>
      <mx:play/>
      <mx:diReference>...</mx:diReference>
      <r:revocationFreshness>
         <r:priorToStart>PT0S</r:priorToStart>
      </r:revocationFreshness>
   </r:grant>
   <r:issuer>
      <dsig:Signature>
```

```
            <dsig:SignedInfo>...</dsig:SignedInfo>
            <dsig:SignatureValue>L16qOQhAprs=</dsig:SignatureValue>
            <dsig:KeyInfo>...Bob...</dsig:KeyInfo>
        </dsig:Signature>
        <r:details>
            <r:revocationMechanism>
                <r:revocationService>
                    <r:serviceReference>
                        <sx:uddi>
                            <sx:serviceKey>
                                <sx:uuid>
                                    B51B7D19-1CD2-3164-04D9-13D3D132CE14
                                </sx:uuid>
                            </sx:serviceKey>
                        </sx:uddi>
                    </r:serviceReference>
                </r:revocationService>
            </r:revocationMechanism>
        </r:details>
    </r:issuer>
</r:license>
```

Example 5.16

This example License states that Bob permits Cathy to play the video provided that the specified revocation service confirms that the License is still fresh at the instant that Cathy begins to play the video.

In other words, if the service states that this License issued by Bob is guaranteed fresh through the following Wednesday at noon, Cathy can begin playing the video at any time up to and including the following Wednesday at noon. This is because the value of r:priorToStart is PT0S (zero seconds). If the value of r:priorToStart had been PT1H (one hour), then Cathy could have begun playing the video at any time up to and including the following Wednesday at 1:00 P.M. If the value of r:priorToStart had been negative two hours, then Cathy could have begun playing the video at any time up to and including the following Wednesday at 10:00 A.M.

5.5.7 CONSTRUCTING DISTRIBUTION AND OFFER LICENSES

This section describes two types of Licenses:

- *Distribution Licenses*: Licenses that a rights holder issues to allow a distributor to include specified r:grant or r:grantGroup elements in Licenses that the distributor issues to end-Users.
- *Offer Licenses*: Licenses that a distributor issues to allow a User to obtain another License from the distributor.

This section describes these two types of Licenses and some of the issues that relate to them.

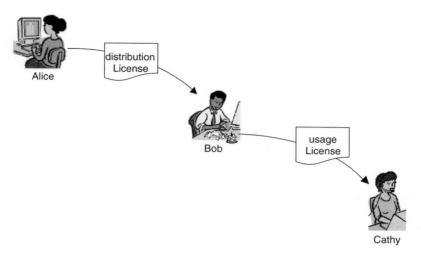

Figure 5.3   Multi-tier distribution

5.5.7.1 Issuing Distribution Licenses

A distribution License is a License that grants the `r:issue` Right. Granting the `r:issue` Right is a basic mechanism used in multi-tier distribution scenarios (scenarios in which content is redistributed or repackaged) such as the one illustrated in Figure 5.3.

With this mechanism, a rights holder, Alice, can allow a distributor, Bob, to include a specified `r:grant` or `r:grantGroup` (the one specified as the Resource inside the `r:grant` in the distribution License) in Licenses that the distributor issues to end-Users, such as Cathy. The `r:issue` Right only applies to `r:grant` or `r:grantGroup` Resources (or elements that can substitute for `r:grant` or `r:grantGroup`); the `r:issue` Right cannot be used with any other type of Resource. This section describes using the `r:issue` Right with `r:grant` elements. For information about using the `r:issue` Right with an `r:grantGroup`, see Section 5.5.7.7.

Alice can specify the `r:grant` that Bob may include in Licenses he issues using any level of detail. The following are some examples:

- Alice can completely specify the Principal, Right, Resource and/or Condition.
- Alice can use variables to represent the Principal, Right, Resource, Condition and/or elements within these elements. The use of variables provides Bob with greater flexibility in issuing Licenses.

The following example illustrates a distribution License using two variables. In this License, Alice grants the `r:issue` Right to Bob.

```
<r:license...>
   <r:grant>
      <r:forAll varName="x"/>
      <r:forAll varName="y"/>
```

```
        <r:keyHolder>...Bob...</r:keyHolder>
        <r:issue/>
        <r:grant>
           <r:keyHolder varRef="x"/>
           <mx:play/>
           <mx:diReference>...</mx:diReference>
           <r:validityInterval varRef="y"/>
        </r:grant>
     </r:grant>
     <r:issuer>
        <dsig:Signature>
           <dsig:SignedInfo>...</dsig:SignedInfo>
           <dsig:SignatureValue>L17qOQhAprs=</dsig:SignatureValue>
           <dsig:KeyInfo>...Ali...</dsig:KeyInfo>
        </dsig:Signature>
     </r:issuer>
  </r:license>
```

Example 5.17

By resolving the variable *x* to an r:keyHolder that identifies Cathy and the variable *y* to an r:validityInterval for 2003, Bob may include the r:grant shown in Example 5.3 in a License he issues as shown in Example 5.5. Bob's distribution License from Alice enables him to issue many different Licenses in addition to the one in Example 5.5 by substituting different r:keyHolder and r:validityInterval elements for the two variables, *x* and *y*. However, Alice identified a specific Right (mx:play) and specific Resource (an mx:diReference with identifier urn:example:001), and Bob cannot change these elements when he issues a License, so all of the r:grant elements that Bob includes in Licenses according to this distribution License from Alice will have the same Right and Resource.

5.5.7.2 Effect of Variable Declaration Location in Distribution Licenses

In Example 5.17, both the variables (*x* and *y*) are declared at the same level as the r:issue Right. Therefore, Bob replaces both of these variables with actual values when he issues a License.

The following example illustrates a distribution License that contains two variables declared at different levels within the License: one that the distributor, Bob, replaces with a value when issuing a License and one that is resolved when the consumer who receives that License plays the DI.

```
<r:license...>
   <r:grant>
      <r:forAll varName="y"/>
      <r:keyHolder>...Bob...</r:keyHolder>
      <r:issue/>
      <r:grant>
         <r:forAll varName="x"/>
         <r:keyHolder varRef="x"/>
         <mx:play/>
```

```
            <mx:diReference>...</mx:diReference>
            <r:validityInterval varRef="y"/>
        </r:grant>
    </r:grant>
    <r:issuer>
        <dsig:Signature>
            <dsig:SignedInfo>...</dsig:SignedInfo>
            <dsig:SignatureValue>L18qOQhAprs=</dsig:SignatureValue>
            <dsig:KeyInfo>...Ali...</dsig:KeyInfo>
        </dsig:Signature>
    </r:issuer>
</r:license>
```

Example 5.18

This License is structured as illustrated in Figure 5.4. For brevity, this figure omits the detail in the r:issuer element.

In Figure 5.4, it can be seen that, although both the variables (*x* and *y*) are *referenced* within the inner r:grant (the one that Bob may include in Licenses he issues), these two variables are *declared* at two different levels in the License.

- The variable *y* is declared at the same level as the r:issue Right. Therefore, Bob replaces this variable with a value (a specific validity interval) when he issues a usage License according to this distribution License.

- The variable *x* is declared within the inner r:grant at the same level as the mx:play Right. For this reason, the r:grant that Bob includes in the usage License he issues will still contain both the variable declaration (<r:forAll varName="x"/>) and the variable reference (<r:keyHolder varRef="x"/>) when he issues it. No particular User's key will be specified in that License and the variable *x* will not be resolved until someone plays the DI using that License, at which point the variable *x* is resolved to identify that person.

5.5.7.3 Issuing Offer Licenses

Offer Licenses can be thought of as machine-readable advertisements. A group of offer Licenses can be likened to a machine-readable product catalogue in which the products are the different kinds of rights available. Since they are just advertisements, offer Licenses can often be bypassed by using different negotiation mechanisms to arrive at the final usage License. However, offer Licenses do facilitate some important applications. Because the offer License has a machine-readable description of the rights available, they enable consumer applications that help the consumer search for rights that meet certain criteria and present the results of those searches in a way the consumer can understand. For instance, a consumer wanting rights to a new song could use such an application to discover that three offers are available: one to preview the first 30 s for free, one that allows playing at the cost of one credit/play, and one that allows unlimited use for one dollar.

An offer License grants the r:obtain Right. By issuing a License containing the r:obtain Right, the issuer is offering Users the opportunity to obtain another License

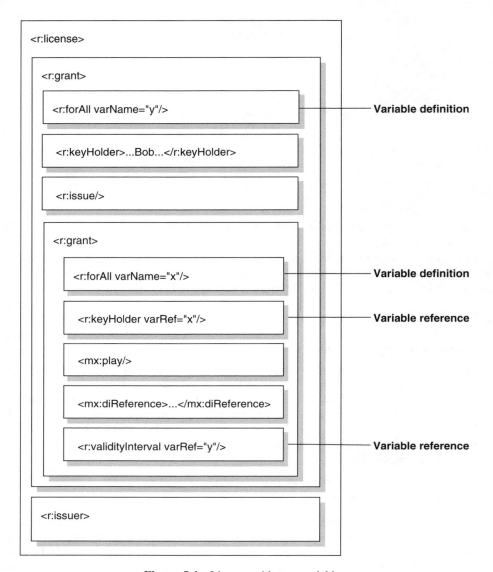

Figure 5.4 License with two variables

containing a specified r:grant or r:grantGroup (the one specified as the Resource inside the r:grant in the offer License). The r:obtain Right only applies to r:grant or r:grantGroup Resources (or elements that can substitute for r:grant or r:grant Group); the r:obtain Right cannot be used with any other type of Resource. This section describes using the r:obtain Right with the r:grant element. For information about using the r:obtain Right with an r:grantGroup, see Section 5.5.7.7.

An offer can specify the r:grant that the end-User may obtain in any level of detail. For instance,

- the offer can completely specify the Principal, Right, Resource and/or Condition.
- the offer can use variables to represent the Principal, Right, Resource, Condition and/or elements within these elements.

The following example illustrates an offer that Bob makes allowing consumers to get a free license to play the video throughout 2003.

```
<r:license...>
   <r:grant>
      <r:forAll varName="x"/>
      <r:keyHolder varRef="x"/>
      <r:obtain/>
      <r:grant>
         <r:keyHolder varRef="x"/>
         <mx:play/>
         <mx:diReference>...</mx:diReference>
          <r:validityInterval>
             <r:notBefore>2003-01-01T00:00:00</r:notBefore>
             <r:notAfter>2004-01-01T00:00:00</r:notAfter>
          </r:validityInterval>
      </r:grant>
   </r:grant>
   <r:issuer>
      <dsig:Signature>
         <dsig:SignedInfo>...</dsig:SignedInfo>
         <dsig:SignatureValue>L19qOQhAprs=</dsig:SignatureValue>
         <dsig:KeyInfo>...Bob...</dsig:KeyInfo>
      </dsig:Signature>
   </r:issuer>
</r:license>
```

Example 5.19

This License is structured as illustrated in Figure 5.5. For brevity, this figure omits the detail in the r:issuer element.

In Figure 5.5, the issuer, Bob, offers any r:keyHolder (the variable *x*) the right to obtain an r:grant containing the following Principal, Right, Resource and Condition:

- The same r:keyHolder (the variable **x**)
- mx:play
- an mx:diReference with identifier urn:example:001
- the r:validityInterval midnight of January 1, 2003, through midnight of January 1, 2004.

Since the variable *x* is referenced in two places in this License, the same r:keyHolder must be substituted in both the places. So, for instance, if Cathy uses the offer to obtain an r:grant, the r:grant that Cathy receives with the mx:play Right must have an r:keyHolder Principal identifying Cathy. In other words, she can only use this offer to obtain a License for herself.

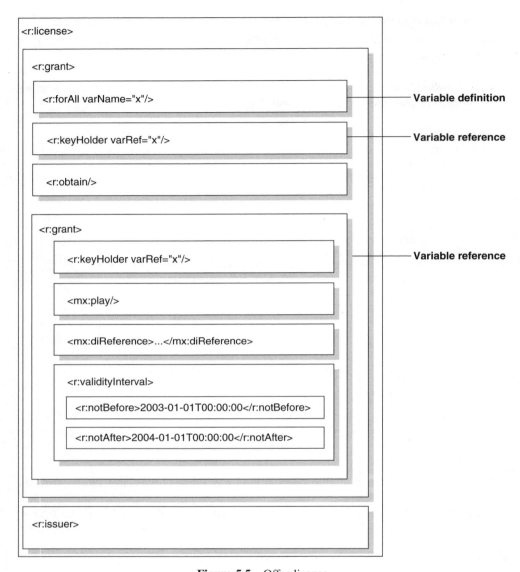

Figure 5.5   Offer license

5.5.7.4 Multi-tier Authorization Walkthrough

The walkthrough in this section illustrates how authorization requests, distribution Licenses and offer Licenses all work together. Three entities are involved in this walkthrough: Alice, Bob and Cathy. Alice is the author of the DI identified by `urn:example:001`. Bob is a distributor of DIs. Cathy is a consumer of DIs. Figure 5.6 illustrates this scenario.

Alice and Bob have reached an agreement allowing Bob to distribute permissions to Alice's DI. This agreement is represented by a License from Alice to Bob (Example 5.17).

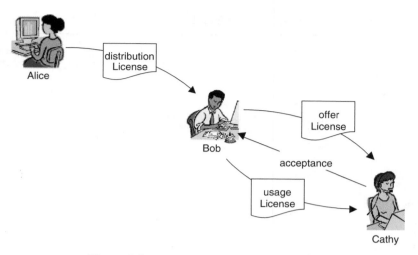

Figure 5.6   Multi-tier distribution using offers

Consequent to this agreement, Bob offers consumers' permission to play Alice's DI. This offer is represented by an offer License from Bob to the consumers (Example 5.19).

Cathy wishes to acquire permission to play Alice's DI. To do this, she elects to contact Bob and engages in a negotiation for the permissions she wishes to acquire. This negotiation can take any number of forms, both automated and manual. One method of automated negotiation is for Cathy simply to select one of the offers already advertised by Bob. If this method of automated negotiation is selected, Cathy informs Bob of the r:grant she desires (Example 5.3) and an authorization request is evaluated to verify that Cathy is indeed permitted to request that r:grant from Bob. This authorization request is demonstrated in the section 'May Cathy Obtain?' If another method of negotiation is selected, the above-mentioned section can be skipped and the corresponding alternate process followed to arrive at the r:grant that Bob and Cathy agree on for inclusion in a License from Bob to Cathy.

Next, Bob verifies that Alice has permitted him to include the r:grant he and Cathy had agreed on into a License. He does this by evaluating an authorization request as demonstrated in the section 'May Bob Issue?'

Finally, Bob issues the License to Cathy (Example 5.5). Cathy now has one additional License that she can use in constructing authorization requests. When Cathy wants to play Alice's DI, she can use this License as one of the Licenses in the set of Licenses in the authorization request she constructs. This authorization request is demonstrated in the section 'May Cathy Play?'

May Cathy Obtain?
When Cathy selects one of the offers that Bob has advertised, an authorization request is evaluated to verify that Cathy is indeed permitted to obtain that r:grant from Bob.

Authorization Request
The authorization request outlined in Table 5.8 asks whether Cathy may obtain from Bob the r:grant shown in Example 5.3. Because Bob is expected to give it to her, 'trustedRootIssuers Bob' is used as the set of root grants.

Table 5.8   Can Cathy obtain a specific grant from Bob?

Component	Value

Principal	`<r:keyHolder>...Cat...</r:keyHolder>`
Right	`<r:obtain/>`
Resource	`<r:grant...>`
	`<r:keyHolder>...Cat...</r:keyHolder>`
	`<mx:play/>`
	`<mx:diReference>...</mx:diReference>`
	`<r:validityInterval>...</r:validityInterval>`
	`</r:grant>`
Interval of time	`...`
Authorization context	`...`
Set of Licenses	`<r:license...>`
	` <r:grant>`
	`  <r:forAll varName="x"/>`
	`  <r:forAll varName="y"/>`
	`  <r:keyHolder>...Bob...</r:keyHolder>`
	`  <r:issue/>`
	`  <r:grant>`
	`   <r:keyHolder varRef="x"/>`
	`   <mx:play/>`
	`   <mx:diReference>...</mx:diReference>`
	`   <r:validityInterval varRef="y"/>`
	`  </r:grant>`
	` </r:grant>`
	` <r:issuer>`
	`  <dsig:Signature>`
	`   <dsig:SignedInfo>...</dsig:SignedInfo>`
	`   <dsig:SignatureValue>L17qOQhAprs=</dsig:SignatureValue>`
	`   <dsig:KeyInfo>...Ali...</dsig:KeyInfo>`
	`  </dsig:Signature>`
	` </r:issuer>`
	`</r:license>`
	`<r:license...>`
	` <r:grant>`
	`  <r:forAll varName="x"/>`
	`  <r:keyHolder varRef="x"/>`
	`  <r:obtain/>`
	`  <r:grant>`
	`   <r:keyHolder varRef="x"/>`
	`   <mx:play/>`
	`   <mx:diReference>...</mx:diReference>`
	`   <r:validityInterval>...</r:validityInterval>`
	`  </r:grant>`

Table 5.8   (*continued*)

Component	Value

	`</r:grant>`
	`<r:issuer>`
	` <dsig:Signature>`
	`  <dsig:SignedInfo>...</dsig:SignedInfo>`
	`  <dsig:SignatureValue>L19qOQhAprs=</dsig:SignatureValue>`
	`  <dsig:KeyInfo>...Bob...</dsig:KeyInfo>`
	` </dsig:Signature>`
	`</r:issuer>`
	`</r:license>`

Set of root grants	trustedRootIssuers Bob

Step 1

As before, the first step in evaluating an authorization request by intuition is to check whether the activity given by the Principal, Right, Resource, interval of time and authorization context components of the authorization request corresponds to one of the root grants given in the authorization request.

In this case, no correspondence is found in this step because the activity 'Cathy obtaining the `r:grant`' does not correspond to the root grants, which state that 'Bob may put whatever he wants in the Licenses he issues'.

Evaluation proceeds to Step 2.

Step 2

The second step in intuitively evaluating an authorization request (if a correspondence for the activity is not found in one of the root grants) is to check for a correspondence in one of the `r:grant` elements in one of the Licenses in the authorization request.

In this example, a correspondence is found in Step 2 because the activity, which is 'Cathy obtaining the `r:grant`', corresponds with the `r:grant` element in the second License, which, when the variable x is resolved to Cathy, states that 'Cathy may obtain the `r:grant`'.

However, Step 2 is not complete because Bob might not be permitted to issue such Licenses. Since a correspondence was found in a License issued by Bob, a new activity comes into question: the activity of Bob putting that `r:grant` (the top-level one with the `r:obtain` Right) in the License he issued. A correspondence must now be found for this activity, starting at Step 1.

Step 1 (Issuance chaining link 1)

When Step 1 (checking whether a correspondence for the activity is found in one of the root grants) is performed with the new activity, a correspondence is found because the activity is 'Bob putting the `r:grant` with the `r:obtain` Right in the License he issued' and the root grants state that 'Bob may put whatever he wants in the Licenses he issues'. The authorization request is answered 'yes'.

Step 2 (Issuance chaining link 1)

Step 2 does not apply for the first link of issuance chaining because a correspondence was found in Step 1.

May Bob Issue?
In order for Cathy to obtain a License containing a particular `r:grant` from Bob, Bob must be authorized to issue the requested `r:grant`.

Authorization Request
The authorization request outlined in Table 5.9 asks whether Bob may include the `r:grant` shown in Example 5.3 in Licenses he issues. Because that `r:grant` applies to Alice's work, 'trustedRootIssuers Alice' is used as the set of root grants.

Table 5.9 Can Bob issue a specific grant?

Component	Value
Principal	`<r:keyHolder>...Bob...</r:keyHolder>`
Right	`<r:issue/>`

Resource	`<r:grant...>`
	`  <r:keyHolder>...Cat...</r:keyHolder>`
	`  <mx:play/>`
	`  <mx:diReference>...</mx:diReference>`
	`  <r:validityInterval>...</r:validityInterval>`
	`</r:grant>`

Interval of time	...
Authorization context	...

Set of Licenses	`<r:license...>`
	`  <r:grant>`
	`    <r:forAll varName="x"/>`
	`    <r:forAll varName="y"/>`
	`    <r:keyHolder>...Bob...</r:keyHolder>`
	`    <r:issue/>`
	`    <r:grant>`
	`      <r:keyHolder varRef="x"/>`
	`      <mx:play/>`
	`      <mx:diReference>...</mx:diReference>`
	`      <r:validityInterval varRef="y"/>`
	`    </r:grant>`
	`  </r:grant>`
	`  <r:issuer>`

Table 5.9 (*continued*)

Component	Value

```
            <dsig:Signature>
             <dsig:SignedInfo>...</dsig:SignedInfo>
             <dsig:SignatureValue>L17qOQhAprs=</dsig:SignatureValue>
             <dsig:KeyInfo>...Ali...</dsig:KeyInfo>
            </dsig:Signature>
           </r:issuer>
          </r:license>

          <r:license...>
           <r:grant>
            <r:forAll varName="x"/>
            <r:keyHolder varRef="x"/>
            <r:obtain/>
            <r:grant>
             <r:keyHolder varRef="x"/>
             <mx:play/>
             <mx:diReference>...</mx:diReference>
             <r:validityInterval>...</r:validityInterval>
            </r:grant>
           </r:grant>
           <r:issuer>
            <dsig:Signature>
             <dsig:SignedInfo>...</dsig:SignedInfo>
             <dsig:SignatureValue>L19qOQhAprs=</dsig:SignatureValue>
             <dsig:KeyInfo>...Bob...</dsig:KeyInfo>
            </dsig:Signature>
           </r:issuer>
          </r:license>
```

Set of root grants	trustedRootIssuers Alice

Step 1

As before, the first step in evaluating an authorization request by intuition is to check whether the activity given by the Principal, Right, Resource, interval of time and authorization context components of the authorization request corresponds to one of the root grants given in the authorization request.

In this case, no correspondence is found in this step because the activity, 'Bob including the `r:grant` with the `mx:play` right in the License he issues', does not correspond to the root grants, which state that 'Alice may put whatever she wants in the Licenses she issues'.

Evaluation proceeds to Step 2.

Step 2

The second step in intuitively evaluating an authorization request (if a correspondence for the activity is not found in one of the root grants) is to check for a correspondence in one of the r:grant elements in one of the Licenses in the authorization request.

In this example, a correspondence is found in Step 2 because the activity, which is 'Bob including the r:grant with the mx:play right in the License he issues', corresponds with the r:grant element in the first License, which, when the variable *x* is resolved to Cathy and the variable *y* is resolved to 2003, states that 'Bob may include that r:grant in the Licenses he issues'.

However, Step 2 is not complete because Alice might not be permitted to issue such Licenses. Since a correspondence was found in a License issued by Alice, a new activity comes into question: the activity of Alice putting that r:grant (the top-level one with the r:issue Right) in the License she issued. A correspondence must now be found for this activity, starting at Step 1.

Step 1 (Issuance chaining link 1)

When Step 1 (checking whether a correspondence for the activity is found in one of the root grants) is performed with the new activity, a correspondence is found because the activity is 'Alice putting the r:grant with the r:issue Right in the License she issued' and the root grants state that 'Alice may put whatever she wants in the Licenses she issues'. The authorization request is answered 'yes'.

Step 2 (Issuance chaining link 1)

Step 2 does not apply for the first link of issuance chaining because a correspondence was found in Step 1.

May Cathy Play?

Having received a License with the requested r:grant with the mx:play right from Bob, a final authorization request must be evaluated in order for Cathy to use that License to play Alice's DI.

Authorization Request

The authorization request outlined in Table 5.10 asks whether Cathy may play Alice's DI on September 1, 2003. Because the DI belongs to Alice, 'trustedRootIssuers Alice' is used as the set of root grants.

Table 5.10 Can Cathy play Alice's digital item on September 1, 2003?

Component	Value
Principal	`<r:keyHolder>...Cat...</r:keyHolder>`
Right	`<mx:play/>`

Resource	`<mx:diReference>`
	`  <mx:identifier>urn:example:001</mx:identifier>`
	`</mx:diReference>`

Interval of time	Midnight of 2003-09-01 through midnight of 2003-09-02

Table 5.10   (*continued*)

Component	Value
Authorization context	...

Set of Licenses

```
<r:license...>
 <r:grant>
  <r:forAll varName="x"/>
  <r:forAll varName="y"/>
  <r:keyHolder>...Bob...</r:keyHolder>
  <r:issue/>
  <r:grant>
   <r:keyHolder varRef="x"/>
   <mx:play/>
   <mx:diReference>...</mx:diReference>
   <r:validityInterval varRef="y"/>
  </r:grant>
 </r:grant>
 <r:issuer>
  <dsig:Signature>
   <dsig:SignedInfo>...</dsig:SignedInfo>
   <dsig:SignatureValue>L17qOQhAprs=</dsig:SignatureValue>
   <dsig:KeyInfo>...Ali...</dsig:KeyInfo>
  </dsig:Signature>
 </r:issuer>
</r:license>

<r:license...>
 <r:grant>
  <r:forAll varName="x"/>
  <r:keyHolder varRef="x"/>
  <r:obtain/>
  <r:grant>
   <r:keyHolder varRef="x"/>
   <mx:play/>
   <mx:diReference>...</mx:diReference>
   <r:validityInterval>...</r:validityInterval>
  </r:grant>
 </r:grant>
 <r:issuer>
  <dsig:Signature>
   <dsig:SignedInfo>...</dsig:SignedInfo>
   <dsig:SignatureValue>L19qOQhAprs=</dsig:SignatureValue>
   <dsig:KeyInfo>...Bob...</dsig:KeyInfo>
  </dsig:Signature>
 </r:issuer>
</r:license>
```

(*continued overleaf*)

Table 5.10   *(continued)*

Component	Value

```
<r:license...>
 <r:grant>
  <r:keyHolder>...Cat...</r:keyHolder>

  <mx:play/>
  <mx:diReference>
   <mx:identifier>urn:example:001</mx:identifier>
  </mx:diReference>
  <r:validityInterval>
   <r:notBefore>2003-01-01T00:00:00</r:notBefore>
   <r:notAfter>2004-01-01T00:00:00</r:notAfter>
  </r:validityInterval>
 </r:grant>
 <r:issuer>
  <dsig:Signature>
   <dsig:SignedInfo>...</dsig:SignedInfo>
   <dsig:SignatureValue>L05qOQhAprs=</dsig:SignatureValue>
   <dsig:KeyInfo>...Bob...</dsig:KeyInfo>
  </dsig:Signature>
 </r:issuer>
</r:license>
```

Set of root grants	trustedRootIssuers Alice

Step 1

As before, the first step in evaluating an authorization request by intuition is to check whether the activity given by the Principal, Right, Resource, interval of time and authorization context components of the authorization request corresponds to one of the root grants given in the authorization request.

In this case, no correspondence is found in this step because the activity, 'Cathy playing Alice's DI on September 1, 2003', does not correspond to the root grants, which state that 'Alice may put whatever she wants in the Licenses she issues'.

Evaluation proceeds to Step 2.

Step 2

The second step in intuitively evaluating an authorization request (if a correspondence for the activity is not found in one of the root grants) is to check for a correspondence in one of the r:grant elements in one of the Licenses in the authorization request.

In this example, a correspondence is found in Step 2 because the activity 'Cathy playing Alice's DI on September 1, 2003', corresponds with the r:grant element in the third License, which states that 'Cathy may play Alice's DI in 2003'.

However, Step 2 is not complete because Bob might not be permitted to issue such Licenses. Since a correspondence was found in a License issued by Bob, a new activity comes into question: the activity of Bob putting that `r:grant` (the one with the `mx:play` Right) in the License he issued. A correspondence must now be found for this activity, starting at Step 1.

Step 1 (Issuance chaining link 1)

When Step 1 (checking whether a correspondence for the activity is found in one of the root grants) is performed with the new activity, no correspondence is found in this step because the activity, 'Bob including the `r:grant` in the License he issues', does not correspond to the root grants, which state that 'Alice may put whatever she wants in the Licenses she issues'.

Evaluation proceeds to Step 2.

Step 2 (Issuance chaining link 1)

When Step 2 (checking whether a correspondence for the activity is found in one of the `r:grant` elements in one of the Licenses) is performed with the new activity, a correspondence is found because the activity, which is 'Bob including the `r:grant` with the `mx:play` right in the License he issues', corresponds with the `r:grant` element in the first License, which, when the variable x is resolved to Cathy and the variable y is resolved to 2003, states that 'Bob may include that `r:grant` in the Licenses he issues'.

However, Step 2 is not complete because Alice might not be permitted to issue such Licenses. Since a correspondence was found in a License issued by Alice, a new activity comes into question: the activity of Alice putting that `r:grant` (the top-level one with the `r:issue` Right) in the License she issued. A correspondence must now be found for this activity, starting at Step 1.

Step 1 (Issuance chaining link 2)

When Step 1 (checking whether a correspondence for the activity is found in one of the root grants) is performed with the new activity, a correspondence is found because the activity is 'Alice putting the `r:grant` with the `r:issue` Right in the License she issued' and the root grants state that 'Alice may put whatever she wants in the Licenses she issues'. The authorization request is answered 'yes'.

Step 2 (Issuance chaining link 2)

Step 2 does not apply for the second link of issuance chaining because a correspondence was found in Step 1.

5.5.7.5 Adding Constraints on Distribution

Sometimes it is desirable to constrain a distribution License. For example, building upon Example 5.17, Alice might wish to stipulate that, when Bob issues a usage License according to the distribution License that she has given him, (1) the usage License must have a validity period of one year and (2) he must report to her the Principal to whom he has issued the usage License.

To accomplish this, two mechanisms are employed, as shown in the following example:

- A pattern is placed inside a variable declaration. The effect of placing a pattern inside a variable declaration is to constrain the set of bindings of that variable, thereby constraining the set of eligible bindings of that variable (the values that variable can take on). Any value that the declared variable takes on must match the pattern specified in the variable declaration.

- A condition is placed in the r:grant with the r:issue Right. The effect of placing a Condition in a r:grant with an r:issue Right is no different than the effect of placing a Condition in any other r:grant: in order for the r:grant to be used, the Condition must be satisfied. In distribution Licenses, a Condition in the outer r:grant needs to be satisfied when the outer r:grant is used to include the inner r:grant in a usage License that gets issued. A Condition in the inner r:grant carries over into the usage License along with the inner r:grant and needs to be satisfied when that r:grant is used to use a DI.

```
<r:license...>
   <r:grant>
      <r:forAll varName="x"/>
      <r:forAll varName="y">
         <sx:validityIntervalDurationPattern>
            <sx:duration>P1Y</sx:duration>
         </sx:validityIntervalDurationPattern>
      </r:forAll>
      <r:keyHolder>...Bob...</r:keyHolder>
      <r:issue/>
      <r:grant>
         <r:keyHolder varRef="x"/>
         <mx:play/>
         <mx:diReference>...</mx:diReference>
         <r:validityInterval varRef="y"/>
      </r:grant>
      <sx:trackReport>
         <r:serviceReference>
            <sx:uddi>
               <sx:serviceKey>
                  <sx:uuid>D1135111-B7DB-D1C2-32CD-3964D3E14049</sx:uuid>
               </sx:serviceKey>
            </sx:uddi>
            <r:serviceParameters>
               <r:datum>
                  <r:keyHolder varRef="x"/>
               </r:datum>
            </r:serviceParameters>
         </r:serviceReference>
      </sx:trackReport>
   </r:grant>
   <r:issuer>
      <dsig:Signature>
         <dsig:SignedInfo>...</dsig:SignedInfo>
         <dsig:SignatureValue>L20qOQhAprs=</dsig:SignatureValue>
```

```
        <dsig:KeyInfo>...Ali...</dsig:KeyInfo>
      </dsig:Signature>
    </r:issuer>
  </r:license>
```

Example 5.20

This example introduces two new elements:

- *sx:validityIntervalDurationPattern*: This pattern matches any r:
  validityInterval with the specified duration. Therefore, when the variable *y* is
  resolved in this case, it must be resolved to some r:validityInterval that represents a specific one-year duration (for instance, midnight of January 1, 2003, through
  midnight of January 1, 2004, or midnight of February 2, 2003, through midnight of
  February 2, 2004). This specific value is placed in the usage License that Bob issues
  and must be satisfied whenever that usage License is used.

- *sx:trackReport*: This Condition is satisfied if a report is made to a specific tracking service. The sx:trackReport Condition contains an r:serviceReference
  element that specifies both the service to which to make the report and the parameters
  required to report to it (for more information on the r:serviceReference element,
  refer to section 5.5.3). In this case, the value of the variable *x* is to be reported to the
  service. Because the variable *x* is also used as the Principal in the inner r:grant and
  because both references to the variable *x* must resolve to the same value, the effect is
  that the service can keep a record of who has been permitted to play Alice's DI.

5.5.7.6 Adding Constraints on Offers

If Bob has a constrained distribution License from Alice like the one in Example 5.20, it
is not wise for him to make unconstrained offers like the one in Example 5.19.

In the constrained distribution License from Alice, Alice only allows Bob to include
r:grant elements with one-year validity intervals in the usage Licenses he issues. Therefore, Bob should not be offering r:grant elements with validity intervals of any length
that the consumer chooses. To constrain his offer to be in accordance with his distribution
License, Bob will want to put the same sx:validityIntervalDurationPattern
in the declaration for the variable *y* in the offer License as is in the distribution License.

In addition, the constrained distribution License from Alice requires Bob to report the
Principals to which he issues usage Licenses. For this constraint, it is not necessary for
Bob to modify his offer at all; however, he may wish to charge a small fee for his trouble
in fulfilling his offers so that he is only obligated to fulfill offers for those who pay his
fee. This is done by placing a Condition in the outer level r:grant (at the same level
as the r:obtain Right).

The following example illustrates Bob's revised constrained offer:

```
<r:license...>
   <r:grant>
      <r:forAll varName="x"/>
      <r:forAll varName="y">
```

```
            <sx:validityIntervalDurationPattern>
                <sx:duration>P1Y</sx:duration>
            </sx:validityIntervalDurationPattern>
        </r:forAll>
        <r:keyHolder varRef="x"/>
        <r:obtain/>
        <r:grant>
            <r:keyHolder varRef="x"/>
            <mx:play/>
            <mx:diReference>...</mx:diReference>
            <r:validityInterval varRef="y"/>
        </r:grant>
        <sx:feePerUse...>
            <sx:rate>
                <sx:amount>1.25</sx:amount>
                <sx:currency>c:USD</sx:currency>
            </sx:rate>
            <sx:to>
                <sx:aba>
                    <sx:institution>123456789</sx:institution>
                    <sx:account>987654321</sx:account>
                </sx:aba>
            </sx:to>
        </sx:feePerUse>
    </r:grant>
    <r:issuer>
        <dsig:Signature>
            <dsig:SignedInfo>...</dsig:SignedInfo>
            <dsig:SignatureValue>L21qOQhAprs=</dsig:SignatureValue>
            <dsig:KeyInfo>...Bob...</dsig:KeyInfo>
        </dsig:Signature>
    </r:issuer>
</r:license>
```

Example 5.21

The `sx:feePerUse` Condition is satisfied if a fee is paid each time the containing `r:grant` is used. In this case, because the fee is on the outer `r:grant`, the fee only needs to be paid each time a User obtains an `r:grant` from Bob, not each time a User plays the DI (a fee that a User pays each time she plays the DI would be specified inside the inner `r:grant`.) The `sx:feePerUse` Condition contains elements that specify the fee amount, the currency and the party to whom the fee must be paid. In this case, US$1.25 must be paid to account number 987654321 at banking institution 123456789.

5.5.7.7 Distributing Multiple Rights in a Package

An `r:grantGroup` is a container for `r:grant` elements. An `r:grantGroup` can appear in a License in many of the same places that an `r:grant` may appear.

If the Resource corresponding to an `r:issue` Right is an `r:grant`, a distributor is allowed to include that `r:grant` in the Licenses he issues. Likewise, if the Resource corresponding to an `r:issue` Right is an `r:grantGroup`, a distributor is allowed to include that `r:grantGroup` in the Licenses he issues. A similar parallelism holds for the `r:obtain` Right.

Because of this behavior of `r:grantGroup` with respect to the `r:issue` and `r:obtain` Rights, a rights holder can use an `r:grantGroup` to package several `r:grant` elements together as a unit in a distribution License. The distributor is allowed to include the whole `r:grantGroup` in Licenses he issues; however, the distributor is not allowed to issue Licenses containing only individual `r:grant` elements from the `r:grantGroup`.

The following example illustrates a distribution License that uses an `r:grantGroup` as the Resource in its top-level `r:grant`.

```
<r:license...>
   <r:grant>
      <r:forAll varName="x"/>
      <r:keyHolder>...Bob...</r:keyHolder>
      <r:issue/>
      <r:grantGroup>
         <r:grant>
            <r:keyHolder varRef="x"/>
            <mx:play/>
            <mx:diReference>...</mx:diReference>
         </r:grant>
         <r:grant>
            <r:keyHolder varRef="x"/>
            <mx:print/>
            <mx:diReference>...</mx:diReference>
         </r:grant>
      </r:grantGroup>
   </r:grant>
   <r:issuer>
      <dsig:Signature>
         <dsig:SignedInfo>...</dsig:SignedInfo>
         <dsig:SignatureValue>L22qOQhAprs=</dsig:SignatureValue>
         <dsig:KeyInfo>...Ali...</dsig:KeyInfo>
      </dsig:Signature>
   </r:issuer>
</r:license>
```

Example 5.22

The `mx:print` element refers to the act of deriving a fixed and directly perceivable representation of a resource, such as a hard-copy print of an image or text.

In Example 5.22, Alice permits Bob to include an `r:grantGroup` in the Licenses he issues. This `r:grantGroup` contains both the `mx:play` Right and the `mx:print` Right. Bob is not permitted to grant only the `mx:play` Right or only the `mx:print` Right. Bob may, however, (in fact he must) resolve the variable x when he issues a License pursuant to the distribution License from Alice. If Bob resolves this variable in the License he issues to Cathy, the License would look as follows:

```
<r:license...>
   <r:grantGroup>
```

```
      <r:grant>
         <r:keyHolder>...Cat...</r:keyHolder>
         <mx:play/>  ·
         <mx:diReference>...</mx:diReference>
      </r:grant>
      <r:grant>
         <r:keyHolder>...Cat...</r:keyHolder>
         <mx:print/>
         <mx:diReference>...</mx:diReference>
      </r:grant>
   </r:grantGroup>
   <r:issuer>
      <dsig:Signature>
         <dsig:SignedInfo>...</dsig:SignedInfo>
         <dsig:SignatureValue>L23qOQhAprs=</dsig:SignatureValue>
         <dsig:KeyInfo>...Bob...</dsig:KeyInfo>
      </dsig:Signature>
   </r:issuer>
</r:license>
```

Example 5.23

It may be noted that an r:grantGroup can also be specified as the Resource corresponding to the r:obtain Right, in which case the end-User can obtain the entire r:grantGroup. However, the end-User cannot use such an offer to obtain individual r:grant elements within the r:grantGroup.

5.5.8 *DELEGATING REVOCATION*

Section 5.5.7 describes how one issuer, Alice, can permit another issuer, Bob, to include an r:grant or r:grantGroup in Licenses he issues and how he can permit other Users to obtain that r:grant or r:grantGroup in a License issued by him. In some cases, Bob may want to permit another User, Ethyl, to revoke a License that he has issued.

To do this, first Bob issues a License like the one in Example 5.16, where Bob permits Cathy to play a video but reserves the right to revoke his permission at a later time. Bob's License to Cathy has an r:revocationMechanism and an r:revocation Freshness Condition.

Then, Bob issues a second License to Ethyl, as shown below:

```
<r:license...>
   <r:grant>
      <r:keyHolder>...Eth...</r:keyHolder>
      <r:revoke/>
      <r:revocable>
         <dsig:SignatureValue>L16qOQhAprs=</dsig:SignatureValue>
      </r:revocable>
   </r:grant>
   <r:issuer>
      <dsig:Signature>
         <dsig:SignedInfo>...</dsig:SignedInfo>
```

```
        <dsig:SignatureValue>L24qOQhAprs=</dsig:SignatureValue>
        <dsig:KeyInfo>...Bob...</dsig:KeyInfo>
      </dsig:Signature>
    </r:issuer>
  </r:license>
```

Example 5.24

This License demonstrates the use of a new Right, `r:revoke`, and a new Resource, `r:revocable`. As defined in ISO/IEC 21000-5, these two elements are meant to be used together, and they are most easily understood together. Although it is possible (usually in the context of an extension to the Rights Expression Language) to use these elements separately, this use is not described in this chapter.

When the `r:revoke` Right is used, the Resource is `r:revocable`, which specifies the License and issuer pair that may be revoked. One way to specify the License and issuer pair is to give a `dsig:SignatureValue` because the `dsig:SignatureValue` is generated by the issuer on the basis of the License and is unique to that License and issuer pair.

To understand exactly what this new License permits Ethyl to do, it is useful to consider Cathy's License from Bob (described in Section 5.5.6.2). Cathy's License contains a revocation mechanism that maps the `dsig:SignatureValue` in that License ('L16...') to a guaranteed fresh-through time. The revocation mechanism keeps updating the guaranteed fresh-through time until Bob tells it to stop. By issuing the new License as shown above, Bob informs the revocation mechanism that it should also allow Ethyl to tell it to stop updating the guaranteed fresh-through time for the 'L16...' `dsig:SignatureValue`. Once Ethyl tells the revocation mechanism to stop, Cathy will no longer be able to get updated guaranteed fresh-through times and will eventually be no longer able to play the video.

5.5.9 USING CROSS-REFERENCES

Cross-references are a convenience feature provided by the Rights Expression Language. The use of this feature can greatly reduce the size of a License by assigning identifiers to commonly used elements and then simply referencing the elements by those identifiers elsewhere in the License.

5.5.9.1 Defining and Referencing Identifiers

An identifier can be assigned to any LicensePart (that is, any element that can substitute for the `r:licensePart` element). Many commonly used elements can substitute for the `r:licensePart` element, including `r:principal`, `r:right`, `r:resource`, `r:condition`, `r:grant`, `r:grantGroup` and `r:forAll`.

All LicenseParts have two attributes that are used to implement cross-referencing:

- `r:licensePartId`: This attribute assigns an identifier to an element. The identifier is required to be unique only within a given License.
- `r:licensePartIdRef`: This attribute references a LicensePart using its assigned identifier.

In a License, an element can have one of these two attributes, but not both.

The following example License illustrates how these attributes may be used to take advantage of the cross-referencing feature.

```
<r:license...>
   <r:grant>
      <r:keyHolder licensePartId="Cat">
         <r:info>
            <dsig:KeyValue>
               <dsig:RSAKeyValue>
                  <dsig:Modulus>Cat7QzxAprs=</dsig:Modulus>
                  <dsig:Exponent>AQABAA==</dsig:Exponent>
               </dsig:RSAKeyValue>
            </dsig:KeyValue>
         </r:info>
      </r:keyHolder>
      <mx:play/>
      <mx:diReference licensePartId="OnlineMagazine">
         <mx:identifier>urn:example:000</mx:identifier>
      </mx:diReference>
   </r:grant>
   <r:grant>
      <r:keyHolder licensePartIdRef="Cat"/>
      <mx:print/>
      <mx:diReference licensePartIdRef="OnlineMagazine"/>
   </r:grant>
   <r:issuer>
      <dsig:Signature>
         <dsig:SignedInfo>...</dsig:SignedInfo>
         <dsig:SignatureValue>L25qOQhAprs=</dsig:SignatureValue>
         <dsig:KeyInfo>...Bob...</dsig:KeyInfo>
      </dsig:Signature>
   </r:issuer>
</r:license>
```

Example 5.25

This License contains two `r:grant` elements, as illustrated in Figure 5.7. For brevity, the figure omits the detail in the `r:issuer` element.

The first `r:grant` contains the `r:keyHolder` and `mx:diReference` elements in their entirety, assigning a LicensePart identifier to each. In the second `r:grant`, the `r:keyHolder` and `mx:diReference` elements do not have any content. These elements simply use the `r:licensePartIdRef` attribute to reference the earlier `r:keyHolder` and `mx:diReference` elements by the identifier.

When this License is interpreted, the content of the `r:keyHolder` that defines the LicensePart identifier 'Cat' is copied into the `r:keyHolder` element that references this LicensePart identifier. A similar replacement takes place for the LicensePart identifier 'OnlineMagazine'. When a License is interpreted, these cross-references are resolved before resolving variables but after constructing and verifying signatures.

An element that uses an `r:licensePartIdRef` attribute must reference an element of the same name. For example, an `r:keyHolder` element that uses an `r:license PartIdRef` attribute must reference another `r:keyHolder` element.

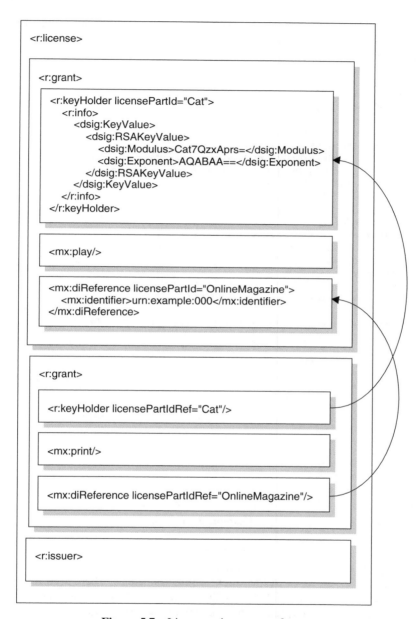

Figure 5.7   License using cross-references

5.5.10 *AGGREGATING IDENTIFIER DEFINITIONS USING INVENTORY*

To make Licenses that use cross-referencing easier to read and process, it may be helpful to use the inventory feature. The `r:inventory` element is simply a container for LicenseParts. The r:inventory element appears at the beginning of a License and provides a convenient place in which to assign identifiers to LicenseParts. An element's inclusion within the r:inventory element does not convey any meaning; it is simply a matter of convenience. The following example License conveys the same meaning as the one in Example 5.25.

```
<r:license...>
   <r:inventory>
      <r:keyHolder licensePartId="Cat">
         <r:info>
            <dsig:KeyValue>
               <dsig:RSAKeyValue>
                  <dsig:Modulus>Cat7QzxAprs=</dsig:Modulus>
                  <dsig:Exponent>AQABAA==</dsig:Exponent>
               </dsig:RSAKeyValue>
            </dsig:KeyValue>
         </r:info>
      </r:keyHolder>
      <mx:diReference licensePartId="OnlineMagazine">
         <mx:identifier>urn:example:000</mx:identifier>
      </mx:diReference>
   </r:inventory>
   <r:grant>
      <r:keyHolder licensePartIdRef="Cat"/>
      <mx:play/>
      <mx:diReference licensePartIdRef="OnlineMagazine"/>
   </r:grant>
   <r:grant>
      <r:keyHolder licensePartIdRef="Cat"/>
      <mx:print/>
      <mx:diReference licensePartIdRef="OnlineMagazine"/>
   </r:grant>
   <r:issuer>
      <dsig:Signature>
         <dsig:SignedInfo>...</dsig:SignedInfo>
         <dsig:SignatureValue>L26qOQhAprs=</dsig:SignatureValue>
         <dsig:KeyInfo>...Bob...</dsig:KeyInfo>
      </dsig:Signature>
   </r:issuer>
</r:license>
```

Example 5.26

In this example, the r:inventory element contains the r:keyHolder element, which is assigned the LicensePart identifier 'Cat', and the mx:diReference element, which is assigned the LicensePart identifier 'OnlineMagazine'. Then, both the r:grant elements in the License simply reference these elements using the assigned identifiers.

When this License is interpreted, the content of the r:keyHolder within the r:inventory element is copied into the r:KeyHolder elements in the two r:grant elements. A similar replacement takes place for the mx:diReference elements.

5.5.11 DELEGATING LICENSES
One particular form of delegation (which will be referred to here as 'one-deep delegation') can be modeled using features that have already been discussed in this chapter. In one-deep delegation, a User is given a package permitting him/her to use a resource and also

permitting him/her to delegate that permission to other Users. The Users to whom he/she delegates are able to use the resource but are not able to delegate to further Users.

To implement one-deep delegation using concepts already introduced, a User is given a License containing an `r:grantGroup` (introduced in section 5.5.7.7) containing two `r:grant` elements: one with `mx:play` (introduced in section 5.3.1.1) for a DI and one with `r:issue` (introduced in section 5.5.7.1) for an `r:grant` having `mx:play` for the same DI. This way, the User is given a hybrid usage/distribution License. The User can use the license to play the DI and/or to issue Licenses to other Users allowing them to play the same DI.

There are other forms of delegation besides the one-deep delegation described so far. These include n-deep delegation, infinite delegation, delegation within a group, and restrictive delegation (where the delegatee has fewer rights or more conditions than the delegator). While it is sometimes best to model simple delegation cases (like one-deep delegation) using features that have already been discussed in previous sections, some of the more complex delegation cases are better modeled using the `r:delegationControl` element.

When `r:delegationControl` appears in an `r:grant`, it signifies that the `r:grant` is delegatable. If the `r:delegationControl` has no children, delegation of the `r:grant` is unconstrained. However, the `r:delegationControl` may have several children, each of which represents a constraint on the delegation of the `r:grant`. ISO/IEC 21000-5 defines several delegation control constraints that can appear as children of `r:delegationControl` to model many forms of delegation, including those mentioned above. Other forms of delegation can be modeled in extensions to the Rights Expression Language.

The following example demonstrates the use of `r:delegationControl` with a simple delegation control constraint, `r:conditionUnchanged`. This constraint requires that the Conditions on the delegatee be the same as those on the delegator.

```
<r:license...>
   <r:grant>
      <r:delegationControl>
         <r:conditionUnchanged/>
      </r:delegationControl>
      <r:keyHolder>...Cat...</r:keyHolder>
      <mx:play/>
      <mx:diReference>...</mx:diReference>
      <r:validityInterval>...</r:validityInterval>
   </r:grant>
   <r:issuer>
      <dsig:Signature>
         <dsig:SignedInfo>...</dsig:SignedInfo>
         <dsig:SignatureValue>L27qOQhAprs=</dsig:SignatureValue>
         <dsig:KeyInfo>...Bob...</dsig:KeyInfo>
      </dsig:Signature>
   </r:issuer>
</r:license>
```

Example 5.27

This example License states that Bob permits Cathy to play a video in 2003. The presence of the `r:delegationControl` element specifies that Cathy's `r:grant` is delegatable. This means that Cathy can issue a License (to Fred, for example) that looks exactly like the one above with four possible exceptions:

- the User to whom the License is issued. Where the License above has Cathy's identity, the new License would have Fred's identity.
- the User who issued the License. The new License would be issued by Cathy instead of by Bob.
- the `r:delegationControl`. In this case, Cathy wants to allow Fred to delegate as well, so she opts to leave the `r:delegationControl` as is, not adding additional constraints to it.
- the Condition. In this case, the exception allowing Cathy to change the Condition is overridden by the `r:conditionUnchanged` in the `r:delegationControl`, which requires Cathy to leave the Condition unchanged when delegating to Fred.

The License that Cathy issues would allow Fred to play the same video, also in 2003. To demonstrate this, Table 5.11 shows an authorization request.

Table 5.11   Authorization request using delegation

Component	Value
Principal	`<r:keyHolder>...Fre...</r:keyHolder>`
Right	`<mx:play/>`

Resource	`<mx:diReference>`
	`  <mx:identifier>urn:example:001</mx:identifier>`
	`</mx:diReference>`

Interval of time	Midnight of 2003-09-01 through midnight of 2003-09-02
Authorization context	...

Set of Licenses	`<r:license...>`
	`  <r:grant>`
	`   <r:delegationControl>`
	`    <r:conditionUnchanged/>`
	`   </r:delegationControl>`
	`   <r:keyHolder>...Cat...</r:keyHolder>`
	`   <mx:play/>`
	`   <mx:diReference>`
	`    <mx:identifier>urn:example:001</mx:identifier>`
	`   </mx:diReference>`
	`   <r:validityInterval>`
	`    <r:notBefore>2003-01-01T00:00:00</r:notBefore>`
	`    <r:notAfter>2004-01-01T00:00:00</r:notAfter>`
	`   </r:validityInterval>`
	`  </r:grant>`
	`  <r:issuer>`
	`   <dsig:Signature>`

Table 5.11   (*continued*)

Component	Value
| | ```
<dsig:SignedInfo>...</dsig:SignedInfo>
<dsig:SignatureValue>L27qOQhAprs=</dsig:SignatureValue>
<dsig:KeyInfo>...Bob...</dsig:KeyInfo>
</dsig:Signature>
</r:issuer>
</r:license>

Example 5.28
<r:license...>
 <r:grant>

  <r:delegationControl>
   <r:conditionUnchanged/>
  </r:delegationControl>
  <r:keyHolder>...Fre...</r:keyHolder>
  <mx:play/>
  <mx:diReference>
   <mx:identifier>urn:example:001</mx:identifier>
  </mx:diReference>
  <r:validityInterval>
   <r:notBefore>2003-01-01T00:00:00</r:notBefore>
   <r:notAfter>2004-01-01T00:00:00</r:notAfter>
  </r:validityInterval>
 </r:grant>
 <r:issuer>
  <dsig:Signature>
   <dsig:SignedInfo>...</dsig:SignedInfo>
   <dsig:SignatureValue>L28qOQhAprs=</dsig:SignatureValue>
   <dsig:KeyInfo>...Cat...</dsig:KeyInfo>
  </dsig:Signature>
 </r:issuer>
</r:license>
``` |
| Set of root grants | trustedRootIssuers Bob |

The authorization request has two Licenses: the License from Bob to Cathy and the License from Cathy to Fred. It also has a set of root grants that say 'Bob can include whatever he wants in the Licenses he issues'. As mentioned in the section 'Evaluating a More Typical Authorization Request by Intuition', (see 5.4.2.2) this scenario involves two links of 'issuance chaining' as shown in Table 5.12.

To determine whether the activity of Fred playing the video on September 1, 2003, is allowed, a correspondence for that activity is found in the License issued by Cathy. Since this correspondence is found in a License (rather than a root grant), a link of 'issuance chaining' is needed to inquire about the activity of Cathy including that r:grant to
```

Table 5.12 Issuance chaining links

Activity	Step 1	Step 2
Fred playing the video on September 1, 2003, (original request)	No correspondence found in root grants	Correspondence found in second License
Cathy including the r:grant to Fred in the License she issues (issuance chaining link 1)	No correspondence found in root grants	Correspondence found in first License
Bob including the r:grant to Cathy in the License he issues (issuance chaining link 2)	Correspondence found in root grants	Does not apply because correspondence was found in Step 1

Fred in a License. The correspondence for this activity is found in the License from Bob because the r:grant in that License contains an r:delegationControl allowing Cathy to delegate. Because this second correspondence is again found in a License (rather than a root grant), a second link of 'issuance chaining' is needed to inquire about the activity of Bob including the r:grant to Cathy in a License. The correspondence for this activity is found in the root grants that say that Bob may put whatever he wants in the Licenses he issues. At this point, the 'issuance chain' is complete, so the answer to the authorization request is 'yes'.

A point worth noting in this scenario is that, in the end, both Cathy and Fred can play the video. In some scenarios, Bob may wish to allow Cathy to delegate to Fred only if, in so doing, Cathy gives up her rights to play the video. The following License represents this scenario.

```
<r:license...>
   <r:grant>
      <r:delegationControl>
          <r:conditionUnchanged/>
      </r:delegationControl>
      <r:keyHolder>...Cat...</r:keyHolder>
      <mx:play/>
      <mx:diReference>...</mx:diReference>
      <sx:transferControl>
         <r:serviceReference>
            <sx:uddi>
               <sx:serviceKey>
                  <sx:uuid>8F0B5111-D1C2-3964-3f47-D3B70132CE14</sx:uuid>
               </sx:serviceKey>
            </sx:uddi>
            <r:serviceParameters>
               <r:datum>
                  <sx:stateDistinguisher>
                      D11350EE-B7DB-D049-827C-06959EE527E9
                  </sx:stateDistinguisher>
               </r:datum>
            </r:serviceParameters>
```

```
            </r:serviceReference>
          </sx:transferControl>
      </r:grant>
      <r:issuer>
        <dsig:Signature>
          <dsig:SignedInfo>...</dsig:SignedInfo>
          <dsig:SignatureValue>L29qOQhAprs=</dsig:SignatureValue>
          <dsig:KeyInfo>...Bob...</dsig:KeyInfo>
        </dsig:Signature>
      </r:issuer>
  </r:license>
```

Example 5.29

The difference between this example License and Example 5.27 is that this License uses an `sx:transferControl` Condition that works in conjunction with the `r:delegation` Control. The `r:delegationControl` allows the `r:grant` containing the `mx:play` Right to propagate from Cathy to Fred without changes to the associated Condition, `sx:transfer` Control. The result is that both Cathy and Fred receive an `r:grant` containing the `mx:play` Right and an identical `sx:transferControl` Condition. The `sx:transferControl` element ensures that either Cathy or Fred can play at any one time, but not both. This is accomplished by specifying a service (which can be anything, including an online third party or an off-line secure baton-passing protocol) and a state distinguisher unique to this `r:grant` (so that the same service can be used for the Cathy–Fred group as well as for any other group, depending on the state distinguisher value). Depending on the service specified, the transfer from Cathy to Fred may be permanent (Cathy cannot ever play again) or temporary (Cathy may play when Fred is not playing, and vice versa).

5.6 CONCLUSION

The Rights Expression Language defined by ISO/IEC 21000-5 provides both the features and the flexibility needed to specify rights, terms and conditions supporting digital content distribution in a wide variety of business applications. This chapter introduced many of the important concepts and features of the Rights Expression Language, providing a basis for further investigation of the language.

There are many more features to take advantage of, and work is ongoing in this area, developing extensions, profiles and reference software. The purpose of the extensions and profiles that are under development is to make the language meet the specific needs of particular industry segments. The reference software under development will help understand the language from the software point of view. Section 5.6.1 describes the ongoing standards work related to the Rights Expression Language. Section 5.6.2 provides an overview of the reference materials available for further reading, including ISO/IEC 21000-5 itself. The reader interested in learning more about the language or how to use it in a deployed system is encouraged to consult these resources (in addition to the references listed in Section 5.8). While reading ISO/IEC 21000-5, it may be helpful to refer back to the examples in this chapter for illustrations of some of the capabilities that the language provides.

5.6.1 ONGOING STANDARDS WORK

Ongoing standards work related to the Rights Expression Language can be classified into four areas: extensions, profiles, conformance and reference software.

5.6.1.1 Extensions

Extensions add elements to the Rights Expression Language that are needed to address the requirements of a particular business segment. As of August 2004, the Rights Expression Language is being extended in a number of forums.

Open eBook Forum (OeBF)
The OeBF [3] is an international trade and standards organization for electronic publishing industries. Its Rights and Rules Working Group is focused on creating an open and commercially viable standard for interoperability of DRM systems providing trusted exchange of eBooks among all parties in a distribution chain. This working group has created a Working Group Draft of an extension to the Rights Expression Language that specifies elements specifically targeted at the eBook domain.

ISO/IEC JTC1/SC29/WG11 (MPEG)
MPEG is the Working Group (specifically, ISO/IEC Joint Technical Committee 1, Subcommittee 29, Working Group 11) of ISO/IEC in charge of the development of standards for coded representation of digital audio and video. MPEG has a Proposed Draft Amendment to ISO/IEC 21000-7 (DI Adaptation) containing an extension to the Rights Expression Language. The amendment specifies Conditions for fine-grained control of the changes permitted when playing, printing, modifying or adapting a DI.

5.6.1.2 Profiles

Profiles identify a subset of the elements in the Rights Expression Language that is sufficient to address the needs of a particular business segment. As of August 2004, the Rights Expression Language is being profiled in a number of forums.

Organization for the Advancement of Structured Information Standards (OASIS)
The Organization for the Advancement of Structured Information Standards (OASIS) Web Services Security Technical Committee [8] is focused on forming the necessary technical foundation for higher-level security services, which are to be defined in other specifications. This committee has a Committee Draft named *Rights Expression Language Token Profile*, a profile of the Rights Expression Language for the web services security domain.

Web Services Interoperability Organization (WS-I)
Web Services Interoperability (WS-I) [9] is an open industry organization chartered to promote WS-I. WS-I is working on a profile of the Rights Expression Language Token Profile (from OASIS) that will guide the creation of sample applications and testing tools and collectively promote WS-I when using the Rights Expression Language.

ISO/IEC JTC1/SC36 – Information Technology for Learning, Education and Training
ISO/IEC JTC1/SC36 [10] is the Subcommittee of the ISO/IEC Information Technology Joint Technical Committee focused on information technology for learning, education and training. ISO/IEC JTC1/SC36 has a Proposed Draft International Standard Profile of the Rights Expression Language for the learning technologies domain.

ISO/IEC JTC1/SC29/WG11 (MPEG)
MPEG is the Working Group (specifically, ISO/IEC Joint Technical Committee 1, Sub-committee 29, Working Group 11) of ISO/IEC in charge of the development of standards for coded representation of digital audio and video. MPEG has an MPEG-21 Profiles Under Consideration document [11] including a draft profile of the Rights Expression Language for mobile devices.

Content Reference Forum (CRF)
The Content Reference Forum (CRF) [12] is a standards group established to develop a universal way to distribute digital content across various media and geographies. The CRF has two candidate specifications containing profiles of the Rights Expression Language, one for offer Licenses and one for usage Licenses.

5.6.1.3 Conformance

As of August 2004, MPEG (ISO/IEC JTC1/SC29/WG11) has a Working Draft of ISO/IEC 21000-14 (Conformance) [11] specifying conformance points for ISO/IEC 21000.

5.6.1.4 Reference Software

As of August 2004, ISO/IEC JTC1/SC29/WG11 (MPEG, the Working Group for ISO/IEC 21000) has a Final Committee Draft of ISO/IEC 21000-8 (Reference Software) containing software implementing the various parts of ISO/IEC 21000. It also includes utility software demonstrating particular nonnormative uses of the various parts of ISO/IEC 21000.

5.6.2 FURTHER READING

Additional reference materials are available that address other aspects of the Rights Expression Language. Readers interested in learning more about the language are encouraged to refer to these materials.

5.6.2.1 The Rights Expression Language Specification

ISO/IEC 21000-5 specifies the syntax and semantics of the Rights Expression Language. It defines all of the Rights Expression Language types and elements, describes the relationships among the elements, and explains the authorization model by which Rights Expressions can be evaluated to determine the authorization they convey.

When reading ISO/IEC 21000-5 after this chapter, it is helpful to start reading ISO/IEC 21000-5 at Clause 4. Clause 4 describes the conventions used in ISO/IEC 21000-5. Clause 1 (Scope) can be skipped because this chapter gives a good understanding of the scope. Clause 2 (Normative references) and Clause 3 (Terms, definitions, symbols and abbreviated terms) can also be skipped initially and referred to later for help in understanding some of the terms and symbols used in the later Clauses.

The information about the authorization model and architectural concepts can be found in Clauses 5 and 6. This is where the details of authorization requests, authorization proofs, variables, license parts, and so on, are defined. The normative specification of equality appears in Clause 6, but Annex C provides some implementation hints for equality.

Table 5.13 Core elements (Clause 7)

allConditions	exerciseMechanism	obtain	revoke
allPrincipals	existsRight	patternFromLicensePart	right
anXmlExpression	forAll	possessProperty	rightPattern
anXmlPatternAbstract	fulfiller	prerequisiteRight	rightPatternAbstract
condition	grant	principal	serviceDescription
conditionIncremental	grantGroup	principalPattern	serviceReference
conditionPattern	grantGroupPattern	principalPatternAbstract	toConstraint
conditionPatternAbstract	grantPattern	propertyAbstract	trustedRootGrants
datum	issue	propertyPossessor	trustedRootIssuers
conditionUnchanged	issuer	resource	trustRoot
dcConstraint	keyHolder	resourcePattern	validityInterval
delegationControl	license	resourcePatternAbstract	
depthConstraint	licenseGroup	revocable	
digitalResource	licensePart	revocationFreshness	

Table 5.14 Standard extension elements (Clause 8)

anonymousStateService	licenseIdPattern	uddi
callForCondition	name	validityIntervalDurationPattern
commonName	propertyUri	validityIntervalFloating
dnsName	rate	validityIntervalStartsNow
emailName	rightUri	validityTimeMetered
exerciseLimit	seekApproval	validityTimePeriodic
feeFlat	stateReferenceValuePattern	wsdlAddress
feeMetered	territory	wsdlComplete
feePerInterval	trackQuery	x509SubjectName
feePerUse	trackReport	x509SubjectNamePattern
feePerUsePrePay	transferControl	

Table 5.15 Multimedia extension elements (Clause 9)

modify	play	diCriteria	resourceSignedBy
enlarge	print	diPartOf	requiredAttributeChanges
reduce	execute	isMarked	prohibitedAttributeChanges
move	install	mark	complement
adapt	uninstall	source	intersection
diminish	delete	destination	set
enhance	diItemReference	helper	
embed	diReference	renderer	

Table 5.16 Elements described in this chapter

Element	Section introducing the element
r:allConditions	5.5.2
r:allPrincipals	5.5.1
r:anXmlExpression	5.5.5.2
r:conditionUnchanged	5.5.11
r:delegationControl	5.5.11
mx:diCriteria	5.5.5.2
mx:diPartOf	5.5.5.3
mx:diReference	5.3.1.1
sx:exerciseLimit	5.5.2
sx:feePerUse	5.5.7.6
r:forAll	5.5.5.1
r:grant	5.3.1
r:grantGroup	5.5.7.7
r:inventory	5.5.10
r:issue	5.5.7.1
r:issuer	5.3.2
r:keyHolder	5.3.1.1
r:license	5.3
r:obtain	5.5.7.3
mx:play	5.3.1.1
r:possessProperty	5.5.4.1
r:prerequisteRight	5.5.4.2
mx:print	5.5.7.7
sx:propertyUri	5.5.4.1
r:revocable	5.5.8
r:revocationFreshness	5.5.6.2
r:revocationMechanism	5.5.6.1
r:revoke	5.5.8
r:serviceReference	5.5.3
sx:trackReport	5.5.7.5
sx:transferControl	5.5.11
r:validityInterval	5.3.1.1
sx:validityIntervalDurationPattern	5.5.7.5

Details about the semantics of specific elements in the core namespace (abbreviated 'r'), standard extension namespace (abbreviated 'sx'), and multimedia namespace (abbreviated 'mx') are found in Clauses 7 (Table 5.13), 8 (Table 5.14), and 9 (Table 5.15), respectively. Details about the syntax of all these elements are found in Annex A, which provides listings of the W3C XML schemas. For reference, tables listing the global elements defined in each namespace are provided below.

Details about the qualified names that are defined for identifying countries, regions and currencies can be found in Annex B.

Additional examples can be found in Annexes D, F, G and J.

A discussion of how to extend the Rights Expression Language to add elements or to profile it for a particular application can be found in Annexes E, F and G.

Some notes about the usage of the Rights Expression Language in conjunction with other parts of ISO/IEC 21000 can be found in Annexes H and I.

5.6.2.2 Online Resources

The website of ContentGuard, Inc. [13] contains additional educational materials for the Rights Expression Language. At the time of writing, these materials include:

- *Feature example pages*: These web pages each describe a specific feature of the Rights Expression Language in detail, including examples of the feature's use. The feature examples include basic features (such as License and Grant) and advanced features (such as delegation and revocation).

- *Use case pages*: These web pages each describe an extended, real-world use case for the Rights Expression Language in a particular business model. In these web pages, each example in the use case builds on the previous examples, gradually increasing the number of parameters used to specify rights.

- *Design guideline papers*: These papers address issues that commonly arise when using the Rights Expression Language in real-world applications, such as creating extensions and profiles.

5.7 ELEMENT INDEX

The elements described in this chapter are listed in Table 5.16.

REFERENCES

[1] ISO/IEC, 21000-5:2004, "Information technology – Multimedia framework (MPEG-21) – Part 5: Rights Expression Language", available at http://www.iso.org/iso/en/CatalogueDetailPage.CatalogueDetail? CSNUMBER=36095&ICS1=35&ICS2=40&ICS3=, 2004.

[2] This book's supplementary materials web page, http://www.wiley.com/go/MPEG-21.

[3] The Open eBook Forum (OeBF), web site, http://www.openebook.org/, accessed 2004.

[4] The Open Mobile Alliance (OMA), web site, http://www.openmobilealliance.org/, accessed 2004.

[5] XML-Signature Syntax and Processing, W3C Recommendation, 12 February 2002, available at http://www.w3.org/TR/xmldsig-core/.

[6] Organization for the Advancement of Structured Information Standards (OASIS) Universal Description, Discovery and Integration (UDDI) Technical Committee (TC) Committee Specification, "Schema Centric XML Canonicalization Version 1.0", 19 July 2002, available at http://uddi.org/pubs/SchemaCentric Canonicalization.htm.

[7] OASIS UDDI, web site, http://uddi.org/, accessed 2004.

[8] OASIS Web Services Security (WSS) TC, web site, http://www.oasis-open.org/committees/tc_home.php? wg_abbrev=wss, accessed 2004.

[9] Web Services Interoperability Organization (WS-I), web site, http://www.ws-i.org/, accessed 2004.

[10] ISO/IEC, JTC1 SC36 web site, http://jtc1sc36.org/, accessed 2004.

[11] MPEG, Working Documents web page, http://www.chiariglione.org/mpeg/working_documents.htm, accessed 2004.

[12] Content Reference Forum (CRF), web site, http://www.crforum.org/, accessed 2004.

[13] ContentGuard, Inc. web site, http://www.contentguard.com/, accessed 2004.

6

The MPEG-21 Rights Data Dictionary and New Approaches to Semantics

Chris Barlas, Martin Dow and Godfrey Rust

6.1 INTRODUCTION

The MPEG Rights Data Dictionary (RDD) is a sister specification to the MPEG Rights Expression Language (REL), designed to provide semantics to the REL and support the interoperability of rights expressions. The scope statement for the specification makes it clear that it has two objectives, but there is a third too: the creation of a methodology for the development of interoperable metadata to support the general trading of intellectual property rights.

This chapter provides an explanation of the MPEG RDD. Section 6.2 set the scene for its specification, including the semantic problems it is intended to address. Section 6.3 explains the specification itself, with sub-sections devoted to the dictionary, to the methodology and to the Registration Authority (RA) that will be established as a repository for Terms as they are developed. Section 6.4 shows how the RDD can be used in combination with the REL to provide Terms for rights expressions and how it is proposed to automate the process. Section 6.5 deals with the expected developments of the RDD, particularly with respect to future semantic interoperability issues in the context of pervasive networked computing and invisible support for extensive, automated rights management on a global scale.

6.2 BACKGROUND

It is sometimes said that a major cause of the slow uptake of DRM technologies is the failure to agree on a specification for interoperable DRM. Yet, while such a specification might be highly desirable, it is not immediately likely. Companies have invested huge

The MPEG-21 Book Ian S Burnett, Fernando Pereira, Rik Van de Walle, Rob Koenen
© 2006 John Wiley & Sons, Ltd

amounts in proprietary technology and they are not about to give up their own technology in favour of a committee-ordained solution.

A viable alternative is to tackle the problem at an infrastructure level. The MPEG RDD was envisaged as part of this infrastructural solution as standards for rights semantics are an essential precursor to releasing the power of entrepreneurship in network delivery of intellectual property. While the development and maintenance of a standardized RDD may appear to be complex and difficult, it is, in reality, an essential tool to bring *coherence* to the otherwise impenetrable complexity of the digital management of rights.

Without the underlying semantic standards, there will be no satisfactory way of exchanging data between different digital rights proprietary management applications; every communication will depend on imprecise and unmaintainable many-to-many semantic mappings. This will be costly and prove to be a non-incentive for rights holders, who will have to bear the cost. In turn, this will damage the interests of DRM companies that will depend on a plentiful supply of interoperable content (interoperable because it is identified and described in a consistent manner). In short, for everyone involved, the cost of maintaining communications between proprietary applications without a standard semantic layer will be unsupportable.

6.2.1 THE <INDECS> LEGACY

The MPEG RDD is founded on a very considerable legacy of semantic knowledge. Perhaps, the most influential input to the RDD was the project [1], funded under the European Commission RTD programme in 1998–2000.

The subject of the <indecs> project was the practical interoperability of digital content identification systems and their related rights metadata within multimedia e-commerce. It was a fast-track project aimed at finding practical solutions to some basic infrastructure issues that affect all types of rights holders.

The project dealt with metadata relating to any type of intellectual property in the form of digital objects (such as MPEG's Digital Items) and also related to packages (such as books and sound carriers) and intangible properties (such as musical and literary works) that contain or are contained in digital objects and are themselves subjects of e-commerce.

The specific problem <indecs> sought to address was the question of the interpretation of identifiers. In the case where an item of intellectual property was already identified by its creator or owner, it identified the steps necessary to ensure that identifiers could be properly interpreted within the e-commerce environment for the purpose of searches, licences, financial transactions and protection. Fundamentally, it was recognized that at that time there were substantial gaps and inconsistencies in the global 'metadata infrastructure' that prohibited secure and automated navigation. In fact, it was opined that there was no such thing as a global infrastructure – only a collection of unconnected or occasionally linked schemes, messages and dictionaries existed.

The project delivered its final report in June 2000 [2]. It was this report that provided the intellectual bones of the RDD. Indeed, the <indecs> report, along with the Imprimatur Business Model [3], had already been the subject of input documents to MPEG and the issues addressed by the project were understood by MPEG experts.

The final <indecs> report identified four axioms and had, as its basic tenet, the following statement:

People make stuff. People use stuff. People do deals about stuff.

The paragraph continued:

The stuff and the deals may come in any order, but neither come before the people. This is the basic model of commerce that underlies the <indecs> framework and models. While the approach described here may be usefully applied in many domains, the main focus of <indecs> is on the use of what is commonly (if imprecisely) called content or intellectual property. The model applies in many contexts, but is particularly useful in the digital and Internet environments where the problems of metadata interoperability are becoming especially acute. Commerce is used here in its broadest sense, not necessarily having financial gain as its object. The model applies equally to cultural transactions in places such as libraries in which people 'make deals' that enable others to have free access to 'stuff'.

The four axioms identified in the report provided the guiding principles for the creations of the dictionary.

6.2.1.1 Axiom 1: Metadata is Critical

Electronic trading depends to a far greater extent than traditional commerce on the way in which things are identified (whether they are people, stuff or deals) and the terms in which they are described (metadata, or data about data).

E-commerce requires the linking of identifiers that connect people with goods and services: stuff. In dealing with intellectual property, these identifiers form complex and dynamic chains. All kinds of metadata elements find their way into them. Where there is a gap or an ambiguity in these elements, it is likely that the chains will be broken or misrouted, and the required transaction will not happen or will have the wrong results. As e-commerce grows, reliance on metadata chains grows with it.

6.2.1.2 Axiom 2: Stuff is Complex

The second axiom on which <indecs> rests is that, when dealing with intellectual property, stuff is complex. The generic <indecs> term for a piece of stuff that may carry intellectual property rights is a creation. While an apple bought at a market stall is a single physical entity owned entirely by one person, a single digital audiovisual creation may contain hundreds or even thousands of separate pieces of content protected by intellectual property law. These may include moving pictures, recorded audio, still photographs, graphics, text and software applications, some only in part or in modified form. Each of these separate manifestations of intellectual property may have rights. These manifestations are normally expressions of abstract works or abstractions in which there may be further rights; and those expressions may come into being through the medium of spatio-temporal performances in which yet further rights may exist. All of these rights may be owned or controlled by different people for different places and different periods of time. The trading of one digital creation may involve rights transactions affecting thousands of people and companies, from whom permissions may be required and to whom payment may be due.

6.2.1.3 Axiom 3: Metadata is Modular

Because stuff is complex, metadata is modular. E-commerce metadata is made up of connecting pieces created by different people. Each of the basic entities (such as parties, creations and transactions) must have its own metadata set if stuff is to be found and used, and rights are to be protected and rewarded. If the rights in a complex creation come from many different people, so inevitably must the metadata. Constraints of cost, time and knowledge ensure that the multimedia producer is dependent on his suppliers of content to also provide the metadata on which future management depends. The same dependency is increasingly true for others in the chain, including non-profit–driven organizations such as libraries and academic institutions. Metadata in the digital environment can therefore be viewed as a set of 'modules', produced in different places and for different purposes, which must link together easily into complex forms to create new metadata modules for different stuff, people and deals. The result can be described as the metadata network, or in a narrower context, the Semantic Web [4].

6.2.1.4 Axiom 4: Transactions Need Automation

The final axiom is that, in an increasing range of cases, transactions need to be highly or completely automated. In physical commerce, much metadata complexity has been dealt with (if at all) in administrative systems within bounded organizations such as publishers or collecting societies, each operating its own local data standards and systems. The scale and nature of e-commerce has made it imperative that these local standards and systems can interoperate with others in automated ways.

For example, in the non-digital environment, securing copyright 'permissions' is a complicated, time-consuming and often unsatisfactory process. Owners and publishers are already often unable to cope with the volume of low-value permission requests made in conventional ways.

In the digital environment, the volume and nature of such uses is increasing exponentially. Because stuff is complex, technology is ingenious and the virtual world does not recognize national boundaries, the number of creations, agreements and potential rights holders and users multiplies rapidly and continually. Without automation, all but the most valuable permissions will become impossible to administer.

The developers of the MPEG RDD followed the <indecs> approach described above, and as a result, the RDD specification, while providing dictionary Terms in accordance with its scope statement, provides a much broader, deeper solution to the requirements published by MPEG at the 57th meeting in Sydney in July 2001 [5]. In particular, as is proposed in the <indecs> framework, the RDD treats all metadata, including rights metadata, alike, on the basis that to achieve the automated processing of rights expressions, elements of both rights and descriptive metadata must be available to those creating rights expressions.

6.2.2 THE MPEG CALL FOR REQUIREMENTS AND PROPOSALS

In February 2001, MPEG issued a call for requirements for a REL and a RDD. This call was based on a protracted requirements-gathering exercise, which had begun with the 55th MPEG meeting held in Pisa in January 2000. These requirements were further refined at the 56th meeting (Singapore) and at an Ad Hoc Group meeting convened in London between the 56th and 57th meetings. This protracted requirements-gathering exercise was set up because it was considered essential to understand what the requirements for such technologies might be and whether such technologies were actually necessary at the standards level. During this requirements-gathering exercise, it became clear that the industry wanted both a language and a dictionary. Both were required because while a language was able to express rights, a dictionary was considered essential, given the complexity of the semantics that were likely to be required in future in an automated rights management environment. However, it also became clear that the semantics of rights were far from agreed upon. Indeed, the central term in the debate – rights – was hotly debated. On the one side were those who claimed the word as a proxy for 'copyright', that is the owner's right to permit or deny someone to do something with a piece of protected content, and on the other side were those who claimed the word as a proxy for 'the right to do something', that is the user's right to act on a piece of content.

After a prolonged period of debate and division, MPEG published a set of requirements for the REL and the RDD in July 2001 at the 57th meeting in Sydney. Many of the requirements were relevant only to the REL, but a few were relevant to both. In particular, the requirements specified that the RDD should be able to deal with all creation types, including abstractions and manifestations.

In anticipation of the call, a Consortium (which went on to be selected to provide the baseline technology) had been assembled to respond to the requirements. The consortium was composed of a number of major rights-holder groups, including both the Recording Industry Association of America, the International Federation of the Recording Industry and the Motion Picture Association, together with a technology company (ContentGuard), a multimedia support company (Melodies and Memories Global) and an infrastructure organization (the International DOI Foundation). These organizations had declared an interest in the development of a RDD based on the <indecs> analysis and supported the development of the proposal during the period of competitive development, which lasted from July to October 2001.

6.2.3 THE SUBMISSIONS

Three proposals were submitted specifically for the RDD. The Open Digital Rights Language and Dictionary was submitted jointly by IPR Systems and Nokia, IPROnto was submitted by the Distributed Multimedia Applications Group of the University of Pompeu Fabra and <indecs>2RDD was submitted by the <indecs>2 Consortium. At the 58th MPEG meeting held in Thailand, there was a 10-day technology assessment. The proposals were carefully considered with the <indecs>2 consortium proposal being eventually selected as the baseline, on the basis of its thoroughness and approach. In particular, the <indecs>2RDD proposal was predicated incorporating existing descriptive metadata schemes already developed by rights owners. As noted earlier, the proposal relied substantially on the previous work carried out in the <indecs> project.

6.2.4 THE <INDECS>2RDD MPEG PROPOSAL

The <indecs>2RDD Consortium proposed a fundamental model for the RDD, designed to create a dictionary that was consistent and extensible, with semantics as granular as required by any particular use case.

In the preamble to the proposal for the specification, the Consortium made it clear that there were three separate, differentiated objectives being recommended to MPEG for the RDD.

Firstly, the proposal aimed at a specification that would 'provide a set of clear, consistent, structured and integrated definitions of Terms that could be used in different schemes'. This was considered essential for an environment in which different content verticals already had extensive metadata schemes and which would therefore require Terms and definitions that could be used across different schemes.

Secondly, the proposal set out to 'provide a comprehensive methodology for the interoperability of metadata necessary for the management of rights and permissions'. This was considered necessary on the basis that different content communities already had their own metadata schemes and it was considered essential to have a method by which existing metadata could be incorporated into a right management environment for multimedia compositions.

Thirdly, reflecting the discussions that had taken place during the requirements period, the proposal made it clear that the RDD should be completely separate from any legal regime for rights. Thus, it was clearly stated that '<indecs>2RDD should accurately *describe* but in no way *prescribe* how rights and permissions operate'. Bearing in mind the reservations expressed by commentators such as Professor Laurence Lessig [6], to the effect that DRM must not encode law, the RDD was intended to be only descriptive, although the meaning of each of its Terms is, in itself, prescriptive. That is to say, the creators of the proposal insisted that the division of law and code be maintained and that the semantics provided by the RDD would serve only to describe intellectual property rights and the content to which they apply and would not have any quasi-legal status.

These three objectives formed the underlying philosophy of the proposal.

6.3 THE RDD SPECIFICATION

6.3.1 SCOPE

The RDD specification sets out to describe how an RDD can be created, what it can contain and how it can be used. The dictionary itself comprises a set of clear, consistent, structured, integrated and uniquely identified Terms (as defined in Clause 5.4 of the specification) to support the MPEG-21 REL, ISO/IEC 21000-5.

To show how the dictionary has been created and how it can be extended, Annex A of the specification sets out the methodology for and structure of the RDD, and specifies how further Terms may be defined under the governance of a RA, whose requirements are described in Annex C of the specification. Use of the RDD is intended to facilitate the accurate exchange and processing of information between interested parties involved in the administration of rights in and use of Digital Items and in particular, it is intended to support ISO/IEC 21000-5 (REL). Section 6.4.1 of this chapter describes how the RDD relates to the REL (ISO/IEC 21000-5) and how the REL can use the RDD.

Besides providing definitions of Terms for use in the MPEG REL, the MPEG RDD is designed to support the mapping of Terms from different namespaces. Such mapping will enable the transformation of metadata from the terminology of one namespace (or Authority) into that of another namespace (or Authority). This is further described in Section 6.3.10. Mapping, to ensure minimum ambiguity or loss of semantic integrity, is the responsibility of the RA, which is also obligated to provide an automated Term look-up service.

The RDD is a *prescriptive* dictionary in the sense that it defines a single meaning for a Term represented by a particular RddAuthorized TermName, but it is also inclusive in that it can recognize the prescription of other Headwords and definitions by other Authorities and incorporate them through mappings. The RDD also supports the circumstance that the same name may have different meanings under different Authorities. The specification describes audit provisions so that additions, amendments and deletions to Terms and their attributes can be tracked.

Finally, and very importantly, the RDD recognizes legal definitions as, and only as, Terms from other Authorities that can be mapped into the dictionary. From a legal perspective, Terms that are directly authorized by the RDD RA neither define nor prescribe intellectual property rights or other legal entities.

6.3.2 ORGANIZATION OF THE SPECIFICATION

The standard is set out in six main clauses and four annexes, three of which are normative. The main clauses contain not only the usual scope statement and definitions, but also the standardized Terms that constitute the current standard dictionary. The main annex, which is normative, sets out the methodology upon which the dictionary is based, providing the means by which the dictionary can be extended. The second normative annex lays down rules for styles to be used within the dictionary. The third normative annex contains the requirements for the RA. The fourth annex, which is informative, contains examples showing how Terms can be specialized (made specific to particular requirements) and how Terms can be mapped to external Terms to facilitate interoperability.

The following technical summary shows how a Term is constructed, how meaning is introduced into the dictionary through the analysis of verbs and how a logical model is used to integrate Terms. There is then an explanation of the scope of the current standardized Terms and how the dictionary can be extended, under the RA.

6.3.3 THE SPECIFICATION DESCRIBED

It is worth remarking that the RDD is not what, at first sight, many people would expect from a RDD, and this is a result of the difference in its first two objectives (see 6.2.1 above). On the one hand, RDD provides a small number (14) of specific 'rights' verbs for immediate use in the REL.

To support these verbs, the RDD provides a substantial ontological infrastructure of hundreds of interlinked Terms, many of which may at first glance appear to have nothing to do with 'rights' at all. It is this substantial 'dictionary' that provides the basis for interoperability, and for the addition of new Terms in RDD which will enable a user, in principle, to write the terms of a licence to any degree of specialization and granularity, while at the same time ensuring that their language is mapped in a computable way to other Terms used in REL licences. Such a goal is not to be achieved simply, though simplicity lies at the heart of the specification.

RDD is based, like the <indecs> framework that preceded it, on *Acts* and the events (called *contexts* in the RDD) in which they occur. The underlying RDD model shows how different Terms all relate in a structured way to an underlying act. So, for example, the act *Adapt* defines the existence of an *Adapter* who makes an *Adaptation* from a *SourceOfAdaptation* at a *TimeOfAdapting* and in a *PlaceOfAdapting*. These 'families' of Terms, based on verbs, provide an ontology capable of supporting descriptions of the most complex content and events.

However, although RDD is very prescriptive in its meaning, it has a flexible trick up its sleeve: it allows people with their own metadata schemes to map their Terms and dictionaries precisely onto a corresponding RDD Term, and if the required Term does not yet exist, it can be created in the right place within the ontology hierarchies or 'genealogies'.

This does not suppose that everyone always means the same thing by the same word: rather that if namespace A has a Term 'JournalArticleWork' that means the same as namespace B's 'SerialArticle', both can be mapped to a common RDD Term; and if, on the other hand, A and B each use the word 'License' but mean something different by it, both can map to a corresponding RDD Term in the MPEG lingua franca. Accurate mapping is not a straightforward task, and this is the reason for the RDD's underlying model and Terms bearing that complexity which, at first glance, may be unexpected of such a standard. The RDD specification is as much a methodology as a dictionary.

The RDD specification outlines:

- the attributes of a Term;
- the initial set of REL verbs;
- the context-based model on which Terms are 'generated' from verbs and related to one another;
- the list of Terms that result from the combination of the REL verbs and the model;
- the means by which the list of Terms can be mapped to and added to; and
- the function of the Registration Authority that is responsible for the administration of the RDD.

6.3.4 TERM

A Term is defined as 'a semantic element with a defined meaning and an RddIdentifier'. Terms can be of any level of granularity or complexity. Each term has a unique identifier within the RDD and a range of Attributes.

6.3.5 TERM ATTRIBUTES

Each Term has attributes as illustrated in Figure 6.1.

The most significant Attributes shown in this table are explained in the following text.

6.3.5.1 RDD Identifier

Each Term has a unique identifier. No identification scheme is mandated by the standard. It will be the responsibility of the RA to select one.

6.3.5.2 Meaning Type

A Meaning is defined in the RDD dictionary as 'an Abstract element of significance represented in the RDD by a Term'. A Term can have an OriginalMeaning (the first Term Act is the only Term with an OriginalMeaning), a PartlyDerivedMeaning, where Meaning is derived through inheritance or a DerivedMeaning, where Meaning is wholly composed of a combination of two or more existing Meanings.

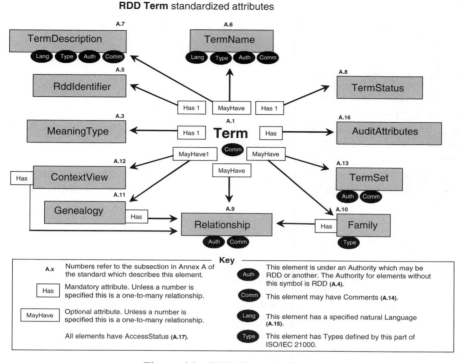

Figure 6.1   RDD Term Attributes

6.3.5.3 Term Status

Using TermStatus, it is possible to show whether the Term is Standardized, Native, Adopted, Mapped or Isolated. Standardized Terms are those that have been incorporated into the specification itself.

Native Terms are those that have a RDD Authorized Headword and Definition, with meaning inherited from another Standardized or Native Term. However, Native Terms are established under the authority of the RDD RA, rather than by the standardization process. It is likely that Native Terms will eventually be incorporated into the specification by amendment.

Adopted Terms are those under the governance of an Authority external to the RDD RA but recognized by the RDD RA and are consequently to be relied upon.

Mapped Terms are those under an external Authority, typically a single Authority responsible for a specific vertical domain. However, they are mapped to RDD Standardized and Native Terms to enable interoperability.

Isolated Terms are those registered with the RDD RA but not mapped into the RDD, hence they do not provide interoperability.

6.3.5.4 Genealogy

Genealogy is defined in the RDD dictionary as 'a group of Relationships that determine the derivation of, and constraints on, Meaning for a Term, and which are true regardless of Context'. Genealogies are an essential part of the ontological structure of the RDD.

6.3.5.5 Audit Attributes

These are used to set out the history of a Term, such as how it entered the dictionary and whether it has been changed.

6.3.6 RIGHTS VERBS

The 'first Term' at the top of the ontology supporting RDD is the verb *Act*. The most important Terms in the dictionary are the fourteen ActTypes (such as *Play, Print* and *Adapt*), which correspond to the standard 'Rights' that may be granted in the REL. These rights are located within a hierarchical ontology as explained in Section 6.3.3.

Table 6.1 is taken from the specification. The left-hand column contains the verbs (called ActTypes in the specification), the second column shows the parentage, the third column contains the normative definition and the fourth column contains information comments.

Table 6.1   The fourteen RDD ActTypes

ActType	Parent(s)	Definition	Comments (Informative)
Adapt	Derive, Change Transiently	To *Change Transiently* an existing *Resource* to *Derive* a new *Resource*.	With *Adapt*, two distinct *Resources* will exist as a result of the process, one of which is the original *Resource* in unchanged form, and one of which is newly made. Changes can include the addition to and removal of elements of the original *Resource*, including the *Embedding* of other *Resources*. Changes can be made temporarily to the original resource in the course of the *Adapt* process, but such changes are not saved in the original *Resource* at the end of the process.
			Specializations of *Adapt* can be differentiated by specific attributes of the *Resource* which are preserved or changed. The specific attributes can be on a list or can be called out by using a list. Lists can be inclusive (for example, 'Attributes a and b must be changed') or exclusive (for example, 'Everything except attributes c and d must be changed'). Attributes that are not constrained in specializations can be changed.

Table 6.1   (*continued*)

ActType	Parent(s)	Definition	Comments (Informative)
			Most *Act Types* that are generally known as 'copying' may be represented in the RDD Dictionary as children of *Adapt*. In most domains 'copy' typically means to *Derive* a new *Resource* which has the same set of specified or implied attributes as its *Source*, a common example being the 'copying' of a Digital Object. However, the concept of 'sameness' is not to be confused with that of identity as two things cannot technically be 'identical' because at the very least they will have different spatial or temporal attributes (that is, they will be located in a different place, or created at a different time), and so a 'copy' with absolutely identical attributes to the original cannot logically exist. Particular interpretations of 'copy' can be defined as specializations of *Adapt* [for further explanation see Annex D of the specification].
Delete	Destroy	*To* Destroy *a* DigitalResource.	*Delete* applies only to *Digital Resources. Delete* is not capable of reversal. After a *Delete* process, an 'undelete' action is impossible.
Diminish	Adapt	To *Derive* a new *Resource* which is smaller than its *Source*.	With *Diminish*, two distinct *Resources* will exist at the end of the process, one of which is the original *Resource* in unchanged form, and one of which is newly made, whose content is *Adapted* from the original *Resource*, and a Measure of which is smaller than that of the original while no Measures of it are larger. Changes can include the removal of elements of the original *Resource*. Changes can be made temporarily to the original *Resource* in the course of the *Diminish* process, but such changes are not saved in the original *Resource* at the end of the process.

(*continued overleaf*)

Table 6.1   (*continued*)

ActType	Parent(s)	Definition	Comments (Informative)
Embed	Relate	To put a *Resource* into another *Resource*.	The *Resource* into which a *Resource* is *Embedded* can be pre-existing or can be created by the act of combining the *EmbeddedResource* with one or more others. *Embed* refers only to the embedding of an existing *Resource* in another: if a 'copy' of an existing *Resource* is to be created and *Embedded* in another, then both *Adapt* and *Embed* would be used.
Enhance	Adapt	To *Derive* a new *Resource* which is larger than its *Source*.	With *Enhance*, two distinct *Resources* will exist at the end of the process, one of which is the original *Resource* in unchanged form, and one of which is newly made, whose content is *Adapted* from the original *Resource*, and a Measure of which is larger than that of the original while no Measures of it are smaller. Changes can include the addition of elements to the original *Resource*, including the *Embedding* of other *Resources*. Changes can be made temporarily to the original *Resource* in the course of the *Enhance* process, but such changes are not saved in the original *Resource* at the end of the process.
Enlarge	Modify	*To* Modify a Resource *by adding to it*.	With *Enlarge*, a single *Resource* is preserved at the end of the process. Changes can include the addition of new material, including the *Embedding* of other *Resources*, but not the changing or removal of existing elements of the original *Resource*.
Execute	Activate	To execute a *DigitalResource*.	*Execute* refers to the primitive computing process of executing. *Execute* applies only to a *Digital Resource*.
Install	UseTool	To follow the instructions provided by an *Installing Resource*.	An *InstallingResource* is a *Resource* that provides instructions which when followed result in one or more *Resources* that are new, or *Enabled*, or both new and *Enabled*.

Table 6.1   (*continued*)

ActType	Parent(s)	Definition	Comments (Informative)
Modify	Change	To *Change* a *Resource*, preserving the alterations made.	With *Modify*, a single *Resource* is preserved at the end of the process (that is, no additional Resource(s) come into existence). Changes can include the addition to and removal of elements of the original *Resource*, including the *Embedding* of other *Resources* within it.
			Specializations of *Modify* can be differentiated by specific attributes of the *Resource* being preserved or changed. The specific attributes can be on a list or can be called out by using a list. Lists can be inclusive (for example, 'Attributes a and b must be changed') or exclusive (for example, 'Everything except attributes c and d must be changed'). Attributes that are not constrained in specializations can be changed.
Move	Modify	To relocate a *Resource* from one *Place* to another.	With *Move*, at least the location of the Resource is *Changed*.
Play	Render, Perform	To *Derive* a *Transient* and directly *Perceivable* representation of a *Resource*.	*Play* covers the making of any forms of *Transient* representation that can be Perceived directly (that is, without any intermediary process) with at least one of the five human senses. Play includes playing a video or audio clip, displaying an image or text document, or creating *Transient* representations that can be touched, or *Perceived* to be touched. When Play is applied to a *DigitalResource*, content can be rendered in any order or sequence according to the technical constraints of the *DigitalResource* and renderer.
Print	Render, Fix	To *Derive* a *Fixed* and directly *Perceivable* representation of a *Resource*.	*Print* refers to the making of a *Fixed* physical representation, such as a hard-copy print of an image or text, that can be *Perceived* directly (that is, without any intermediary process) with one or more of the five human senses.

(*continued overleaf*)

Table 6.1   (*continued*)

ActType	Parent(s)	Definition	Comments (Informative)
Reduce	Modify	To *Modify* a *Resource* by taking away from it.	With *Reduce*, a single *Resource* is preserved at the end of the process. Changes can include only the removal of existing elements of the original *Resource*.
Uninstall	UseTool	To follow the instructions provided by an *Uninstalling Resource*.	An *UninstallingResource* is a Resource that provides instructions which when followed result in one or more *Resources* that had previously been Installed being *Disabled* or *Destroyed*.

6.3.7 CONTEXT MODEL

RDD supports a range of common ontological axioms such as type inheritance and disjunction, but enhances them with an additional core structure. Terms are organized into *Families* according to the roles they play in *Contexts*. Contexts come in two varieties – Events (through which change happens) and Situations (which are unchanging) – which conform to the rules of the RDD *ContextModel*. This model (Figure 6.2) defines a group of five Terms that form the 'BasicTermSet' – *Context, Agent, Resource, Time* and *Place*. These Terms and their associated Classes and Relators form the core semantic architecture of the RDD – that is, the mechanisms by which meanings are derived from one another.

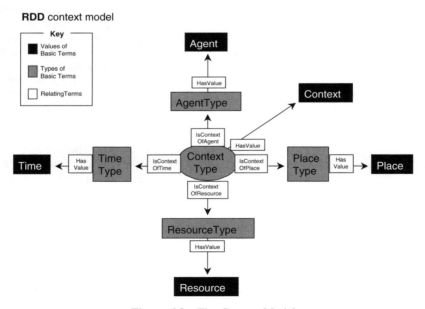

Figure 6.2   The Context Model

Through the ContextModel, when a new Act (say, *Modify*) is defined, a full range of related Terms embodying some aspect of the Act can also be defined, and each new Term placed in its logical node of the Dictionary. This Modify 'Begets' the Terms *Modifier, Modification, TimeOfModification* and *PlaceOfModification.*

The Model also provides all the specific Relationship Types that exist between any two elements of a Context (for example, *IsModifierOf* or *HasPlaceOfModification*). This ready proliferation of derived Terms creates mapping points for potentially any Term in any Schema.

6.3.8 STANDARDIZED TERMS

The RDD standard includes all the Terms required to support the ontology of the fourteen REL ActTypes and their related Families. These Terms are defined with their Term Attributes, including all their hierarchical and Family Relationships.

6.3.9 EXTENSIBILITY

As noted above, the StandardizedTerms provide only the basic semantics of RDD. However, as one of the primary purposes of the RDD is to provide a means of mapping existing Terms from other ('Alien') schemes into the dictionary for the purpose of creating more specialized ActTypes and Conditions for use in the REL, several different types of Terms are defined. This means that within the dictionary there is a kind of hierarchy of Terms, ranging from those owned and standardized by the RDD to those that are merely incorporated and cannot be vouched for by the RDD RA.

NativeTerms are those that are derived from StandardizedTerms and are under the Authority of the RDD, but which have not yet been standardized. AdoptedTerms are those whose definition is under another Authority but is one recognized by the RDD Authority (such as ISO Territory and Language codes). MappedTerms are defined by external Authorities that are not formally recognized by RDD, but which are mapped to other Terms in the dictionary. Finally, IsolatedTerms are those that are incorporated into the dictionary but are not mapped to any Term in the dictionary.

New Terms can be defined at any time through mapping by the RDD RA (see the following text).

The primary purpose of the MPEG RDD is to support the MPEG REL. To achieve this, the fourteen RDD ActTypes (Figure 6.1) are normatively referenced by the REL specification. These 'rights verbs' are the baseline actions that can be used in REL grants. The fourteen ActTypes, with their supporting ontology, are the result of a very careful analysis and cover the most common expected actions a user might wish to undertake with respect to digital content. However, it is fully expected that in future the meaning of these ActTypes will be extended by specialization (explained in Section 6.4.2). While such specializations can be used in a vacuum, it was recognized that to be useful they also needed to be added to the dictionary. This could be achieved either by amendments to the Standard or be the establishment of a RA that would keep a register of such specializations and any other new Terms.

6.3.10 THE RDD REGISTRATION AUTHORITY

During the development of the RDD, it was decided that the dictionary in the published standard should be restricted to those Terms necessary to support the first version of the

REL. Without such a restriction, it was judged that the dictionary would be too large to manage in a paper version. Furthermore, without such a specific restriction, its extent would have been arbitrary, which could have been confusing to users. However, it was also recognized that the dictionary would have to be substantially enlarged in the course of time to fulfill the requirements of its first users. In the normal course of standards making, this would be managed through cumulative amendments and new versions of the standard. But, in the case of a dictionary, which is by nature a living document, formal amendments of the standard would be simply too slow a method.

In this way, it became clear that, if constant amendments were to be avoided, some kind of RA would be required for this function. However, it was also clear that any RA for the RDD would be very different from the registration authorities set up for other specifications. For instance, the MPEG-4 RA for IPMP systems merely requires that owners and operators of IPMP systems register with the Authority. The MPEG-21 Digital Item Identification Registration Authority works on a similar principle, requiring those organizations managing an identification system to register with the Authority. Registration is not onerous but the benefits of registration are clear.

When it comes to the RDD, the situation is somewhat different. While the registration of a Term and its definition might be useful for the purpose of querying the Authority's database to discover the meaning of a Term, mere registration does nothing for interoperability. And it is interoperability of Terms that the specification implicitly promises, in as much as it proposes a methodology by which this might be achieved.

It was therefore decided that a RA should be established, with requirements laid out in an annex to the standard. This RA, to be set up in accordance with JTC1 directives [7], will be established with a very specific mandate. Its primary duty will be to manage the orderly development of the dictionary. This will require the development of a set of procedures for the acceptance of new Terms submitted to the RA, which will then be incorporated into the dictionary. A secondary, but equally important, activity will be the Authority's work on mapping submitted Terms to standardized, native, adopted and mapped Terms already in the dictionary. This mapping is vital as it will ensure that the primary purpose of the dictionary, the supply of consistent, interoperable semantics for the purposes of rights management across all content domain using MPEG technologies, is achieved.

Hence, the requirements for a RA, as set out in the Annex C of the specification, are directed towards the creation of what might be termed an *active* RA, that is, one that does more than simply register.

It is worth considering these requirements in some depth, for they provide a clear outline of what the future RDD RA will do and what its place in the secure content trading environment might be. The major requirements upon the organization establishing the RA are set out in two sections of Annex C of the specification.

The first set of requirements concerns the setting up of the Authority. The requirements call for processes to register and identify Terms. This is entirely in line with a conventional MPEG RA. However, the next requirement to '*establish administrative procedures for the introduction and mapping of Terms and TermSets in the RDD System*' means that the RA will have to provide actual services to support the creation of a dynamic dictionary, in which newly registered Terms can be mapped to already registered Terms. This is essential in terms of interoperability if Terms from one domain are to be capable of semantically sound interpretation in another domain that uses a different controlled vocabulary.

Clearly, this requirement puts the onus on the RA to employ semantic specialists who can create the appropriate mappings in accordance with the methodology contained in Annex A of the specification. Furthermore, it implicitly requires the Authority to acquire agreement to the mappings, as one Term mapped to another without agreement could not be called *reliable*. So, besides the semantic specialists, the RA will also require managers to negotiate agreement to the mappings once they are created. Thus, the RA is required to collaborate with external authorities managing alien metadata schemes to extend the dictionary beyond the standardized Terms.

This activity is important as it will provide interoperability between different metadata schemes by mapping alien Terms to Terms already in the dictionary. For instance, the publishing industry relies on ONIX message metadata, which is managed by EDItEUR [8] on behalf of the industry. It is likely that publishers will want to see ONIX Terms incorporated into the RDD, not on a Term by Term basis, but rather by wholesale inclusion. This will involve the RDD RA negotiating with EDItEUR for the inclusion of ONIX Terms into the RDD. It is hoped that such a course will be followed by metadata Authorities in other verticals, such as the music and film industries. By the incorporation of whole Term sets, the RDD will be rendered increasingly useful, providing a central mapping service to players in the content industry.

A third requirement in this first section calls for the RA operator to *'establish procedures for the management and coordination of applications for Terms and Term Sets that may be made on a sectoral, geographic or linguistic basis, including the appointment (and revocation) of sectoral, geographic or linguistic agencies as necessary'*. This is in recognition of the fact that the trade in intellectual property rights and content is carried on at a national and regional level and that there may be specific issues that can only be dealt with on a distributed basis. Such a process is increasingly being adopted for networks, where rigid centralized control is superseded by much more localized management. It also suggests that Terms in the dictionary may not all be in English, the baseline language of the dictionary. For instance, if a Korean Term were to be added to the dictionary, it is clear that a linguist would be required to ensure that the Korean meaning is preserved in the English version.

The remaining requirements in the first set deal with the issues such as making a look-up service available and ensuring that there are processes for the recognition of authorities. Together, this first set of requirements clearly implies that the RA will be (i) an active organization; (ii) a distributed organization, with regional and perhaps national centres; and (iii) capable of setting up and maintaining relationships with other authorities that maintain metadata sets.

The second set of requirements deals with the ongoing management and maintenance of the RA. These requirements cover issues such as the ability of the RA to maintain registers of Term identifiers and attributes, to amend or delete Terms in the dictionary and to keep the system well documented. Also included is a requirement for promotion of the RDD, on the basis that uptake of the RDD process is an important step towards interoperability for the management and protection of content online.

One objection that might be raised to the way it is proposed to set up and run the RA is that the activity will be costly. This is of course true, but under its rules ISO does allow those running Registration Authorities to operate on a cost recovery basis. That said, it will clearly be incumbent on the organization chosen to manage the RA to ensure that costs do not discourage potential registrants or the entire purpose of the RDD will be destroyed.

It is not yet known (July 2005) which organization will eventually be chosen as the RA, although WG11 has recommended the International DOI Foundation to SC29.

6.4 USING THE RDD

6.4.1 USING THE RDD IN ASSOCIATION WITH THE REL

The immediate application of the RDD is the simple linking of the fourteen basics Act-Types to their corresponding REL 'rights'. RDD provides the standardized semantics for interpreting these rights in an REL license. When an REL license grants the right to 'Play', it means 'Play' as defined by RDD (that is, 'to Derive a Transient and directly Perceivable representation of a Resource'). Each of the Terms with capital letters is also defined within the RDD ontology, so further look up will enable the user to understand the meaning as precisely as possible. The verb definitions are also supported by commentaries to remove further ambiguity.

6.4.2 SPECIALIZING THE VERBS FOR MORE COMPLEX USES

However, 'Play' in the highly generalized sense will often be inadequate to constrain a licensed use. A rights owner may wish to grant the right to Play his resource on a specific kind of computer, or using specific kind of software, or in a particular environment, or for a particular purpose (for example, commercial or educational). They may want to make quantitative restrictions on bandwidth or resolution or image cropping.

What could be even more problematic is that the rights owner may need to issue REL licenses for content that is not digital: the MPEG-21 requirements cover the licensing not only of digital content per se, but of physical carriers, recorded performances and abstract works contained within or referenced by a Digital Item. The use of a particular verb (like Adapt) will take on a quite different meaning in relation to an abstract work than for a digital work.

The RDD approach to this is to allow for the specialization of verbs. For example, Adapt may be specialized to AdaptPerformance or TranslateTextualContent or (to take an extreme but not quite absurd case to illustrate the point) AdaptByRemovingNoMoreThan 15PercentOfAudioVisualContentUsingAnAcmeComputerOnAWednesdayInBulgaria. The mechanism of RDD enables the rights owner (or any other interested party) to define and map all data elements relevant to a particular Act and create a specific verb or ActionFamily of Terms that can be deployed into an REL grant.

This process will result in the addition of a wide range of different Terms – including nouns, adjectives and prepositions as well as verbs – that may be deployed into REL grants with a standard meaning.

Such specialization will not, though, be the activity of the rights owner only. Device manufacturers or data processors are likely to want to map the semantics of their own technology into the RDD, for this will enable them to map their specifications to the specialized Terms in users' licenses to automatically determine whether a Digital Item may be processed in the way specified.

6.5 AUTOMATING THE RDD PROCESS

The RDD specification does not mandate any particular implementation. In fact, during the drafting of the standard, it was even decided not to mandate any particular representation of the RDD. This decision was taken because it was believed that implementation could be achieved in many ways. What follows is just one approach to implementation, which is set out here to demonstrate some of the issues that might be encountered by a RA or other users of the specification.

6.5.1 REQUIREMENTS FOR AUTOMATION

Providing automated support for the RDD process will entail the RA maintaining a system to service the needs of the different categories of users. These broadly consist of users participating in processes within the RA and users who constitute clients that will use RDD Terms and TermSets. Any user will need to be identified and authenticated, and the policies that apply to a user will determine the RDD content they have access to and the necessary workflows applicable according to the relevant agreement with the RA.

The broad set of requirements on RDD automation are dictated by the elements set out in the previous section and the need to support the axioms of <indecs>2RDD set out at the beginning of the chapter.

An implementation of the RDD would likely be satisfied by the following architectural choices:

- A repository where RDD Terms can be created, updated, versioned and deleted. Assigning unique identifiers to Terms and the ability to interrelate them in accordance with the Context Model in a managed environment would be a function of the repository. The repository is thus the component underlying the ability to relate disparate schemes together through the mapping process and thus support transformation of data between schemes. The repository should support a discovery mechanism and workflow for RA staff using system-defined metadata designed for the purposes of administering the RDD. Access controls and authorization are required for controlled management.

- A registry interface for client systems to submit Terms to the RA for mapping. The registry also serves as a discovery mechanism for various mapped Terms and should also support Term discovery through search, browse and query mechanisms. These functions would serve metadata content according to a policy for the user, supported by fine-grained access control and authorization functionality.

- Both an inbuilt human user interface and a machine service interface are required to support human tasks and allow for solely machine-based interactions with the RDD. This provides for component-driven extensibility in the future, such as customized user interfaces and notification services. All interfaces are required to operate in a distributed and scalable environment, so various Internet-based access methods (HTML, Web service, email) would be appropriate.

- Open integration standards should be used wherever possible to simplify integration of system interfaces with the maximum number of types of existing systems, thereby making it easy to adopt the RDD Terms and methods into client systems of the RDD. Relevant examples include XML for data interchange, SOAP for Web service interactions and RDF (Resource Description Framework) for object–object declaration of relationships, as well as support for appropriate MPEG-21 standards such as the DID [9] and REL.

- To support ongoing distributed installation requirements of RDD administration software for use by registration authorities, software components compatible with the maximum number of base computing platforms would become important.

- For the RDD to support data transformations between schemes, a function-centric view on computing with TermSets is required to ensure reusability of functions in various contexts free of side effects.

- The development and exposure of explicit links between business concepts and process definitions and technical concepts, algorithms and process models registered with the RDD. These links must have a clear model and method of access and processing for client systems so that the precise semantics are both available and preserved unambiguously when client systems come to process them, deriving an interpretation of data, such as evaluating whether the conditions of a specific license have been met.

- To support the capability for unambiguous interpretation (i.e., evaluation of Terms and data), the RDD semantics must be capable of being expressed as stand-alone data rather than relying on certain specific embedded code to carry this out. This is to ensure that the semantics derived from the RDD are transmittable independently of any code or applications that use the data. Typically, this relies upon the definition of a schema for RDD Terms and their relationships using a well-defined schema language; using an underlying triples data model, such schemata can be defined using the W3C's Web Ontology Language ('OWL') [10] or RDF-Schema language [11].

6.5.2 TERM REGISTRATION AND IDENTIFICATION

6.5.2.1 Registration

In accordance with the procedures set out in the standard, it will be possible to register Terms. The procedures specify a process based on form submission. To automate this element, online forms may be used. If any of the XML-based forms methods are used, such as the Adobe's smart forms or the W3C's XForms standard [12], then an electronic forms repository can be built as standardized declarative templates of explicit data definitions providing look-ups into the registry. Electronic form submission forms only part of a broader workflow or overall business process that would also be likely to include location of the forms, review of content, possible requests for clarification, subsequent Termregistration and notifications.

Accuracy and complete data acquisition by using standardized electronic forms is essential to populating the dictionary with consistent, 'valid' entries and satisfying subsequent process requirements.

6.5.2.2 Identity

Sharing semantics between authoritative sources requires an effective means to uniquely label individual artefacts, or 'resources'. Treating Terms and TermSets themselves as artefacts and assigning to each a *unique identifier* allows Terms and TermSets to be used at any lifecycle point to exchange precise semantics. This identifier would need to be globally unique, rather than an identifier local to a Termrepository, so that these precise semantics may be shared. This then allows for the RDD registry to be an effectively federated one, at a national level under the auspices of the RDD RA, and indeed amongst partner authorities.

The fact that a Term may also be used at different levels of granularity means that it may be used to refer to any artefact, and the Term's global identifier must uniquely identify it. Examples of what a Term refers to are as follows:

- A RDD native concept in an ontology schema language such as Web Ontology Language (OWL)
- A business rule expressed as a logical implication
- A relational database table or column
- A business process definition
- A protocol definition
- A Web service function definition
- A format definition, such as an XML Schema or Relax NG [13]
- A grammar definition in Extended Backus–Naur Form [14] (EBNF)
- An electronic form definition
- An explicit information node within an electronic form
- A codelist or controlled set of values, such as territory codes or currencies
- An agreement between parties (RDD users or authorities), or any sub-part of the agreement, such as an obligation
- A collection of information about an Authority
- An 'alien' concept in an ontology language that is linked to an external dictionary or thesaurus, thus providing a link to Semantic Web ontology–based systems
- Some isolated element known to the RDD and therefore linked to a user's data dictionary that is not administered by the RDD.

Terms are annotated with their relationship with the RDD itself. This allows users to cross-reference metadata sets using the RDD as a starting point, and allowing TermSets to combine mapped Terms with unmapped Terms. This flexibility will allow users to specialize in RDD Terms, use them and submit them for registration at a later date.

6.5.3 IMPLEMENTING A MODEL FOR INTER-TERM RELATIONSHIPS
6.5.3.1 Using XML and RDF

The key to implementing the RDD and therefore assigning unique identifiers to Terms and to collections of Terms, TermSets, is a successful implementation of the underlying

RDD Context Model. This model essentially takes values denoted by single Terms and the process of RDD contextualization turns them into information according to the RDD Context Model. This is in direct contrast to, say, XML's non-contextual approach, where, in the simplest case, each unit of information has a single, fixed (in the non-RDD sense) context described by its schema (usually a DTD or XML Schema definition).

The need to allow for interoperability of information exchange with the RDD RA would suggest that an XML format is required at its interfaces. This does, however, mean that flexibility to define context at this point is largely lost. XML namespaces, which divide sets of Terms in elements and attributes into specific vocabularies with fixed definitions, and XML modularization both allow a degree of contextualization by allowing multiple schemata to be mixed in a single document.

The W3C RDF standard takes the concept of composing models requiring a high degree of context to its logical conclusion by disposing of the notion of a document and instead using a collection of statements. The context of this collection of statements, expressed as node−arc−node 'triples' that correspond to RDF 'assertions' consisting of a subject, predicate and an object, cannot be known beforehand. The semantics are completely defined by the triple statements, and the whole network of relationships between the nodes constitutes a data model, often termed a *graph*. The fact that the RDD itself represents complex relationships between Terms expressed as triples means that constructing a graph using RDF is a natural, standards-based implementation choice. It affords the benefit of allowing various operations to be performed on a complex network of Terms that have well-known properties, both for adding new Terms and querying for subsets of these on the basis of certain constraints, and mixing Terms with data values. It further allows the expressive schema languages OWL and RDF-S to constrain the model to various combinations of triples, enabling classification and validation of both Terms and data, just as a relational schema allows classification of data arranged in table structures of tuples rather than triples.

Using RDF, a unique identifier scheme can be implemented by assigning unique node IDs encoded as a URI to uniquely identify a RDF 'resource', where a resource may be an RDF triple itself. This, in the context of RDF, is known as 'reification'. If graphs themselves need to be identified as a graph node in their own right to support certain aspects of RDD management, there are strategies but these are non-standard, such as constraining the statements in a RDF file on a file system. Research into areas such as RDF named *graphs* may eventually address these issues by being able to identify both each triple and a graph fragment itself, so that operations may be issued against multiple federated graphs of Terms in a consistent manner.

In addition, within RDF, other identifier schemes such as the Digital Object Identifier may be applied to uniquely identify nodes, as long as a URI encoding can be achieved.

The ability to uniquely identify Terms within the registry means that these identifiers, or their unique name equivalent, may be mixed into a message, thereby linking precise semantics without recourse to both sender and receiver requiring an identical context encoding in XML around the Term. The context is provided precisely and authoritatively with a look up against the RDF by either party on the RDD, the normal method being a query in a RDF query language such as SPARQL [15]. This is an extremely effective method, combining the best of the fixed non-contextual XML approach for scenarios requiring data exchange, and a semantic expression capability somewhat akin to the vision of Semantic Web practitioners but injecting the consistent application of context as the means of ensuring precise semantic interoperability in differing usages. It also ensures there is a low overhead in terms of sharing the Terms among various metadata authorities, and is simple for practitioners to implement since there is no need for complex XML schema namespace combinations within a document.

6.5.4 ADMINISTRATIVE PROCEDURES AND THE DYNAMIC DICTIONARY

The RA appointed to run the RDD is obliged to establish administrative procedures for the introduction and mapping of Terms and TermSets in the RDD System. Semantic experts will review submitted Terms and TermSets, and use the mapping methods outlined in the standard as part of the workflow process. The use of tools will partially automate this process, ensuring consistency and validity. These tools are likely to rely on a mapping of distinct Term subsets to at least one ontology language such as OWL-DL to ensure that mapping operations are achievable and consistent with respect to each other. Examples might be traversing a hierarchy of Terms or adding a constraint in a relationship between Terms.

As discussed earlier, the benefits of using the W3C family of ontology languages (RDF-S, OWL-Lite, OWL-DL and OWL-Full) may be realized by an expression of RDD triples as RDF triples, since this family is based on formal RDF syntax and semantics. Terms that represent Type, such as AgentType or ResourceType, can be represented as nodes in a graph, where these nodes can be ascribed as being nodes that denote a class in an ontology language. This allows machines using languages such as RDF-S to 'understand' the RDD Term hierarchies, and the mappings to the language could be authorized by the RA. Furthermore, ontology languages such as OWL-DL can similarly 'understand' constraints expressed using Arbitrary Values. This 'understanding' is only in the formal mathematical or logical sense – the mapping relates sets of RDD Terms to a formalism with known mathematical properties, and it is this that allows the fully automated computation of the semantics present in the RDD to be realized. Since the RDD specification is itself agnostic to any specific mathematical model, and indeed a single formalism of logic is unlikely to cover the entire term set without having undesirable computational properties, a single Term may eventually acquire an authorized mapping to many mathematic or logical 'domains of interpretation'. This feature has the advantage of allowing flexibility as to how an Authority should formally process Terms. Simple interpretations can be achieved using a Term in the context of the logic of a relational schema, and more complex ones may call for a logic such as that used by OWL-DL or ISO's SCL.

The mapping to an ontology language has the further advantage that any instance data classified using the RDD can be collected into a fully featured knowledge base system and analysed. Such facilities are akin to powerful OLAP tools; 'interesting' questions can be asked of the data, trends discovered, inconsistencies and conflicts exposed, and drill-downs performed over a huge combination of categories provided by the RDD.

6.5.5 TERM DISCOVERY AND RETRIEVAL

Once registered and processed, Terms are available to be 'discovered' in an automated way. Fully mapped Terms allow 'intelligent' discovery based on the Context Model to be performed, allowing the links and groups to be viewed in a highly structured and meaningful way within TermSets and a user interface tailored to the permissions granted to that user. Structured search and query tasks may be performed, both using this interface and with an automated look-up service. Mappings ensure that various representations of registered Term data can be returned, possibly supporting standard metadata harvesting protocols so that RDD metadata may be accessed using external tools. Such services may be automated via a Web service invocation. The consistency of the Context Model allows automated computing machinery to group together Terms meaningful to the user's queries, so that at a user interface level understanding the RDD structure becomes simplified and relates to user's information requirements around the interpretation and processing of rights data. The relationship types a user will primarily navigate are those of the Context Model framework provided by ContextFamily and ActionFamily instantiations. Users will be able to decompose Terms using the links, the user interface implementation providing guidance throughout.

6.5.6 MAINTENANCE OF THE RDD

A system that introduces Terms, imports schema containing TermSets, encapsulates procedures set out in the standard and with users and Authorities related in a submit-review-authorize-publish workflow might be characterized as a content management system and content repository, where the content is data about the RDD Terms and the mappings. The fact that the Terms and mappings provide a rich source of metadata means that this content management system also has the features of a general-purpose metadata editor for entities of any sort associated with rights metadata. A system that provides a facility for a user to introduce Terms and validate them according to an available mapping to an ontology or rules language might be characterized as an integrated ontology editor with reasoning capabilities. A system that provides preservation, discovery and look-up services might best be characterized as a registry and repository. A combination of all these features on a single platform might be termed an 'Ontology Management System' (OMS).

6.5.7 SUPPORT FOR USING THE RDD

Design-time tool support for users of the RDD that would allow users to build, register and thereby share models based on the RDD Terms might require both a version of the OMS and also support for relating it to instance data. This instance data refers to records of specific RDD Acts and their related specific Resources for subsequent look up and reuse. Using the terminology of Knowledge Management, this would be thought of as building and maintaining a knowledge base available to a user's systems. If the instance data are considered digital assets, such a system would also be thought of as a Digital Asset Management (DAM) system. The knowledge base implementation itself would remain stable over time relative to the various business processes and technical products and systems that evolve within an organization, while its content evolves to reflect this continuous process of change. A knowledge base composed of key digital assets is capable of driving the 'business model' implementations normally embedded in code within multiple systems. Examples of usage are likely to include the construction, distribution and interpretation of licenses, conflict resolution and rules- and content-based access mechanisms to specific functionality. Furthermore, if a knowledge base is connected to the

RDD's 'dynamic dictionary', then its value would also grow alongside the growth of the dictionary.

Integration of a knowledge base requires that it must also be compatible with software engineering practices. There is, however, one major element that must be borne in mind – the use of the RDD Terms and a knowledge base approach is essentially data driven, rather than being an approach focused around coding styles in imperative languages such as Java or C. While a data-driven approach is extremely valuable in terms of systems robustness and the ability to control the evolution of system behaviour over time, there are implications for an integration approach. Typical approaches to integration are considered below:

Data tier integration approaches in a relational model are often appropriate for systems sharing bulk data where the exact 'meaning' of all data structures is known and is unlikely to change. Data tier integration is likely to fail when the data might be consumed in different contexts or if the systems are expected to deliver new functionality. Complex and brittle query strategies are required when there is a significant complexity of metadata. Simple RDD aspects such as lists of controlled values like territory codes are candidates for incorporation into a relational model.

Middle-tier integration would involve some Terms and their relationships being implemented as class encapsulations and associations in an object-oriented design, and behaviours described by attributes in the RDD built-in methods on classes. Although design-time use of a modelling language to represent RDD Terms helps to reduce ambiguity, the major drawback here is that object-oriented implementations have no declarative semantics, so the precise response of a given computing result will ultimately depend on the depth of understanding of the programmer of the problem domain and empirical testing of the delivered functionality.

Enterprise Application Integration ('EAI') typically means that each software component transforms its own semantic elements down to a 'lowest common denominator', and messages are passed between information components on the basis of an information broker that understands the lowest common denominator. Although a degree of semantic mediation using the RDD does become possible, the major drawback is that the semantics encoded as metadata within the messages sent via the broker remain largely static and therefore are unable to derive benefit from the rich metadata extensibility inherent within the RDD Context Model.

Integration using Service-oriented Architecture (SOA) requires a registry to store and cross-reference metadata for both RDD concepts and metadata that describe the interfaces to concrete services that claim to implement some RDD Act. These metadata are exposed via a look-up or 'query' interface to compose collections of these services (even to compose them 'on-demand') and to construct and decode messages sent between components on the system. 'Meaningful' messages may be defined by embedding combinations of Terms sourced from the Registry into messages. The registry might therefore not only issue Terms related to semantic concepts involving rights and their inter-relationships but also constrain Terms involved in representational forms, string encodings and message structure, for example. An SOA makes inherent use of the characteristics of a network and is thus capable of linking services between radically different types of systems, from legacy mainframe databases down to embedded device firmware. Therefore, by linking to concrete service description metadata, a RDD-based metadata strategy may be devised between components across heterogeneous technologies, regardless of physical location. SOAs can scale very effectively – from fulfilling the RDD RA's interoperability requirements across multiple organizations and even industry domains, right down to the use of the very same basic SOA pattern within a small organization or private network.

From the appraisals above it can be seen that SOA integration offers the clearest leveraging of the value of the RDD, with a clear alignment of approach through the use of a registry/repository sitting at the heart of various functional components. Defining each service invocation as a RDD 'Act Type' whose parameters are typed Resources (in the RDD sense of the word Resource) offers a route to orchestrating the behaviour of complex systems coherently by using the precise semantics of the RDD. It gives a coherent ontological basis for the exact characterization and multi-purposing of system behaviour. This aspect is a crucial factor for building robust yet complex system behaviours from discrete software components that can be 'trusted' to behave predictably against digital rights and policy definitions in a machine–machine automated environment.

6.5.8 TERM UPDATE MECHANISM

Registries containing RDD Terms may operate in a 'federated' network where updates are managed according to agreements and policy, as set out by the RDD RA's procedures in accordance with the requirements in Annex C of the specification. For instance, this would allow a registry present in a system managing rights-holder information to share only specific information about their catalogue with a registry belonging to a mobile telecommunications network operator through the exchange of messages. Present at the 'top' of the Terms authority would be the RDD RA, representing a broad constituency of organizations requiring the management of rights metadata, and the links to other Authorities set up for different but complementary aims. It is likely that the RA would be maintaining these links by building systems that use the various established domain-specific schemes and protocols available for sharing and disseminating metadata. Versioned updates to the RDD can then be propagated to client systems either via notifications and feeds or via on-demand queries to the RDD registry system.

The mechanism by which Terms are updated also needs to maintain a register of party identity to ensure that the claims or assertions held by the registry can be shared consistently, has an authentication mechanism to check user's credentials that establishes they are who they claim to be, and has an encryption mechanism to protect the data itself.

The registries belonging to organizations within this constituency all benefit from these upper-level agreements by having the opportunity to participate in a wider network of information exchange. The benefits are myriad: repositories of digital assets can be opened to trade with organizations unforeseen when those repositories were originally created; enforcement of content protection (often called 'DRM') based on a complex variety of deals and agreements, and decoupled from the protection mechanism itself; digital content can be aggregated and repackaged for new markets while still respecting the rights of the stakeholders in the constituent contents.

In today's ubiquitous Internet environment, the readiest route to achieving the federated registry environment is via a Web services environment. The development of related products and standards has progressed to the extent that they are usable, although there is debate over the overlapping standards developed by consortia and other organizations. For instance, OASIS's ebXML [16] standards framework for e-commerce clearly has potential for provision of the basic underlying aspects of a multi-party infrastructure but it is not the only choice.

As a means of federating identity across domains, the Liberty Alliance specifications [17] might also provide a Web service–based solution. Liberty has the concept of Identity Providers alongside multiple Service Providers, which is used to broker and authenticate requests, an approach that, in principle, could allow the mixing of identifier schemes between Authorities.

An alternative to using a Web service approach is to use autonomous agents deployed over the Internet to specify interactions based on a common RDD ontology. However, despite the FIPA (Foundation of Intelligent and Physical Agent) standards [18] being reasonably mature, there is a lack of requisite skills base and vendor implementations in the broad commercial arena. Committing Authorities and organizations to developing FIPA-compliant agents and messages could, in all likelihood, prove a barrier to interoperability at an implementation level.

6.6 BUILDING ON THE RDD

The development of RDD has not happened in isolation from the more general work in the deployment of ontologies, which are emerging as key components in the management of metadata of all kinds, within and between enterprises.

The semantic interoperability work of the <indecs> framework, which provided information for the development of the RDD, is now being carried forward by the work of the Ontologyx initiative [19] in the UK-based Rightscom consultancy.

6.6.1 ONTOLOGYX – A SEMANTIC APPROACH TO MANAGING RIGHTS

Ontologyx has grown from the development of 'Ontology X', an 'ontology of ontologies' based on the Context Model and now supporting several major standard media data dictionaries: for the International DOI (Digital Object Identifier Foundation, the international music and copyright industry's MI3P project, and the publishing industry's ONIX metadata message standards as well as incorporating the MPEG-21 RDD and supporting several proprietary commercial metadata schemes.

The convergence of these schemes demonstrates that the requirement for multimedia metadata interoperability is becoming a marketplace reality. While DOI, MI3P and ONIX all emerged from traditional 'vertical' markets (journal publishing, the music industry and book publishing, respectively), all three have, in a short time, identified the need for supporting metadata of all media types – audio, images, audiovisual, text and software specifically.

Two other developments are highly significant in the context of MPEG-21 REL/RDD. The first is the widespread and quite abrupt acceptance of the importance of 'abstract' works as well as their manifestations. While the music industry has always recognized the value of rights in the 'abstract' composition underlying a performance, this has been largely ignored by other media sectors who have not until now generally seen a need to 'unbundle' an abstract work from its manifestation, as all rights have tended to be controlled by the same publishers, producers or creators.

However, the development of digital commerce in content, with the multiplicity of possible forms of representation and location for a 'creation', has led to an explosion in the recognition of 'abstract' works, exemplified in the recent standardization of the ISTC (International Standard Textual Work Code) and in the remarkable statistic that all of the first 10 million DOIs have been issued to identify abstract works (principally journal articles) rather than specific manifestations per se.

The second significant development is that all these 'vertical' schemes have begun by issuing descriptive or simple commercial metadata, and then moved on to the need for more complex rights expressions, which must be integrated with the original descriptive metadata.

Taken together, these two developments have greatly escalated the urgency and the complexity of requirement for an REL and its supporting data dictionary.

6.6.2 THE 'COA FRAMEWORK' AND RIGHTS MODEL: CREATING DOMAIN-SPECIFIC DATA DICTIONARIES AND MESSAGES

Each of the Ontologyx-supported domain dictionaries described above is based on a common ontology framework (the Contextual Ontologyx Architecture or 'COA'), which is a distillation of the Context Model and its key supporting Terms. This framework might be said to combine the functionality of the Context Model, Dublin Core and OWL (the Web Ontology Language standard for the Semantic Web), and as such provides both a foundation for a domain or enterprise ontology and also a head start on interoperability.

The COA also has a specific extension for rights (the COA Rights Model). It describes the complete flow of rights from their origination in law to their exploitation in licences (or violation without them) as a series of eight types of Context, with semantics from the Context Model.

By combining this contextual framework with the tools of specialization and mapping described earlier in relation to the RDD, Ontologyx has begun to create or support specialized yet fully interoperable domain ontologies for content descriptions and rights (such as those described in Section 6.1)

The initial requirement driving the development of these ontologies is for the support of consistent XML Message suites for transporting metadata between and within enterprises. The new dimensions are (a) the integration of the ontology with the message – the ontology contains the specification of the messages themselves as ontology Terms, and (b) the use of the ontology to support message transformation.

6.6.3 INTEGRATING AND LINKING SEMANTIC DOMAINS

Underlying all of this development is a simple assumption: meaning should be managed and communicated automatically through changes in semantic elements, not changes in computer code: 'the smarts are in the data'. MPEG-21 RDD and its relatives such as 'Ontology X' are predicated on the <indecs> axiom that metadata is complex, and the companion axiom that 'permanent change is here to stay', and so the mapping and development of semantic relationships must be flexibly parameterized to have any chance of success.

The Ontologyx technology assumes the paradox of multiple and inconsistent metadata standards as a given. Domains and enterprises will have, and continue to develop, their own semantics. Far from metadata standards such as Dublin Core or Onix leading to fewer standards, in fact they only spawn more variants. As a consequence, MPEG-21 RDD does not limit the scope of rights expressions, but encourages their fullest diversity. The integration of such semantics depends upon

- mappings that respect and retain the names, authorities and attributes of the originals;
- a 'many-to-one' hub-and-spoke mapping process: that is, a central ontology (such as RDD or Ontologyx) to support pairwise mapping of any two participating schemes;
- the richest possible model in the central ontology, so that domain Terms can be mapped and transformed without semantic loss; and
- a mechanism for formal agreement of the mappings.

Each of these requirements is met by the RDD and (on a wider scale) the Ontologyx approach.

The critical element in this is the central model – in this case, the Context Model. The automatic proliferation of Terms described briefly above (ref) creates the possibility of transforming all incoming metadata, in whatever semantic form, into a common contextual representation, and thence output again into a different form.

The key here is in the verbosity of the Context Model which supports the relatively easy generation of specific Relators (the equivalent of 'properties' in RDF or OWL) that precisely determine the attributes of their domains and ranges. For example, an Ontologyx or RDD statement such as 'X IsHostSoundCarrierPrincipalTitleOf Y' may be cumbersome as a human-readable construction, but in semantic processing it can convey precisely the semantic 'DNA' of both X and Y, which in combination with other arbitrary statements can support radical possibilities of inference.

6.7 CONCLUSION

Using the techniques of semantic interoperability outlined throughout the chapter, it is possible to perform automated mediation, or transformations, between data held by the different parties involved in complex transactions, which is crucial to the effective processing of all but the most trivial digital rights and policy. Accuracy of these data is driven by the precision of the semantics contained within message exchanges, which drives the accuracy of interpretation that can be derived from the message data by computational processes.

In common practice, this means a message constrained not only to a certain format (in XML) can be transformed to data not only in another XML format, but also using a different vocabulary of Terms, given certain overall requirements are met:

- The RDD Terms and relationships are maintained coherently and that semantic accuracy can be largely supported using a formal ontology representation of the RDD, over a common data model consisting of sets of triples
- The RDD maintains interfaces to allow a machine-based query and look up in addition to the human user interface.
- The vocabularies that comprise the XML elements and attributes in a message consist of RDD mapped Terms.

- The functions or services that are required to process the data are fully described in RDD Terms.
- The functions or services introduce no side effects into data, as is the case with a good service-oriented architecture implementation.

In addition, RDD clients may leverage RDD semantics in their own business systems as well as with external parties, linking together previously stovepiped databases and applications via a specialized RDD 'knowledge base' approach.

REFERENCES

[1] www.indecs.org 2000.
[2] M. Bide, G. Rust, "The <indecs> Metadata Framework – Principles, Model and Data Dictionary," (www.indecs.org) June 2000.
[3] Imprimatur, a previous European Commission funded research project which led to the <indecs> project, developed an abstract model illustrating the flows of rights and content that could take place in DRM systems. The model was published in the final report of the Imprimatur Project. Unfortunately, the original Imprimatur web site is no longer in existence, 1998.
[4] T. Berners-Lee, J. Hendler, O. Lassila, Also see The Semantic Web: www.w3c.org/2001/sw/, Scientific American May 2001.
[5] MPEG-21, "Requirements for a Rights Data Dictionary and a Rights Expression Language," (W4336), Sydney, July 2001.
[6] "Code and other Laws of Cyberspace," Basic Books, 1999.
[7] ISO/IEC Joint Technical Committee 1 Procedures and Directives.
[8] www.editeur.org. Onix Version, 2000.
[9] ISO/IEC 21000-2 Digital Item Declaration.
[10] www.w3c.org/TR/owl-features, "OWL Web Ontology Language," 10 February 2004.
[11] www.w3c.org/TR/rdf-schema/, "RDF Vocabulary Description Language 1.0: RDF Schema," W3C Recommendation 10 February 2004.
[12] www.w3c.org/TR/xforms/, "XForms 1.0," W3C Recommendation 14 October 2003.
[13] www.relaxng.org, Oasis Relax NG Technical Committee, Relax NG is being developed into an International Standard (ISO/IEC 19757-2) by ISO/EIEC JTC1/SC34/WG1.
[14] ISO/IEC 14977, 1996 (E).
[15] www.w3c.org/TR/WD-rdf-sparql-query-20041012, "SPARQL Query Language for RDF," W3C Working Draft 12 October 2004.
[16] www.ebxml.org, 2001.
[17] www.projectliberty.org, Phase 1 2002, Phase 2 2003
[18] www.fipa.org 1st Draft of FIPA Policies and Procedures Document, 2002.
[19] www.ontologyx.com.

7

Digital Item Adaptation – Tools for Universal Multimedia Access

Anthony Vetro, Christian Timmerer and Sylvain Devillers

7.1 INTRODUCTION

Access devices of all shapes and forms that can connect to different networks and can be used for a myriad of applications are becoming an integral part of our everyday lives. For better or worse, these devices keep us connected at all times to our families, friends and the office. They allow us to share our experiences and emotions, conduct business, or just say 'Hello'. With multimedia, the communication is much more powerful. However, we face the serious problem of heterogeneity in our terminals, in our networks and in the people who ultimately consume and interact with the information being presented to them. Chapters 7 and 8 describe the work developed by the MPEG standardization committee to help alleviate some of the burdens confronting us in connecting a wide range of multimedia content with different terminals, networks and users, which requires the content to be adapted to the context where it is consumed.

Digital Item Adaptation (DIA) is Part 7 of MPEG-21 [1, 2] and has been developed to address the general problem of Universal Multimedia Access (UMA). This chapter provides a detailed introduction of the problem domain and includes a general overview of DIA. It also describes the DIA tools that provide information about the usage environment as well as DIA tools that enable metadata adaptation, session mobility and the configuration of DIA engines. These tools are mainly, but not exclusively, used within the domain of Digital Item consumers. In the next chapter, a collective set of DIA tools that enable the construction of adaptation engines that are independent of media coding formats are covered in detail. These tools are mainly, but are not exclusively, used within the domain of Digital Item producers and providers.

This chapter is organized as follows. In Section 7.2, the concept of Universal Multimedia Access (UMA) is introduced followed by an overview of DIA in Section 7.3. We

The MPEG-21 Book   Ian S Burnett, Fernando Pereira, Rik Van de Walle, Rob Koenen
© 2006 John Wiley & Sons, Ltd

highlight the rich set of tools offered by the standard to assist with the adaptation of Digital Items including tools for the description of usage environments, tools to create high-level descriptions of the bitstream syntax to achieve format-independent adaptation, tools that assist in making trade-offs between feasible adaptation operations and the constraints, tools that enable low-complexity adaptation of metadata, as well as tools for session mobility and configuration. In the sections that follow, tools for describing the usage environment (Section 7.4) and tools for metadata adaptation (Section 7.5), session mobility (Section 7.6), and the configuration of DIA engines (Section 7.7) will be described in more detail. The tools enabling format independence, that is, bitstream syntax description, adaptation Quality of Service (QoS), universal constraints description as well as tools for linking those tools together are described in Chapter 8. We conclude the chapter with some final remarks in Section 7.8.

7.2 UNIVERSAL MULTIMEDIA ACCESS

Enabling access to any multimedia content from any type of terminal or network is a central aspect of the MPEG-21 vision, that is, to achieve interoperable and transparent access to multimedia content. Over the years, many technologies and standards have been developed that aim toward achieving this vision. To start with, compression standards have been considering scalable content representation formats for years, which is significant since a scalable format enables a rather simple means of adaptation to various terminals and networks. Furthermore, the description of multimedia content provides important tools to support Universal Multimedia Access (UMA). For instance, MPEG-7 has standardized tools for summarizing of media resources, tools that provide transcoding hints about the media resources, and tools that indicate the available variations of a given media resource. Additionally, it is critical to know the media format of the content, which is also specified by MPEG-7. Before describing the MPEG-21 tools that have been specified for Digital Item Adaptation (DIA), we first review some of the recent trends in multimedia and introduce the basic concept of UMA.

7.2.1 MULTIMEDIA TRENDS

Three major trends have emerged over the past five years in the way communication, information and entertainment services are provided to consumers: wireless communication, Internet technologies and digital entertainment (audio/video/games). Mobile telephony and Internet technologies have seen tremendous global growth over this period. The scale of this growth combined with market forces made it possible for basic Internet services to be offered over mobile telephone lines, including voice communications, email and text messaging. The third major trend, digital entertainment, can be seen in the growing demand for digital television (TV), Digital Versatile Disc (DVD) and high-end audio products. With more sophisticated and capable devices entering the home, advanced services such as multimedia-on-demand, videoconferencing and interactive TV could be offered to consumers.

In addition to the aforementioned trends in communications, we are also seeing significant changes with regard to the multimedia content being produced and made available. So far, digital multimedia content over the Internet has mainly been dominated by digital music, but streaming video and online video games are quickly catching up. Owing to the success of digital TV and DVD, an increasing amount of high-quality audiovisual content is being created, where personal multimedia repositories that store one's favorite movies, photos and music are becoming increasingly popular. Furthermore, content and service providers are now offering a wide variety of content and services directly to consumers.

In summary, there is an increasing number of ways for us to communicate and a growing amount of digital multimedia content to be accessed. While these trends present great opportunities to improve our interactions and experiences, there are significant technical challenges to overcome before we can reap such benefits.

7.2.2 CONCEPT OF UMA

The concept of UMA is illustrated in Figure 7.1. At the source side, rich multimedia content is made available through various means. This content is likely represented with various content-encoding formats such as MPEG-2, MPEG-4, H.26x, JPEG, AAC, AC-3, and so on, at various quality levels. At the receiving side, a diverse set of terminal devices exists; these are connected through dynamic and heterogeneous communication networks. The terminals have different content playback capabilities, and the users of these terminals will have different preferences that affect the type of content being sought. Likewise, the communication networks are characterized by different bandwidths, error rates and delay, depending on the network infrastructure and current load. The network characteristics, together with the sending and the receiving terminal capabilities and their current state, are important factors to be considered during content delivery.

The mismatch between rich multimedia content and the usage environment is the primary barrier for the fulfillment of the UMA promise. Adaptation is the process that bridges this mismatch by either adapting the content to fit the usage environment or adapting the usage environment to accommodate the content. However, descriptions of the content and usage environment are also needed and this must be combined with negotiation among devices.

The usage environment can consist of a number of network elements such as routers, switches, (wired-to-wireless) gateways, relay servers, video server proxies, protocol converters and media translators. On its way to the receiver terminal, it is likely that the

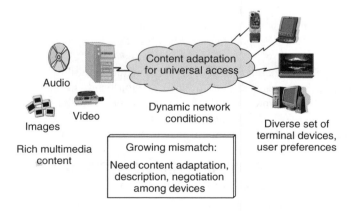

Figure 7.1   Concept of Universal Multimedia Access (UMA)

transmitted multimedia content passes through one or more of these elements. At various stages of the transmission, the content may be adapted, for example, by an application or network service provider, so that the content can be consumed with acceptable quality of service (QoS). The adaptation that is performed can be (i) content selection, (ii) content adaptation and (iii) usage environment adaptation.

The selection of the desired multimedia content plays a crucial role within the multimedia content delivery. An alternative to real-time transcoding is that of creating content in multiple formats and at multiple bitrates and making an appropriate selection at delivery time. This solution is not compute intensive, but is highly storage intensive. Smith *et al.* have proposed an Information Pyramid approach to enable UMA [3]. The approach is based on creating content with different fidelities and modalities and selecting the appropriate content on the basis of a receiver's capabilities. A tool that supports this functionality has been adopted as part of the MPEG-7 specification [4] (cf. VariationSet Description Scheme); it should also be noted that Part 2 of MPEG-21, Digital Item Declaration (DID) [5], also achieves this kind of functionality through the Choice/Selection mechanism. However, the intention of the former is to describe relationships between the source multimedia resource and the different off-line generated variations, whereas the latter has a much broader scope. In particular, it is possible to use tools as described in this chapter for online generation of adapted multimedia resources according to the usage environment. Nevertheless, with so many encoding formats, bitrates and resolutions possible, storing content in multiple formats at multiple bitrates becomes impractical for certain applications. However, for cases in which the types of receiving terminals are limited, storing multiple variants of the media is sufficient. For the general case, where no limitations exist on the terminal or access network, real-time media transcoding is necessary to support a flexible content adaptation framework.

In general, content adaptation refers to the adaptation of content to fit the usage environment. The nature of the content determines the operations involved in the actual adaptation. We distinguish between signal-level and semantic-level adaptation. Signal-level adaptation is achieved by modifying the bitrate, frame rate, resolution or the content representation format. This type of adaptation primarily involves modifications due to terminal and network constraints. On the other hand, semantic-level adaptation modifies the content with a much deeper understanding of the significance and meaning that is embedded in the original data. Obviously, this is a much more difficult problem, but if it accounts for the semantics of the content, as well as the usage environment including human factors and preferences, personalization of the content to particular users could be effectively achieved.

Finally, adapting the usage environment involves acquiring additional resources to handle the content. The resources acquired could be session bandwidth, computing resources at the sending, receiving or intermediate terminals, decoders in the receiver terminals, or improvement in the network delay and/or error rate. The usage environment can change dynamically and content adaptation should match this changing environment.

7.3 OVERVIEW OF DIGITAL ITEM ADAPTATION

7.3.1 SCOPE AND TARGET FOR STANDARDIZATION

DIA specifies tools that are designed to assist the adaptation of Digital Items in order to satisfy transmission, storage and consumption constraints, as well as QoS management by the various Users. In general, an MPEG-21 DIA tool is defined by a syntax, written in XML, and its corresponding semantics. A possible instantiation of one or more tools is referred to as an MPEG-21 DIA description.

DIA addresses the MPEG-21 framework element known as 'Terminals and Networks' described in [6], which aims to achieve interoperable and transparent access to possibly distributed multimedia content. This access should be achieved without end users having to be concerned with terminal and network installation, configuration, implementation and management issues.

The concept of DIA is illustrated in Figure 7.2. A Digital Item is subject to the adaptation process, where the DIA Engine produces the Adapted Digital Item based on DIA descriptions. This adaptation engine usually implements a Resource Adaptation Engine and a Description Adaptation Engine. The former is responsible for adapting the resource normally referenced by means of the MPEG-21 DIDL `Resource` element, whereas the latter adapts the metadata contained within the Digital Item, for example, metadata included in the MPEG-21 DIDL `Descriptor` element. The MPEG-21 DIA specification [1] defines tools that guide the entire adaptation process.

From this, it follows that the scope of standardization is limited to the DIA tools and that the adaptation engines themselves are not normatively specified by the standard. However, descriptions and format-independent mechanisms that provide support for DIA in terms of resource adaptation, description adaptation and/or QoS management are within the scope of the standardization, and are collectively referred to in Figure 7.2 as DIA Tools.

7.3.2 CLASSIFICATION OF TOOLS

The DIA tools are clustered into eight categories as depicted in Figure 7.3. The categories are clustered according to their functionality and use for DIA around the schema tools and low-level data types. The schema tools provide uniform root elements for all DIA descriptions as well as some low-level and basic data types, which can be used by several DIA tools independently.

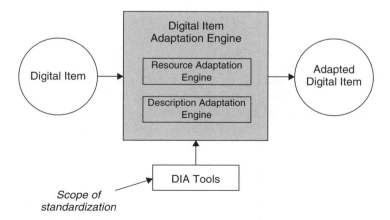

Figure 7.2 Concept of MPEG-21 Digital Item adaptation [1]. Copyright ISO. Reproduced by permission

Figure 7.3 Overview of the Digital Item adaptation tools [1]. Copyright ISO. Reproduced by permission

7.3.2.1 Digital Item Adaptation Tools

The root elements provide a wrapper for complete DIA descriptions, as well as the possibility to generate the so-called DIA description units. A complete DIA description wraps several DIA descriptions, for example, a description of both terminal capabilities and User characteristics. Referencing of external descriptions is also enabled at this level to allow for distributed descriptions. On the other hand, a DIA description unit is used to represent partial information from a complete description within a uniform wrapper element, for example, only the resolution of a terminal.

In addition to the root elements, low-level data types are specified to provide commonly used data types such as a set of unsigned integers, single, vector and matrix data types as well as the base stack function and its possible argument types. The base stack function describes a mathematical expression and is represented by means of arguments and operators in a serialized way, that is, in the form of a stack.

7.3.2.2 Usage Environment Description

The first category, the Usage Environment Description (UED) tools, is further subdivided into four major groups.

- The first group is referred to as *User Characteristics*, which define tools related to the User's content preferences and presentation preferences, as well as accessibility characteristics, mobility characteristics and the destination of a User.
- The second group defines *Terminal Capabilities* in terms of coding and decoding formats supported by a terminal and associates codec-specific parameters to those formats. Furthermore, device properties such as power and storage characteristics and input–output capabilities, for example display capabilities, are defined within this group of tools.
- The third group of tools comprises *Network Characteristics* considering the static capabilities of a network, for example, the maximum capacity of a network, as well as dynamic conditions of a network including the available bandwidth, delay and error characteristics.
- The fourth group provides means for describing the *Natural Environment Characteristics* including location and time, as well as characteristics that pertain to audiovisual aspects, that is, audio environment and illumination characteristics.

7.3.2.3 Bitstream Syntax Description Link

The Bitstream Syntax Description Link (BSDLink) tool provides a simple referencing mechanism to many of the tools specified within DIA. In particular, reference is made to (i) tools that are capable of steering the adaptation, (ii) the bitstream subject to the resource adaptation, (iii) the Bitstream Syntax Description (BSD) describing the structure of the bitstream, and (iv) several transformations including appropriate parameterization. Owing to its flexibility and extensibility, this tool enables the design of a rich variety of adaptation architectures, and provides the connection between the decision-taking and actual adaptation tools. Details about this tool can be found in Chapter 8.

7.3.2.4 Bitstream Syntax Description

The third major category of DIA tools targets the adaptation of binary media resources in a Digital Item. BSDs enable the adaptation of binary media resources in an interoperable and format-independent way. Interoperability in this context refers to the possibility of automatically generating descriptions from a bitstream and vice versa. Therefore, a language – Bitstream Syntax Description Language (BSDL) – which extends XML Schema Language [7–9] has been developed. Format independence is useful when performing the adaptation of Digital Items at network nodes that do not have knowledge of the actual coding format of the resource. Therefore, the generic Bitstream Syntax Schema (gBS Schema) enables User's to construct resource format–independent BSDs, which are also referred to as generic Bitstream Syntax Descriptions (gBSDs). Details about this tool can be found in Chapter 8.

7.3.2.5 Terminal and Network Quality of Service

The Terminal and Network Quality of Service tool forms the fourth category of tools. It describes the relationship between constraints, feasible adaptation operations satisfying these constraints, and associated qualities. It provides a means to make trade-offs among various adaptation parameters so that an optimal adaptation strategy can be formulated. Details about this tool can be found in Chapter 8.

7.3.2.6 Universal Constraints Description Tool

The fifth category of tools is referred to as Universal Constraints Description (UCD) Tool, which can be used for specifying additional constraints on the provider side as well as on the consumer side of a Digital Item. Constraints can be formulated as limitations or optimization constraints and can be differentiated with regard to individual parts of a resource, for example, a scene of a movie, a group of pictures, or a region of interest (a part of an image). Details about this tool can be found in Chapter 8.

7.3.2.7 Metadata Adaptability

Metadata Adaptability, the sixth category of tools, provides tools for filtering and scaling of XML instances (e.g., MPEG-7 descriptions) with reduced complexity compared to having only the XML instance as well as tools for the integration of several XML instances.

7.3.2.8 Session Mobility

The seventh category of tools is referred to as *Session Mobility*, where the configuration-state information that pertains to the consumption of a Digital Item on one device is transferred to a second device. This enables the Digital Item to be consumed on the second device in an adapted way.

7.3.2.9 DIA Configuration Tools

The eighth and last category of tools is called *DIA Configuration tools*. These tools provide the information required for configuration of a DIA Engine, including the identification of how `Choice/Selections` within a DID should be processed, as well as the location of the actual adaptation, that is, the receiver side, the sender side or either side.

7.3.3 SCHEMA DESIGN AND EXTENSIBILITY FEATURES

The MPEG-21 DIA schema, expressed in XML Schema Language [7–9], is designed to meet the following design criteria in an interoperable way: *Extensibility, Reusability, Modularization, Versioning, Compatibility* and *Namespaces*. These design criteria are further elaborated in [10] and [11].

Extensibility: The MPEG-21 DIA schema is written using XML where the 'X' in XML stands for 'eXtensible'. The goal of extensibility within XML should enable the amendment of an existing schema without changing the original schema or breaking legacy applications built upon the original schema. One approach to achieve this goal is by using the type derivation mechanism provided by XML. In this case, the new type needs to be indicated in the instance using the `xsi:type` attribute. This approach has been widely adopted throughout MPEG-21 DIA in combination with the usage of abstract types. Abstract types cannot be used in instances, that is, the type derivation is forced by definition. The abstract type is solely a placeholder for their derived types and can be compared with the usage of abstract types within programming languages such as C++, Java or C#. Its main advantage in standardization is that abstract types provide a basis for future amendments without the need for changing the original specification.

Reusability: The benefits of reusability of XML schema components lie in the reduction of development and maintenance time. Therefore, XML schema components within the MPEG-21 DIA schema have been declared using named types (both simple and complex), because anonymous types cannot be reused. These components can be reused by means of referencing and type derivation as mentioned earlier.

Modularization: When talking about XML schemas, and in particular of the MPEG-21 DIA schema, we tend to think about individual documents. In XML terminology, these individual documents are referred to as *schema documents* and a complete XML schema, like the MPEG-21 DIA schema, can comprise one or more such schema documents. Dividing a schema into several schema documents eases the reuse and maintenance of the individual schema documents dramatically. For instance, the reuse of small and focused components is easier than always including/importing one big schema document. The MPEG-21 DIA core schema as illustrated in Figure 7.3 (i.e., schema tools and low-level data types) is included or imported by the various MPEG-21 DIA tools respectively. In other words, the MPEG-21 DIA tools are built around the *DIA core schema document*. This core schema document is included in DIA tools having the same target namespace as the core schema document; otherwise they are imported. The *unsigned integer's schema document* forms a special case because this schema document has no target namespace. This kind of schema document borrows the namespace from the including schema document and is sometimes called a *neutral building block* [10] or a *chameleon component* [11].

Versioning: Each schema document within MPEG-21 DIA comprises a version attribute to indicate the version of this schema document. Therefore, the official name of the standard has been chosen as the version number, i.e., ISO/IEC 21000-7. It should be mentioned that from an XML Schema point of view this attribute is for documentation purposes only and is ignored by legacy XML processors.

Compatibility: Compatibility of XML schemas can be compared at three different levels of compatibility, namely, *schema, application* and *conversion* compatibility. Backward compatibility at the schema level means that all instances conforming to the previous version of the schema are also valid according to the new version; this could be achieved if, and only if, the previous version is a subset of the new version. This means that only new optional components (elements, attributes and types) can be added and (structural) constraints can be made less restrictive. At the application level, one requirement should be that well-designed applications do not crash when processing instances of a new version. Ignoring any elements or attributes that are not expected and avoiding unnecessary overdependence on the document structure could achieve this. For example, only the name of the actual element one is interested in should be considered without maintaining its parents or grandparents. In practice, however, sometimes compatibility could be neither achieved at the schema level nor at the application level. In this case, appropriate upgrade information should be provided, for example, by style sheets, which allow an automatic upgrade of old instances or clear information on how to handle added or removed components when converting between versions is needed.

Namespaces: The usage of different namespaces plays a crucial role during the development of XML schemas and has been intensively discussed during the development of the MPEG-21 DIA schema. DIA tools that do not implicate any normative processing remain in the same namespace as the namespace for the DIA core schema document, that is, schema tools, Usage Environment Description, BSDLink, Terminal and Network QoS, Universal Constraints Description and Metadata Adaptability tools. On the other hand, tools that implicate normative processing such as Bitstream Syntax Description, Session Mobility and DIA Configuration tools are defined within their own namespace. This enables the implementation of standard conformant processors realizing each normative processing separately without overloading it with tools defined in other namespaces.

7.4 USAGE ENVIRONMENT DESCRIPTION TOOLS

This section introduces the UED tools, which include the description of terminal capabilities and network characteristics, as well as User characteristics and characteristics of the natural environment. The descriptive elements addressed by each category are discussed, which comprise normative aspects of the standard. Following this description, an informative example that exhibits the use of a selected subset of tools within each category is provided. Further readings about the UED tools and how they can assist adaptation engines can be found in [12–14].

7.4.1 TERMINAL CAPABILITIES

While many would think that terminal capabilities are the capabilities of a receiving device in which resources are ultimately consumed, this is not true in the context of DIA. Instead, the notion of a terminal in DIA is much more generic and represents all devices regardless of their location in the delivery chain. In other words, in addition to the typical consumer electronic devices such as mobile phones, televisions, audio players and computers, a terminal could also be a server, proxy or any intermediate network node. A key point is that terminals may have both receiving and transmitting capabilities. Also, the terminal may be defined as a cluster of devices [15].

Given the above, the description of a terminal's capabilities is primarily required to satisfy consumption and processing constraints of a particular terminal. Terminal capabilities are defined by a wide variety of attributes. In particular, the following categories have been considered:

- *Codec capabilities*: encoding and decoding capabilities
- *Device properties*: power, storage and data I/O characteristics
- *Input–output characteristics*: display and audio output capabilities.

The attributes under each of the above categories are described in further detail in the following text.

7.4.1.1 Codec Capabilities

Encoding and decoding capabilities specify the format that a particular terminal is capable of encoding or decoding, for example, an MPEG profile@level. Given the variety of different content representation formats that are available today, it is necessary to be aware of the formats that a terminal is able to deal with. Note that a terminal may be capable of both encoding and decoding, and may also be capable of multiple formats. The syntax of the codec capabilities tool has been designed to instantiate such possibilities.

To describe the format that a terminal is capable of encoding or decoding, it is noted that Part 5 of the MPEG-7 standard [4] specifies the format of multimedia content; this is achieved by means of a controlled term list and classification schemes. Since one of the general goals of DIA is to provide a match between the usage environment and the content, it is natural to use a similar and corresponding structure to describe the usage environment. In this particular case, the structure of the terminal capability should be aligned with the corresponding MPEG-7 tool for describing multimedia formats.

One important point to consider in achieving this alignment is that most MPEG-7 tools are used for content description, where the semantics often explicitly mention how the particular description relates to multimedia content. Consequently, DIA has adopted the same syntax as MPEG-7 for describing multimedia formats, but has redefined the semantics so that it relates to the encoding and decoding capabilities of a terminal. In this way, the classification schemes as defined by MPEG-7 that list the various formats may also be reused.

The classification schemes defined in MPEG-7 include controlled term lists for visual, audio and graphics formats. DIA has additionally defined a list for scene coding formats, such as for MPEG-4, Virtual Reality Modeling Language (VRML), and humanoid animation scenes. For the most part, the terms defined in these lists refer to specific conformance points as defined by the respective standards. For example, in the classification scheme that corresponds to visual coding formats, there exist terms to identify the various MPEG-4 Visual profiles and levels. These lists may be used as is, but may also be extended by means of a registration authority so that coding formats that are yet to be defined could be accounted for.

In addition to the format that a terminal is capable of encoding or decoding, it is sometimes important to know the limits of specific parameters that affect the operation of the codec. In MPEG standards, the level definition often specifies such limits. However, it is possible that some devices are designed with further constraints, or that no specification of a particular limit even exists. Therefore, the codec parameters as defined by DIA would provide a means to describe such limits. Currently, DIA specifies the following parameters:

- minimum buffer size required by a codec in bits,
- average and maximum bitrates that a decoder is capable of decoding and that an encoder can produce,
- maximum memory bandwidth of a codec in units of bits per second,
- maximum vertex processing rate for a graphics codec, and
- maximum fill rate of a graphics codec in units of pixels per second.

As an example, consider a video decoder used in a DVD player that has been designed to decode an MPEG-2 Video bitstream with syntax conformant to the Main Profile. If the device is compliant with the Main Profile @ Main Level (MP@ML), it should be able to decode a stream with a maximum bitrate of 15 Mbps. However, if it is only capable of decoding a maximum bitrate of 8 Mbps, in which the device is no longer compliant to MP@ML, DIA provides a means to describe this limit.

7.4.1.2 Input–Output Characteristics

There are three aspects that we consider as part of the input–output characteristics of a terminal, all of which have significant influence on the adaptation of Digital Items:

- Display capabilities
- Audio output capabilities
- User interaction inputs.

Display Capabilities

Describing the capabilities of a display is obviously very important as certain limitations that impact the visual presentation of information must be taken into consideration. One key aspect to account for in describing a display is that it may have several modes of operation. For instance, a display may be capable of multiple display resolutions, each resolution with a corresponding refresh rate. Subsequently, a `DisplayMode` element has been defined to group this type of dynamic information. Additionally, there are a variety of static attributes that characterize a display. Among these static attributes are the screen size, dot pitch, gamma factor, contrast ratio and maximum brightness. Some color-related attributes include the color bit depth for each RGB component, the color characteristics, which are described using the chromaticity values of the three primaries and the white point, as well as a Boolean flag to simply indicate if the display is color capable or not. In some applications, it may also be very useful to know whether the display is stereoscopic or not, and also what the rendering format is, for example, interlace or progressive. In the case that multiple displays are described, there is also an attribute to indicate which are currently active.

Audio Output

As with displays, audio outputs may also have different modes of operation as defined by the sampling frequency and bits per sample. An analogous element for audio, `AudioMode`, to group such information has also been defined. In addition to this, the lower and upper bounds of the frequency range, the power output, the signal-to-noise ratio, and the number of output channels that the speakers support characterize the audio outputs.

User Interaction

User interaction inputs define the means by which a User can interact with a terminal. With such information available, an adaptation engine could modify the means by which a User would interact with resources contained in a Digital Item. In the following, two instances describing the need for such tools are discussed.

In the first instance, consider an interactive scene that depends on the number of buttons a terminal has. In particular, assume that two scenes with equivalent purposes are designed. The first depends on the presence of a 3-button mouse, while the second is usable with a 1-button mouse. Given this information, the choice between the deliveries of the two scenes can be made. Following this example, the presence of a mouse as a means of input is described by the standard, as well as its properties including the number of buttons the mouse has and whether it has a scroll wheel or not. Other types of inputs such as trackballs, tablets, pens and microphones are also specified as part of the standard.

In a second instance, interaction based on key and string input is considered. Key input is the ability for scene interaction to deal with single keystrokes, thus allowing the author to bind keystrokes to the execution of some commands. String input is the ability for scene interaction to deal with the input of Username, passwords and such multikeystroke input. If the device is capable of key input, services with simple key bindings can be offered. Otherwise, if a mouse or pen is present, services with menus can be offered. Information pertaining to the key input may allow the adaptation engine to choose between two appropriate scenes along this User interaction dimension. In the same way, further information about the string input capability might allow the adaptation engine to choose between a scene requiring input of strings and an equivalent scene making use of a voice-activated server. Following this second example, the capability to enter strings, that is, words or sentences, as well as single character input, is described. As a final note, the syntax of this tool has been designed in such a way as to allow further extensions of additional inputs to be easily defined.

7.4.1.3 Device Properties

There are several tools that have been specified under terminal capabilities that could be classified as device properties. The following are some of those included in this class of tools:

- Device class
- Power characteristics
- Storage characteristics
- Data I/O characteristics.

Additionally, benchmark measures that provide an indication of computational performance as well as tools that describe the IPMP (Intellectual Property Management and Protection) support on a terminal are specified. These device properties are further elaborated in the following text.

Device Class
The device class indicates the type of terminal. A classification scheme of controlled terms has been specified that enumerates these various types. This list of types is extensible and includes terms and identifiers for about a dozen different types of terminals varying from Personal Computers (PCs), Personal Digital Assistants (PDAs) and mobile phones, to set-top boxes, televisions, gateways and routers. The purpose of this description is to provide a high-level description of the terminal that provides some context of the consumption environment.

Power Characteristics
The power characteristics tool is intended to provide information pertaining to the consumption of battery, battery capacity remaining, and battery time remaining. The consumption is specified as the average ampere consumption. With such attributes, a sending device may adapt its transmission strategy in an effort to maximize the battery lifetime.

This may be especially important in ad hoc networking environments where there may exist multiple routes by which data could be transferred and maintaining a minimum level of infrastructure is critical.

Storage Characteristics
Storage characteristics are defined by the input and output transfer rates, the size of the storage and a Boolean variable that indicates whether the device can be written to or not. Data I/O describes bus width and transfer speed, as well as the maximum number of devices supported by the bus and the number of devices that are currently on the bus.

CPU Benchmark
To gauge computational performance, DIA has adopted a benchmark-based description. Descriptions for both Central Processing Unit (CPU) performance and graphics performance have been specified. The CPU performance is described as the number of integer or floating-point operations per second, while the graphics performance is described by the mean value of all benchmark results from a 3D graphics performance test. In both the cases, the name of the particular benchmark is specified. With such measures, the capability of a device to handle a certain type of media or a media encoded at a certain quality could be inferred.

IPMP Support
IPMP tools are modules that perform functions such as authentication, decryption and watermarking. The support for such tools is a device property that is described by DIA to facilitate the adaptation of protected content. Specific characteristics that are specified include the `IPMP_Tool` class descriptions included in MPEG-2 (ISO/IEC 13818) and MPEG-4 (ISO/IEC 14496) as well as `IPMPS_Type` descriptions included in MPEG-4. If, for instance, a terminal cannot support a particular type of IPMP function, it is useless to deliver content protected with such a tool. In that case, the sending device may adapt the protection scheme to something that is supported.

7.4.1.4 Use Case: Format Compatibility

As an example, consider a high-quality full frame-rate video with resolution of 720×480 luminance pixels that is encoded according to the MPEG-2 MP@ML format at a bitrate of 5 Mbps. The MPEG-7 description corresponding to this resource is given in Example 7.1.

Assume that a mobile terminal wants to access this video, but is only capable of decoding JPEG images and video encoded in the MPEG-4 Visual Simple Profile @ Level 1 format. Furthermore, it only has a display resolution of 176×144 luminance pixels. The DIA description of this terminal is given in Example 7.2.

For the mobile terminal to render this high-quality video material on its screen, adaptation from the original MPEG-2 Video format to the destination MPEG-4 Visual format must be performed. The Simple Profile of MPEG-4 Visual indicates the syntax that the adapted bitstream must conform to, while the Level indicates the defined limits on various aspects of the stream, such as the maximum number of macroblocks per second (MBs/s) and the maximum bitrate. In the case of Level 1, the maximum number of MBs/s is 1485, which corresponds to a typical spatial resolution of 176×144 pixels and temporal resolution of 15 Hz, and the maximum bitrate is 64 kbps.

As we see from this simple example, the terminal capability description is a key tool to enable format compatibility. As heterogeneity of devices and the number of source coding formats increases matching of the content to the terminal becomes quite a significant task.

```
<Mpeg7>
  <Description xsi:type="ContentEntityType">
    <MultimediaContent xsi:type="VideoType">
      <Video>
        <MediaInformation id="clip0238">
          <MediaProfile>
            <MediaFormat>
              <VisualCoding>
                <Format href=
                  "urn:mpeg:mpeg7:cs:VisualCodingFormatCS
                    :2001:2.2.2">
                  <Name xml:lang="en">
                    MPEG-2 Video Main Profile @ Main Level
                  </Name>
                </Format>
                <Frame height="720" width="480" rate="30"/>
                <BitRate>5000000</BitRate>
              </VisualCoding>
            </MediaFormat>
          </MediaProfile>
        </MediaInformation>
      </Video>
    </MultimediaContent>
  </Description>
</Mpeg7>
```

Example 7.1   MPEG-7 description of a high-quality MPEG-2 video bitstream.

7.4.2 NETWORK DESCRIPTIONS

Networks are often thought of as this vague entity that transfers data from one point to another. Generally speaking, data passes through many nodes in a network. However, simple point-to-point and peer-to-peer connections are also common. The description of a network in DIA accounts for both ends of this spectrum. As such, DIA takes an abstract view of the network as a connection that conveys information between Users and specifies the description of network characteristics between the Users. Two main categories are considered in the description of network characteristics:

- Capabilities, which define static attributes of a network
- Conditions, which describe more dynamic behavior of a network.

These descriptions primarily enable multimedia adaptation for improved transmission efficiency and are described in further detail in the following text.

7.4.2.1 Capabilities and Conditions

For network characteristics, static and dynamic information is taken into consideration. The former is referred to as network capabilities, whereas the latter is referred to as network conditions.

Network capabilities include attributes that describe the maximum capacity of a network and the minimum guaranteed bandwidth that a network can provide. Attributes that indicate if the network can provide in-sequence packet delivery and how the network deals with erroneous packets, that is, does it forward, correct or discard them, are also specified.

```
<DIA>
  <Description xsi:type="UsageEnvironmentPropertyType">
    <UsageEnvironmentProperty xsi:type="TerminalsType">
      <Terminal>
        <TerminalCapability xsi:type="CodecCapabilitiesType">
          <Decoding xsi:type="ImageCapabilitiesType">
            <Format
              href="urn:mpeg:mpeg7:cs:VisualCodingFormatCS:2001:4">
              <mpeg7:Name xml:lang="en">JPEG</mpeg7:Name>
            </Format>
          </Decoding>
          <Decoding xsi:type="VideoCapabilitiesType">
            <Format href=
              "urn:mpeg:mpeg7:cs:VisualCodingFormatCS
               :2001:3.1.2">
              <mpeg7:Name xml:lang="en">
                MPEG-4 Visual Simple Profile @ Level 1
              </mpeg7:Name>
            </Format>
          </Decoding>
        </TerminalCapability>
        <TerminalCapability xsi:type="DisplaysType">
          <Display>
            <DisplayCapability xsi:type="DisplayCapabilityType">
              <Mode>
                <Resolution horizontal="176" vertical="144"/>
              </Mode>
            </DisplayCapability>
          </Display>
        </TerminalCapability>
      </Terminal>
    </UsageEnvironmentProperty>
  </Description>
</DIA>
```

Example 7.2   DIA description of a terminal that is capable of decoding JPEG and certain MPEG-4 formats.

Network conditions specify attributes that describe the available bandwidth, error and delay. The error is specified in terms of packet loss rate and bit error rate. Several types of delay are considered, including one-way and two-way packet delay, as well as delay variation. Available bandwidth includes attributes that describe the minimum, maximum and average available bandwidth of a network. Since these conditions are dynamic, time stamp information is also needed. Consequently, the start time and duration of all measurements pertaining to network conditions are also specified.

7.4.2.2 Use Case: Efficient and Robust Transmission

To illustrate the use of network descriptions for the efficient and robust transmission of multimedia, consider the video bitstream described in Example 7.1. As indicated in the description, the video has been encoded at a relatively high bitrate of 5 Mbps. For the purpose of this example, assume that the quality of the encoded bitstream has been optimized independently of any network conditions.

Given that the same mobile terminal described in Example 7.2 wishes to access this stream, it is not enough to just know the formats the terminal is capable of decoding and the display resolution. It is also critical to know the network characteristics. Consider the description of a network given in Example 7.3. From this description, we see that the network is characterized by a maximum capacity of 384 kbps and a minimum guaranteed bandwidth of 32 kbps. Over a period of 2 s beginning at 15:22:08 on January 31, 2004, in a time zone that is 1 h different from UTC (Coordinated Universal Time), the average bandwidth was measured at 44 kbps with a packet loss rate of 10 %.

To ensure an efficient and robust transmission of the video resource within the adaptation process, both the available bandwidth and packet loss rate need to be accounted for. So as not to exceed the available bandwidth of the network, the target bitrate of the output stream should be less than the average bandwidth indicated by the DIA description. To maximize the reconstructed video quality at the receiver given the relatively high packet loss rate that has been measured, the adaptation engine may consider several options to increase the robustness of the transmitted stream.

One method of increasing the robustness of the adapted bitstream is to protect important data using some form of error correcting codes, for example, Reed–Solomon codes. Thus, different portions of the bitstream could be partitioned and organized according to their importance. This data-partitioning feature is part of most video coding standards. In this way, bits corresponding to header information, texture data of intracoded frames, and motion vectors of intercoded frames would have higher priority than texture data of intercoded frames. The adaptation engine could then apply forward error correction coding to essentially protect the higher priority data. Of course, there is some redundancy that would be introduced and an overhead that needs to be accounted for in the rate of the video payload data.

Another adaptation strategy that could be employed to maximize reconstructed video quality in the presence of network errors is to reduce the effects of error propagation. As most video coding standards utilize predictive coding methods, a lost packet could impair the quality not only of the current frame but also of the subsequent frames. To spatially localize the propagation of errors, resynchronization markers could be periodically inserted into the bitstream, and to temporally localize the errors, the percentage of intracoded blocks in intercoded frames could be increased. As with forward error correcting codes,

```
<DIA>
    <Description xsi:type="UsageEnvironmentPropertyType">
        <UsageEnvironmentProperty xsi:type="NetworksType">
            <Network>
                <NetworkCharacteristic xsi:type="NetworkCapabilityType"
                    maxCapacity="384000" minGuaranteed="32000"/>
                <NetworkCharacteristic xsi:type="NetworkConditionType"
                    startTime="2004-01-31T15:22:08+01:00" duration="PT2S">
                    <AvailableBandwidth average="44000"/>
                    <Error packetLossRate="0.10"/>
                </NetworkCharacteristic>
            </Network>
        </UsageEnvironmentProperty>
    </Description>
</DIA>
```

Example 7.3   DIA description of a narrow-band network.

these techniques also add redundancy to the output stream and the overhead in rate needs to be accounted for. Further information on the principles of such adaptation strategies, as well as techniques to relate the packet loss rate to the coding parameters to satisfy overall rate constraints, may be found in [16–18].

7.4.3 USER CHARACTERISTICS

The broad notion of a User in MPEG-21 as any entity in the end-to-end delivery chain opens the door to describe many possible characteristics and preferences. In the DIA specification, some of the User characteristics are focused on the end-consumer, for example, characteristics that express audiovisual impairments or presentation preferences, while others could be applied more generally, for example, general information about a User, where a User may represent a person, a group or an organization. We categorize the tools into five subcategories:

- User Info
- Usage Preferences and Usage History
- Presentation Preferences
- Accessibility Characteristics
- Location Characteristics.

The tools included as part of these subcategories will be described in further detail in the following text. After the tools are introduced, we highlight a simple use case for adaptive selection of resources based on user characteristics.

7.4.3.1 User Info

General information about a User is specified as part of DIA under User Info. For this purpose, the standard has not specified anything new, but rather has adopted the Agent Description Scheme (DS) specified by MPEG-7 [4]. The Agent DS describes general characteristics of a User such as name and contact information, where a User can be a person, a group of persons or an organization.

7.4.3.2 Usage Preferences and Usage History

As with User Info, corresponding tools specified by MPEG-7 [4] define the Usage Preference and Usage History description tools in DIA. The Usage Preferences description tool is a container of various descriptors that directly describe the preferences of a User. Specifically, these include descriptors of preferences related to

- the creation of Digital Items (e.g., created when, where, by whom),
- the classification of Digital Items (e.g., form, genre, languages),
- the dissemination of Digital Items (e.g., format, location, disseminator) and
- the type and content of summaries of Digital Items (e.g., duration of an audiovisual summary).

The Usage History description tool describes the history of actions on Digital Items by a User (e.g., recording a video program, playing back a music piece); as such, it describes the preferences of a User indirectly. With the Usage Preferences specified directly or obtained indirectly through observation of User actions, a service provider could easily personalize the set of Digital Items delivered to a User.

7.4.3.3 Presentation Preference

Presentation Preferences are a class of description tools that define a set of preferences related to the means by which Digital Items and their associated resources are presented or rendered for the User. Within this category, DIA has specified a rather rich set of description tools that include preferences related to the following:

- audiovisual presentation or rendering;
- format/modality a User prefers, including the priority of the presentation; and
- focus of a User's attention with respect to audiovisual and textual media.

Audiovisual Presentation

For audio, descriptions of the preferred volume, frequency equalizer settings and audible frequency ranges are specified. Such attributes may affect the way in which the delivered audio resource is encoded, for example, allocating more bits to specific components in the given frequency range. Additionally, for limited capability devices that may not have equalization functionality, equalization may be performed prior to transmission given the designated preferences.

For visual, graphics and display presentations, preferences have been considered. As part of the graphics presentation preferences, attributes related to the User's preferred emphasis of geometry, texture and animation are described. In constrained resource environments, a suitable trade-off considering such preferences could be determined considering the available network bandwidth or computational resources. As part of the display presentation preferences, descriptions of the preferred color temperature, brightness, saturation and contrast are specified. Also, there is a description tool that specifies preferences related to 2D to 3D stereoscopic conversion, and vice versa [19].

Format/Modality Preferences

In the resource adaptation process, various types of conversions may be carried out when a terminal or network cannot support the consumption or transport of a particular modality or format. For each resource, there may be many conversion possibilities. To narrow down these possibilities, a tool that indicates the relative orders for possible conversions of each original modality or format is specified. The orders help an adaptation engine find the destination modality or format when the original needs to be converted under a given constraint. Furthermore, a User can also specify numeric weights for conversions to indicate the relative preference on a given conversion of one modality or format to another. Along similar lines, a tool that allows the User to describe different priorities for different resources is also specified. Further details on how this tool is used to achieve modality conversion of resources may be found in [20].

Focus of Attention Preferences

Adaptation based on a specified region of interest (ROI), or more generally, the focus of attention, is also considered by DIA as a presentation preference. With regard to audiovisual media, the User may describe a ROI encompassing a semantic object in a video program. This ROI can be described using description tools specified in Part 5 of

MPEG-7 [4], which can then be used to personalize the presentation accordingly. It is emphasized that the ROI description itself is not specified by DIA, but the description is linked to DIA as a presentation preference through a Universal Resource Identifier (URI).

With regard to text-based media, the focus of attention is specified in terms of the User's preferred keywords, presentation speed, font types and sizes. As with audiovisual media, text can be summarized, for example, on the basis of a User's preferred keywords. When a User presents the preferred keyword(s), a server can adapt and deliver more meaningful text information.

7.4.3.4 Accessibility Characteristics

Accessibility characteristics provide descriptions that would enable one to adapt content according to certain auditory or visual impairments of the User. For audio, an audiogram is specified for the left and right ear, which specifies the hearing thresholds for a person at various frequencies in the respective ears. For visual impairments, color vision deficiencies are specified, that is, the type and degree of the deficiency. For example, given that a User has a severe green-deficiency, an image or chart containing green colors or shades may be adapted accordingly so that the User can distinguish certain markings. Such descriptions would also be very useful to simply determine the modality of a media resource to be consumed, but may also be used for more sophisticated adaptations, such as the adaptation of color in an image [21].

7.4.3.5 Location Characteristics

There are two description tools standardized by DIA that target location-based characteristics of a User:

- Mobility Characteristics
- Destination.

The first of these tools, mobility characteristics, aims to provide a concise description of the movement of a User over time. In particular, directivity, location update intervals and erraticity are specified. Directivity is defined to be the amount of angular change in the direction of the movement of a User compared to the previous measurement. The location update interval defines the time interval between two consecutive location updates of a particular User. Updates to the location are received when the User crosses a boundary of a predetermined area (e.g., circular, elliptic, etc.) centered at the coordinate of its last location update. Erraticity defines the degree of randomness in a User's movement. Together, these descriptions can be used to classify Users, for example, as pedestrians, highway vehicles, and so on, in order to provide adaptive location-aware services [22].

The second tool, destination, is a tool to indicate, as the name implies, the destination of a User. The destination itself may be specified very precisely, for example, with geographic coordinates, or more conceptually using a specified list of terms. In conjunction with the mobility characteristics descriptor, this tool could also be used for adaptive location-aware services.

7.4.3.6 Use Case: Adaptive Selection of Resources

With the exception of presentation preferences, one of the key applications for User characteristics descriptions is to enable the adaptive selection of resources. In the following, we illustrate this application in three different ways: with usage preferences, with accessibility characteristics and with location characteristics.

Example 7.4 provides a description of usage preferences in which the User has preference for sports and entertainment. In general, such a description enables automatic and personalized discovery, selection, recommendation and recording of media resources. The preferences may be entered by the User directly or automatically inferred from prior viewing or listening habits. A standardized format enables Users to enter or update their preferences using one device and then import them into other devices for instant customization. Users may carry a representation of their preferences together with some general information such as name or age in a secure smart card or other portable storage. Furthermore, if the User allows, these descriptions may be communicated to other Users, for example, content and service providers. In turn, effective filtering and recommendation services could be provided. Further information on personalized selection of content on the basis of preferences may be found in [23, 24].

In addition to having usage preferences guide the type of media resources that are directed to a User, accessibility characteristics could direct the format of media resources. For instance, consider the accessibility characteristics description for two Users as given in Example 7.5. This description indicates that User_A is blind in both eyes and that User_B has auditory impairments in the left ear. Obviously, the delivery of visual media to User_A would make no sense; the implied preference is for audio data or data that could be translated and consumed by touch. On the other hand, for User_B, the description

```
<DIA>
    <Description xsi:type="UsageEnvironmentPropertyType">
        <UsageEnvironmentProperty xsi:type="UsersType">
            <User>
                <UserCharacteristic xsi:type="UsagePreferencesType">
                    <UsagePreferences>
                        <mpeg7:FilteringAndSearchPreferences>
                            <mpeg7:ClassificationPreferences>
                                <mpeg7:Genre>
                                    <mpeg7:Name>Sports</mpeg7:Name>
                                </mpeg7:Genre>
                                <mpeg7:Genre>
                                    <mpeg7:Name>Entertainment</mpeg7:Name>
                                </mpeg7:Genre>
                            </mpeg7:ClassificationPreferences>
                        </mpeg7:FilteringAndSearchPreferences>
                    </UsagePreferences>
                </UserCharacteristic>
            </User>
        </UsageEnvironmentProperty>
    </Description>
</DIA>
```

Example 7.4 DIA description of usage preferences indicating preference for sports and entertainment.

```
<DIA>
  <Description xsi:type="UsageEnvironmentPropertyType">
    <UsageEnvironmentProperty xsi:type="UsersType">
      <User id="User_A">
        <UserCharacteristic xsi:type="VisualImpairmentType">
          <Blindness eyeSide="both"/>
        </UserCharacteristic>
      </User>
      <User id="User_B">
        <UserCharacteristic xsi:type="AuditoryImpairmentType">
          <RightEar>
            <Freq250Hz>0.0</Freq250Hz>
            <Freq500Hz>5.5</Freq500Hz>
            <Freq1000Hz>-0.2</Freq1000Hz>
            <Freq2000Hz>-2.0</Freq2000Hz>
            <Freq4000Hz>1.5</Freq4000Hz>
            <Freq8000Hz>5.5</Freq8000Hz>
          </RightEar>
          <LeftEar>
            <Freq250Hz>9.0</Freq250Hz>
            <Freq500Hz>-1.5</Freq500Hz>
            <Freq1000Hz>9.0</Freq1000Hz>
            <Freq2000Hz>9.0</Freq2000Hz>
            <Freq4000Hz>9.0</Freq4000Hz>
            <Freq8000Hz>10.0</Freq8000Hz>
          </LeftEar>
        </UserCharacteristic>
      </User>
    </UsageEnvironmentProperty>
  </Description>
</DIA>
```

Example 7.5 DIA description of accessibility characteristics for two different Users, where User_A has visual impairments and User_B has auditory impairments.

could be used to compensate for the auditory impairment if the impairment is a minor. If the impairment is severe, the description may be interpreted so that only visual media is directed to the User.

As a final example of adaptive resource selection, consider the DIA description of location characteristics in Example 7.6. This description includes information related to both mobility characteristics and destination. In providing location-based adaptation, we assume a network service provider (NSP) and an application service provider (ASP). The NSP is responsible for keeping track of the User locations, extracting the DIA descriptions and ultimately delivering them to the ASP. After receiving the descriptions, the ASP determines the User's mobility profile, and then performs adaptation according to this profile. A method to extract the mobility profile from the mobility characteristics description has been presented in [22], where the User is classified as a pedestrian, an urban vehicle or a highway vehicle. In some select instances, the resources that a pedestrian may be interested in could be quite different from the resources that a User driving a car may be interested in. Considering a more complete location-aware service, adaptation to additional circumstances such as the User's destination, time and current location should also be accounted for.

```
<DIA>
  <Description xsi:type="UsageEnvironmentPropertyType">
    <UsageEnvironmentProperty xsi:type="UsersType">
      <User>
        <UserCharacteristic xsi:type="MobilityCharacteristicsType">
          <Directivity>
            <Mean>6</Mean>
            <Variance>2</Variance>
            <Values>
              0.3 0.2 0.2 0.1 0.1 0.1 0.0 0.0
              0.0 0.0 0.0 0.0 0.0 0.0 0.0 0.0
            </Values>
          </Directivity>
        </UserCharacteristic>
        <UserCharacteristic xsi:type="DestinationType">
          <DestinationClass>
            <StereotypedClass
              href="urn:mpeg:mpeg21:2003:01-DIA-PlaceTypeCS-NS:12.2">
              <mpeg7:Name xml:lang="en">Supermarket</mpeg7:Name>
            </StereotypedClass>
          </DestinationClass>
          <DestinationName xml:lang="en">Acme</DestinationName>
        </UserCharacteristic>
      </User>
    </UsageEnvironmentProperty>
  </Description>
</DIA>
```

Example 7.6   DIA description of location characteristics.

7.4.4 NATURAL ENVIRONMENT CHARACTERISTICS

The final property of the usage environment is the natural environment, which is an important characteristic that may impact the adaptation of Digital Items. The natural environment pertains to the physical environmental conditions around a User such as lighting condition or noise level, or a circumstance such as the time and location that Digital Items are consumed or processed. In the following text, these characteristics are discussed further.

7.4.4.1 Location and Time

In DIA, the Location description tool refers to the location of usage of a Digital Item, while Time refers to the time of usage of a Digital Item. Both are specified by MPEG-7 description tools, namely, the Place DS for Location and the Time DS for Time. The Place DS describes existing, historical and fictional places and can be used to describe precise geographical location in terms of latitude, longitude and altitude, as well as postal addresses or internal coordinates of the place, for example, an apartment or room number, the conference room, and so on. The Time DS describes dates, time points relative to a time base, and the duration of time. Tools that describe location and time are referenced by both the mobility characteristics and destination tools and may be used to support adaptive location-based services.

7.4.4.2 Visual Environment

With respect to the visual environment, illumination characteristics that may affect the perceived display of visual information are specified. It has been observed that the overall illumination around a display device affects the perceived color of images on the display device and contributes to the distortion or variation of perceived color. By compensating for the estimated distortion, actual distortion caused by the overall illumination can be decreased or removed.

The type of illumination and the illuminance of the overall illumination, both of which could be used to estimate the effect on the perceived color, describe the illumination characteristics. Either the correlated color temperature or the chromaticity of the overall illumination represents the type of the illumination. The value of the correlated color temperature is obtained from a nonuniform quantization of the value in Kelvin units, as defined in Part 3 of MPEG-7 [25], while the $x-y$ chromaticity coordinates are specified in the range of [0,1]. In addition to these descriptions, the overall illuminance of the illumination is specified directly in units of lux.

7.4.4.3 Audio Environment

For audio, the description of the noise levels and a noise frequency spectrum are specified. Such descriptions could be acquired by processing the noise signal input from a microphone of the User's terminal. The noise level is represented as a sound pressure level in decibels, while the noise spectrum are the noise levels for 33 frequency bands of 1/3 octave covering the range of human audible bandwidth. An adaptation engine would enhance the perceived quality of the adapted audio signal by modifying the frequency attenuation of the original audio signal according to the noise characteristics.

7.4.4.4 Use Case: Adaptation to Audiovisual Environment

In this use case, assume that an end user is located in an area in which the natural environment characteristics are given by the DIA description shown in Example 7.7. For the visual environment, this description indicates that the correlated color temperature of the illumination is 159 and the illuminance is 500 lx. For the audio environment, noise level is given as 20 dB, and the 30 values representing the noise frequency spectrum give the noise power for 30 1/3-octave frequency bands in decibels.

Given the above natural environment characteristics, assume that the User requests image and audio data. In this case, the adaptation engine is responsible for providing an image to the illuminated display with minimal color distortion, and an audio signal with improved perceptual quality given the noise environment that has been described. In [26], a method to determine a shift ratio that indicates the difference in illumination under the current condition to that of a reference condition is presented. On the basis of this shift ratio, colors of an image frame could be mapped in such a way that the image is perceived under the reference illumination condition. For audio, enhancement of the perceived quality could be achieved by masking or attenuating selected frequencies during the adaptation based on the noise characteristics.

```
<DIA>
  <Description xsi:type="UsageEnvironmentPropertyType">
    <UsageEnvironmentProperty xsi:type="NaturalEnvironmentsType">
      <NaturalEnvironment>
        <NaturalEnvironmentCharacteristic
           xsi:type="IlluminationCharacteristicsType">
           <TypeOfIllumination>
             <ColorTemperature>159</ColorTemperature>
           </TypeOfIllumination>
           <Illuminance>500</Illuminance>
        </NaturalEnvironmentCharacteristic>
        <NaturalEnvironmentCharacteristic
           xsi:type="AudioEnvironmentType">
           <NoiseLevel>20</NoiseLevel>
           <NoiseFrequencySpectrum>
             40 30 20 10 10 10 10 10 10
             10 40 40 40 30 30 30 20 20 20
             10 10 10 10 10 10 10 10 10 10
             10 10 10
           </NoiseFrequencySpectrum>
        </NaturalEnvironmentCharacteristic>
      </NaturalEnvironment>
    </UsageEnvironmentProperty>
  </Description>
</DIA>
```

Example 7.7   DIA description of natural environment including audiovisual characteristics.

7.5 METADATA ADAPTABILITY

Metadata is widely used within the information technology for associating additional information to all kinds of content to increase the quality of the information stored within the content. One major goal of this additional information is making any audiovisual content as searchable as text. The MPEG-7 standard [27], and especially Part 5 – Multimedia Description Schemes [4], provides a comprehensive set of so-called Description Schemes (DSs) for associating metadata to audiovisual resources. These audiovisual resources might be subject to some kind of adaptation as already elaborated in the former sections. As a result, the corresponding metadata becomes outdated and must be adapted accordingly. In practice, however, this is not the only case when metadata needs to be adapted. There are various use cases where metadata has to be adapted necessarily, which is in the context of MPEG-21 DIA referred to as *metadata adaptation*.

The MPEG-7 description for a complete movie could be very comprehensive, depending on the level of detail. For some applications, only parts of this description might be interesting – to get to these interesting parts requires *filtering* the metadata. On the other hand, *scaling* the metadata refers to reducing the size of the metadata in order to meet the constraints of the network, the receiving terminal or its metadata processing capabilities [28, 29]. Finally, this tool facilitates the *integration* of different descriptions of the same resource by integrating complementary information [30]. This should result in a unique and richer description that matches a specific need.

7.5.1 SCALING AND FILTERING OF METADATA

Multimedia resources in general have various aspects to be described, which might result in comprehensive and sometimes quite complex hierarchical descriptions. On the other hand, resource providers may be interested in offering metadata describing the actual media resource for various different application areas without distributing metadata over several documents, which may lead to undesired duplications. In order to provide only the metadata required by the consuming application, extraction or filtering of the complete set of metadata is needed.

The high-level view of this process in the context of DIA is depicted in Figure 7.4. The metadata within the Digital Item is subject to the description adaptation engine, whereas the media resource (if present) is not necessarily adapted. The adapted Digital Item is constructed using the adapted metadata and any corresponding media resource. The actual metadata is adapted within the description adaptation engine taking into account the User's context, for example, terminal and network capabilities, by means of the information stored in the metadata adaptation description.

In certain instances, metadata represents additional overhead with regard to the actual media resource, while, in other cases, the metadata could be considered a resource itself, for example, it may be rendered as a menu that is used to navigate the content or make selections. Either way, the metadata may be quite large in size, making it difficult to deliver it through a network with low delay. Additionally, its large size may make it difficult to be consumed on a terminal with limited memory for such information. Thus, another major goal of this tool is reducing the size of the metadata by scaling it down using the information stored in the metadata adaptation description.

The metadata adaptation tool provides prior knowledge (meta-metadata) about the metadata, which helps an adaptation engine to determine if and how much relevant information is contained within the metadata without parsing and analyzing the entire description. For example, the metadata adaptation description may contain information about which components and how many components, that is, element names, are available within the

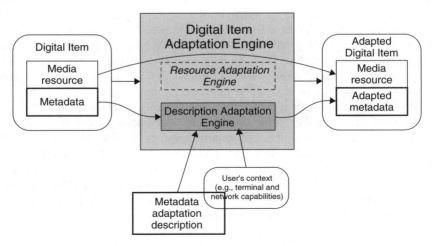

Figure 7.4   Scaling and filtering of metadata

actual metadata and the depth of the hierarchical structure among others. In other words, it provides hints about the complexity of the metadata instance.

Figure 7.5 illustrates how information within a metadata adaptation description can be interpreted for adapting an MPEG-7 description describing an ice hockey game, which is decomposed into multiple scenes with several hierarchies. Additionally, each scene at the lowest level is described using the PointOfView DS and the VisualDescriptor DS.

The metadata adaptation description as shown in the previous text indicates which components (element names) are contained within the MPEG-7 description and which not. For the `TemporalDecomposition` element, the number of successively nested elements is given. Finally, the PointOfView element contains further elements and attributes for which the actual values are given by the metadata adaptation description. Additionally, it is possible to describe, for each component, whether it is complete in terms of possible subcomponents (subelements). A possible interpretation of the metadata adaptation description is also given.

7.5.2 INTEGRATING METADATA

In contrast to filtering and scaling, the integration of metadata is crucial when actually one resource is described by multiple descriptions. In this case, two independent issues need to be taken into account. First, redundant information should be discarded and, second, complementary information should be integrated. This integration process itself follows two goals: the integrated description should provide a unique and richer description as well as match a specific need of the application. An integration process that aims to achieve the aforementioned objectives can be divided into several phases as shown in Figure 7.6.

Phase 1 incorporates the comparison of two descriptions at a structural level. If the two descriptions are not equal, for example metadata types and/or their values are different, a new course and redundant structure is generated, which basically merges the two

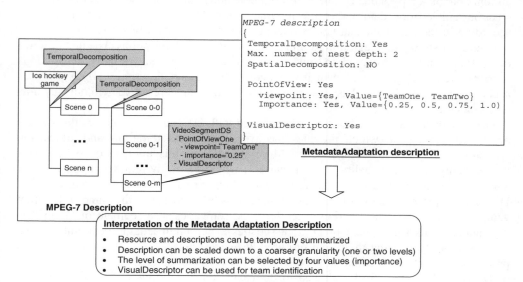

Figure 7.5 Metadata adaptation example [29]

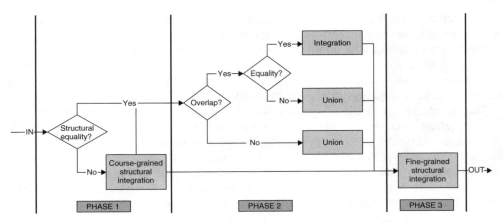

Figure 7.6   Metadata integration [30]

descriptions into a single one. Phase 2 investigates if there is no overlapped information within the description given by Phase 1. In this case, simply the union of the two original descriptions is generated. In order to avoid tedious and complex computations, information stored within the metadata adaptation description could be used for immediately determining the overlapping between the two descriptions. However, if an overlap is found, the descriptions have to be compared regarding the equality of their values. If values are equal, they need to be integrated, which is a complex process and depends on the specific type of metadata. Two specific approaches for MPEG-7 metadata integration have been proposed in [30]. On the other hand, if the values are not equal, their union is generated.

Finally, Phase 3 provides a finer integration at the structural level using both the preintegrated descriptions from Phase 1 and 2. The finer integration mainly concerns the elimination of all possible common information about the described resource such as creation and location data and/or the modification of the structure created in Phase 1, according to the new metadata values obtained by the integration of Phase 2.

Figure 7.7 shows an integration example of two MPEG-7 descriptions. The first description (D1) divides the resource into semantically coherent segments. Normally, descriptions such as these are generated manually, whereas the second description (D2) provides a complete automatic shot-based segmentation of the same resource without any further semantics.

The metadata adaptation description associated with D1 and D2 comprises the `Aver-ageValues` element, which provides the average segment durations and allows the speeding up of the adaptation process. More specifically, this information permits one to eventually skip the equality tests in Phase 1, which generally requires a considerable amount of computational time. The `InvariantProperties` element, which describes D1, enables the expression of the conditioned possibility to apply the same description value to subparts of a description. Therefore, the semantics of a segment of D1 can be propagated to its subsegment of D2 that satisfies a minimum length constraint without processing and analyzing the actual resource. It may be noted that within the integration process the possible inaccurate time boundaries between the manually annotated segments of D1 could be automatically aligned to the exact shot transitions of D2. The result of the integration process is a description (D1+2) comprising the semantic annotations of

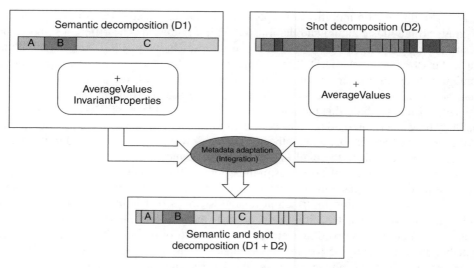

Figure 7.7 Metadata integration example; adapted from [31]

D1 as well as the shot-based decompositions of D2. The use of the integrated description instead of the original ones gives several advantages. From the user's perspective, it is easier to deal with a reduced number of high-quality descriptions, especially in browsing and retrieval, and from the system point of view, less data have to be transmitted through the communication network.

7.6 SESSION MOBILITY

In a world where peer-to-peer networking and redistribution of content among Users is becoming commonplace and secure, the means by which Digital Items are transferred from one device to another device is an important consideration. This section details the Session Mobility tool standardized by DIA that aims to preserve a User's current state of interaction with a Digital Item [32]. In addition to describing what the standard has specified to achieve this, an example of how the tool is used is also discussed.

7.6.1 THE STATE OF INTERACTION

As discussed in Chapter 3, the DID specifies a uniform and flexible abstraction and interoperable schema for declaring the structure and makeup of Digital Items. As a basic functionality, we can declare a Digital Item by specifying its resources, metadata and their interrelationships. Further, the DID can be configured by `Choice` and `Selection` elements that are part of the declaration. We refer to this instantiation of particular `Choice` and `Selection` elements in the DID as the *configuration state* of a Digital Item.

In DIA, session mobility refers to the transfer of configuration-state information that pertains to the consumption of a Digital Item on one device to a second device. This enables the Digital Item to be consumed on the second device in an adapted way. During this transfer, *application-state* information, which pertains to information specific to the application currently rendering the Digital Item, may also be transferred.

To make the session mobility concepts more concrete, consider the following example of an electronic music album and the illustration in Figure 7.8. The different songs of this album are stored at a content server and expressed as individual Selections within a particular Choice of the DID. There may also be other Choices within the DID that configure the platform, acceptable media formats, and so on, for a given User. In this example, assume that the first User is listening to the second song on Device A, and this User would like to share this particular song with a second User on Device B. Assuming that they have the right to do this, the configuration state of the DID on the first device, that is, the Selection that pertains to the second song, would be captured and transferred to the second device. Since the platform and acceptable media formats may be different on the second device, potential Choices in the original DID concerning those Selections would still need to be made there. Supplemental application-state information may also be transferred, such as timing information related to songs or layout information if images and video are involved. One important point to note is that DIDs are used to convey references to content, as well as state or contextual information. For the purpose of the explanation, we refer to the former as a Content DI and the latter as a Context DI.

As we see from the above example, the User's current state of interaction with a Digital Item is completely described by both the configuration-state and the application-state information. In the following section, we describe the means by which a session is captured through such state information.

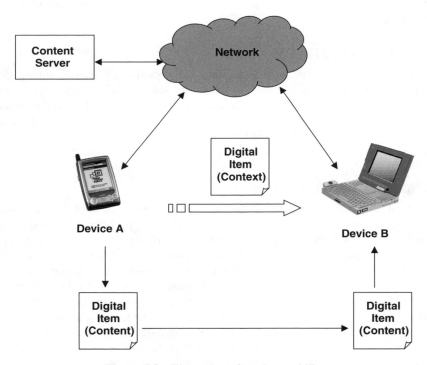

Figure 7.8   Illustration of session mobility

7.6.2 CAPTURING A SESSION

The important points to keep in mind regarding the capture of a session is that the information being transferred must be sufficient to (i) allow configuration of the Digital Item on the second device that is consistent with the configuration state of the first device, while still allowing the reconstructed session to meet the requirements of the second device, and (ii) provide any necessary application-state information to assist in the rendering of the Digital Item on the second device.

Considering the above points, we examine select elements provided by the DIDL and discuss how they are used for the targeted purpose in DIA.

- `Assertion`: This element allows the predicates within a `Choice` of the DID to be asserted as true or false. Therefore, the session mobility tool in DIA to describe the configuration-state information uses this element. In particular, the state of `Selections` made within `Choices` of the DID are indicated.

- `Descriptor`: This element allows metadata from any schema to be added to the Digital Item. As a result, it serves as a container for application-state information specified as part of the session mobility tool in DIA.

- `Annotation`: This element associates `Descriptor` and `Assertion` elements with a Digital Item. For session mobility in DIA, its target attribute is used to identify the Item that is an ancestor to the `Choice` indicated by the `Assertion`.

Given the above, the format of a Context DID that includes both configuration-state and application-state information has been specified by the standard.

The first important piece of information the Context DID must include is a reference to the Content DID. This is simply specified as a URI. With this, the association between content and context information can be made.

The second important piece of information to specify is the configuration state. As indicated above, this is achieved with the `Annotation` and `Assertion` elements of the DIDL. For each partially or fully resolved `Choice` element in the Content DI, an `Annotation` containing an `Assertion` that captures the current configuration state of the `Selections` in the `Choice` of the Content DI is added to the Context DI. The target attribute of the `Annotation` element is used to identify the Item in the Content DI that is an ancestor of the `Choice` to which the `Assertion` applies, while the target attribute of the `Assertion` element identifies the `Choice` in the Content DI to which the `Assertion` applies. Since the state of each partially or fully resolved `Choice` must be captured in the Context DI, only those `Choice` elements for which all `Selections` are in an undecided state are to be omitted.

The final piece of information that may be specified as part of the Context DI is the application-state information. In contrast to the configuration-state information, this information is optional. Two forms of application-state information could be specified: information that pertains to the Content DI as a whole, or information that pertains to just the particular target Item in the Content DI. Either way, the specifics of the application-state information are not defined by the DIA standard. The standard only specifies that it be specified within a `Descriptor` element of the DID and that the description be valid XML, which is qualified using XML namespaces. Examples of application-state information include the position on the screen in which a video contained in the DI is being rendered, the track currently being rendered in a music album, or the view of the Digital Item that the application is presenting to the User.

7.6.3 RECONSTRUCTING A SESSION

The information described by the Context DI is used to reconstruct a session on a second device. First, for each `Assertion` in the Context DI, the corresponding `Choice` in the Content DI is identified. Given this `Choice`, the `Assertion` further identifies which `Selections` are asserted as true and/or which are asserted as false. `Selections` not identified in the `Assertion` are undecided by default.

Once the configuration-state information transferred by the Context DI has been successfully applied to the Content DI, any additional `Choices` that need to be configured before the Digital Item is consumed should be considered at this time. For instance, usage environment characteristics of the second device may be used to automatically configure a `Choice` in the Content DI. Additionally, the User may be expected to configure certain `Choices` manually. After such configurations have been done, the final step of the reconstruction process would be to account for any application-state information that has been transferred within the `Descriptor` element of the Context DI.

7.6.4 SESSION MOBILITY EXAMPLE

To demonstrate the concept of session mobility in more detail, consider the sample Content DI shown in the example below, which references an audio resource as well as a video resource. In this simple example, the Content DI contains a `Choice` element with `Selections` that correspond to the audio and video resources.

```
<DIDL>
  <Item id="CONTEXT_DEMO">
    <Choice choice_id="EXAMPLE_CHOICE" maxSelections="1">
      <Selection select_id="AUDIO">
        <Descriptor>
          <Statement mimeType="text/plain">Audio</Statement>
        </Descriptor>
      </Selection>
      <Selection select_id="VIDEO">
        <Descriptor>
          <Statement mimeType="text/plain">Video</Statement>
        </Descriptor>
      </Selection>
    </Choice>
    <Component>
      <Condition require="AUDIO"/>
      <Resource mimeType="audio/mpeg" ref="audio.mp3"/>
    </Component>
    <Component>
      <Condition require="VIDEO"/>
      <Resource mimeType="video/mpeg" ref="movie.mpg"/>
    </Component>
  </Item>
</DIDL>
```

Example 7.8 Content Digital Item.

To configure this Digital Item, either the AUDIO or VIDEO selection must be made. For the purpose of this example, assume that VIDEO has been selected. The corresponding Context DI that captures this configuration-state information is given in the following example.

From this Context DI, we see that the `SessionMobilityTargetType` identifies the Content DI. Also, we see that the `Assertion` target identifies the `Choice` element within the Content DI and that the VIDEO selection has been asserted. Finally, as part of the `Descriptor` element, application-state information pertaining to the current playback status and media time has been provided. It may be noted that such application-state information is not specified by DIA.

In case the Content DI does not provide any `Choice/Selections` to the Users, the configuration-state information of the Digital Item is assumed to be static. For instance, if the Content DI contains only one `Resource` element, the configuration state is defined as this resource only. The corresponding Context DI looks similar to Example 7.9 without the `Annotation` element. For the sake of completeness, it should be mentioned that a Digital

```
<DIDL>
  <Item>
    <Descriptor>
      <Statement mimeType="text/xml">
        <dia:DIADescriptionUnit
            xsi:type="sm:SessionMobilityAppInfoType">
            <sm:ItemInfoList>
              <sm:ItemInfo
                  target="urn:mpeg:mpeg21:content:smexample#CONTEXT_DEMO">
                  <myApp:CurrentPlaybackStatus>
                    PLAYING
                  </myApp:CurrentPlaybackStatus>
                  <myApp:CurrentMediaTime>
                    4.35s
                  </myApp:CurrentMediaTime>
              </sm:ItemInfo>
            </sm:ItemInfoList>
        </dia:DIADescriptionUnit>
      </Statement>
    </Descriptor>
    <Component>
      <Resource mimeType="text/xml">
        <dia:DIADescriptionUnit
            xsi:type="sm:SessionMobilityTargetType"
            ref="urn:mpeg:mpeg21:content:smexample"/>
      </Resource>
    </Component>
    <Annotation target="urn:mpeg:mpeg21:content:smexample#CONTEXT_DEMO">
      <Assertion
          target="urn:mpeg:mpeg21:content:smexample#EXAMPLE_CHOICE"
          true="VIDEO"/>
    </Annotation>
  </Item>
</DIDL>
```

Example 7.9   Context Digital Item.

Item Adaptation Engine might also adapt the DI according to the usage environment of the device to which the session is transferred.

7.7 DIA CONFIGURATION

The DIA Configuration (DIAC) tools are used to help guide the adaptation process considering the intentions of a DID author. One method by which this is achieved is by allowing authors and providers of Content DIs the ability to specify useful DIA descriptions that would help to either configure the DID or adapt the resources according to the usage environment in which they will be consumed. Additionally, the DIAC tools comprise tools that have been specified for guiding the DID configuration process.

7.7.1 CONFIGURATION TOOLS

There are essentially two types of tools that have been specified under DIAC. The first is used to suggest the particular DIA descriptions that should be employed for resource adaptation, while the second is used to suggest the means by which `Choice/Selections` in a DID should be processed, for example, displayed to Users or configured automatically according to DIA descriptions.

For the first type of tool, suggested DIA descriptors are specified as an XPath expression that points to fragments of a DID or DIA document. For instance, the author or provider (at some point of the delivery chain) may suggest the `Format` element of the `VideoCapabilitiesType`, which is part of the UED descriptions tool, be used for configuration or adaptation. Its corresponding value should be retrieved from a DID or DIA document. The expression itself would be represented as follows:

```
//dia:Decoding[@xsi:type=dia:VideoCapabilitiesType]/dia:Format
```

In addition to indicating the suggested DIA description, this tool also indicates the location where the suggested descriptions should be considered. The possible locations are specified as receiver side, sender side or either side.

The second type of tool under DIAC provides guidance on how the DID `Choice/Selections` should be processed. This is achieved with the definition of two elements, `UserSelection` and `BackgroundConfiguration`. Both elements are meant to appear within the DID as a `Descriptor` of the `Choice` element. For example, in the `Choice` for video quality given in Example 7.10, the presence of the `UserSelection` tag indicates that the User should make the selection manually.

7.7.2 USAGE OF DIA CONFIGURATION

The following use case scenario shows how all the DIAC tools, together with the usage environment descriptors, can be used to facilitate dynamic resource adaptation. Figure 7.9 illustrates the steps involved in the media streaming adaptation and updating process in an end-to-end approach within the MPEG-21 framework.

In this example, the terminal capability description given in Example 7.2 is assumed, while the sample DID is given below in Example 7.11. The suggested DIA descriptions on the receiver side include the `Format` of the `VideoCapabilitiesType`. On the sender side, `Resolution` of the `DisplayCapabilitiesType` could be used for

```
...
  <Choice choice_id="VideoQuality" maxSelections="1" minSelections="1">
     <Descriptor>
        <Statement>
           <diac:UserSelection/>
        </Statement>
     </Descriptor>
     <Selection select_id="VideoQuality_Low">
        <Descriptor>
           <Statement mimeType="text/plain">Low Quality</Statement>
        </Descriptor>
     </Selection>
     <Selection select_id="VideoQuality_High">
        <Descriptor>
           <Statement mimeType="text/plain">High Quality</Statement>
        </Descriptor>
     </Selection>
  </Choice>
...
```

Example 7.10   Fragment of a DID illustrating Choice for video quality.

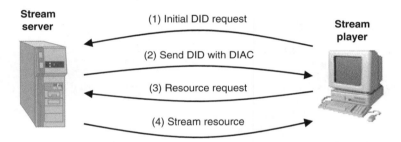

Figure 7.9   Architecture for the media streaming process using DIA Configuration

adaptation and it is suggested that this description be sent from the stream player to the stream server. The UserSelection tag, which appears as a Descriptor of the VideoQuality Choice, indicates that the User manually selects the video quality, while the BackgroundConfiguration tag, which appears as a Descriptor of the SupportedFormat Choice, indicates that the terminal should automatically configure the format. In this case, the terminal supports an MPEG-4 format, so this format will be chosen accordingly by matching the terminal capability description with that of the multimedia format description of the resource, which may be included in the DID.

Given this Content DI, the adaptation scenario is described as follows. First, the stream player makes an initial request for the Content DI. It may be noted that the DIAC information is included as part of the Content DI to provide the stream player with instructions on how to perform the adaptation and how Choice/Selection should be configured, that is, whether a Choice should be configured automatically or manually. Upon receiving the DID, the stream player performs the necessary DID configuration according to the suggested means to configure the DID. DIA UEDs to be applied at the receiver side, as well as direct input from the User, would be taken into account at this step to configure the DID.

```
<DIDL>
  <Item id="foreman_Reply">
    <Descriptor>
      <Statement mimeType="text/plain">
        <dia:DIA>
          <dia:Description
             xsi:type="diac:SuggestedDIADescriptionsType">
             <diac:ReceiverSide>
               <diac:Target xsi:type="diac:TargetType"
targetDescrip-
   tion="//dia:Decoding[@xsi:type=dia:VideoCapabilitiesType]/dia:Format"/>
             </diac:ReceiverSide>
             <diac:SenderSide>
               <diac:Target xsi:type="diac:TargetType"
                   targetDescription="//dia:Display[@xsi:type=dia:>
   <DisplayCapabilitiesType]/dia:Resolution"/>
             </diac:SenderSide>
          </dia:Description>
        </dia:DIA>
      </Statement>
    </Descriptor>
    <Choice choice_id="SupportedFormat" maxSelections="1"
       minSelections="1">
      <Descriptor>
        <Statement>
          <diac:BackgroundConfiguration/>
        </Statement>
      </Descriptor>
      <Selection select_id="SupportedFormat_AVI">
        <Descriptor>
          <Statement mimeType="text/xml">
            <!-- Description of AVI format -->
          </Statement>
        </Descriptor>
      </Selection>
      <Selection select_id="SupportedFormat_MPEG4">
        <Descriptor>
          <Statement mimeType="text/xml">
            <!-- Description of MPEG4 format -->
          </Statement>
        </Descriptor>
      </Selection>
    </Choice>
    <Choice choice_id="VideoQuality" maxSelections="1"
       minSelections="1">
      <Descriptor>
        <Statement>
          <diac:UserSelection/>
        </Statement>
      </Descriptor>
      <Selection select_id="VideoQuality_Low">
        <Descriptor>
          <Statement mimeType="text/plain">Low Quality</Statement>
        </Descriptor>
      </Selection>
```

```
        <Selection select_id="VideoQuality_High">
            <Descriptor>
                <Statement mimeType="text/plain">High Quality</Statement>
            </Descriptor>
        </Selection>
    </Choice>
    <Component>
        <Condition require="SupportedFormat_AVI VideoQuality_Low"/>
        <Resource mimeType="video/streaming" ref="foreman_low.avi"/>
    </Component>
    <Component>
        <Condition require="SupportedFormat_MPEG4 VideoQuality_Low"/>
        <Resource mimeType="video/streaming" ref="foreman_low.mpeg"/>
    </Component>
    <Component>
        <Condition require="SupportedFormat_AVI VideoQuality_High"/>
        <Resource mimeType="video/streaming" ref="foreman_high.avi"/>
    </Component>
    <Component>
        <Condition require="SupportedFormat_MPEG4 VideoQuality_High"/>
        <Resource mimeType="video/streaming" ref="foreman_high.mpeg"/>
    </Component>
  </Item>
</DIDL>
```

Example 7.11   DID with DIA Configuration information.

After the configuration is complete, the stream player requests the designated resources from the stream server. As part of this request, the DIA information according to the suggested DIA descriptors is sent from the player to the stream server. The stream server then determines the appropriate resource adaptation based on the DIA information it receives and begins streaming the adapted resource to the stream player. Of course, if the usage environment changes during transmission, adaptation according to the new constraints or conditions should be performed dynamically.

7.8 CONCLUDING REMARKS

In summary, DIA offers a rich set of tools to assist in the adaptation of Digital Items. In this chapter, we discussed means for describing the fundamental input to every adaptation engine, namely, standardized tools for the description of usage environments. Additionally, we described the tools that enable low-complexity adaptation of metadata, as well as tools for session mobility and tools that help guide the configuration and adaptation process from the perspective of a DID author or provider. The tools for creating high-level descriptions of the bitstream syntax to achieve format-independent adaptation and tools that assist in making trade-offs between feasible adaptation operations and the constraints are described in the next chapter.

Moving forward, amendments to the specification are being considered. Extensions of the DIA technology that is currently in place continue to be investigated, such as tools that provide further assistance with modality conversion and tools related more specifically to the adaptation of audio and graphics media. Looking more toward the big picture, being able to express the rights one has to perform adaptation and how this integrates within a system that governs those rights is also under investigation.

While the basic technology enabling UMA is in place in terms of recent multimedia (metadata) standards, it is still open whether or not and how these standards will be adopted by industry to create UMA-ready content and applications. Given today's heterogeneity of multimedia content, devices and networks, fast and wide deployment of this technology would be highly desirable in order to make future multimedia systems and applications easy and enjoyable to use.

7.9 ACKNOWLEDGMENTS

This chapter summarizes the work and ideas of many people who have actively participated in the process of developing requirements, proposing technology and cooperating to help stabilize and mature the specification. Many thanks are due to all of those who have been involved and helped to bring this part of the MPEG-21 standard to completion.

REFERENCES

[1] ISO/IEC 21000-7:2004(E), "Information Technology – Multimedia Framework – Part 7: Digital Item Adaptation", 2004.

[2] A. Vetro, C. Timmerer, "Digital Item Adaptation: Overview of Standardization and Research Activities", *IEEE Transactions on Multimedia*, vol. **7**, no. 3, pp. 148–426, June 2005.

[3] J. Smith, R. Mohan, S.-S. Lee, "Scalable Multimedia Delivery for Pervasive Computing", *Proc. ACM Multimedia*, pp. 131–140, http://portal.acm.org/citation.cfm?doid=319463.319480, November 1999.

[4] ISO/IEC 15938-5, Information technology – Multimedia Content Description Interface – Part 5: Multimedia Description Schemes, 2003.

[5] ISO/IEC 21000-2:2003(E), "Information Technology – Multimedia Framework – Part 2: Digital Item Declaration", 2003.

[6] ISO/IEC TR 21000-1:2001(E), "Information technology – Multimedia framework (MPEG-21) – Part 1: Vision, Technologies and Strategy", 2001.

[7] World Wide Web Consortium (W3C), "XML Schema Part 0: Primer", Recommendation, Available at http://www.w3.org/TR/xmlschema-0, May 2001.

[8] World Wide Web Consortium (W3C), "XML Schema Part 1: Structures", Recommendation, Available at http://www.w3.org/TR/xmlschema-1, May 2001.

[9] World Wide Web Consortium (W3C), "W3C, XML Schema Part 2: Datatypes", Recommendation, Available at http://www.w3.org/TR/xmlschema-2, May 2001.

[10] E. van der Vlist, "XML Schema", O'Reilly & Associates, June 2002.

[11] P. Walmsley, "Definitive XML Schema", Prentice Hall PTR, Dec 2001.

[12] J. Nam, Y.M. Ro, Y. Huh, M. Kim, "Visual Content Adaptation According to User Perception Characteristics", *IEEE Transactions of Multimedia*, vol. **7**, no. 3, pp. 435–445, June 2005.

[13] B. Feiten, I. Wolf, E. Oh, J. Seo, H.-K. Kim, "Audio Adaptation According to Usage Environment and Perceptual Quality Metrics", *IEEE Transactions of Multimedia*, vol. **7**, no. 3, pp. 446–453, June 2005.

[14] M. van der Schaar, Y. Andreopoulos, "Rate-Distortion-Complexity Modeling for Network and Receiver Aware Adaptation", *IEEE Transactions of Multimedia*, vol. **7**, no. 3, pp. 471–497, June 2005.

[15] I. Burnett, L. De Silva, "MPEG-21 Hierarchical Device Aggregation", ISO/IEC JTC1/SC29/WG11 M8341, Waikoloa, December 2003.

[16] K. Stuhlmuller, N. Farber, M. Link, B. Girod, "Analysis of video transmission over lossy channels", *IEEE Journal of Select Areas of Communication*, vol. **18**, no. 6, pp. 1012–1032, June 2000.

[17] G. de los Reyes, A.R. Reibman, S.-F. Chang, J.C.-I. Chuang, "Error-resilience transcoding for video over wireless channels", *IEEE Journal of Select Areas of Communication*, vol. **18**, no. 6, pp. 1063–1074, June 2000.

[18] S. Dogan, A. Cellatoglu, M. Uyguroglu, A.H. Sadka, A.M. Kondoz, "Error-resilient video transcoding for robust inter-network communications using GPRS", *IEEE Transactions on Circuits System Video Technology*, vol. **12**, no. 6, pp. 453–464, June 2002.

[19] M. Kim, H. Kim, D. Hong, "Stereoscopic Video Adaptation in MPEG-21 DIA", *Proc., 5th Int'l Workshop on Image Analysis for Multimedia Interactive Services (WIAMIS)*, Lisboa, Portugal, April 2004.

[20] T.C. Thang, Y.J. Jung, J.W. Lee, Y.M. Ro, "Modality Conversion For Universal Multimedia Services", *Proc., 5th Int'l Workshop on Image Analysis for Multimedia Interactive Services (WIAMIS)*, Lisboa, Portugal, April 2004.

[21] J. Song, S. Yang, C. Kim, J. Nam, J.W. Hong, Y.M. Ro, "Digital Item Adaptation for Color Vision Variations", *Proc. SPIE Conf, on Human Vision and Electronic Imaging VIII*, Santa Clara, CA, January 2003.

[22] Z. Sahinoglu, A. Vetro, "Mobility characteristics for multimedia service adaptation," *Signal Processing: Image Communication*, vol. **18**, no. 8, pp. 699–719, September 2003.

[23] P. van Beek, J.R. Smith, T. Ebrahimi, T. Suzuki, J. Askelof, "Metadata-driven multimedia access," *IEEE Signal Processing Magazine*, vol. **20**, no. 2, pp. 40–52, March 2003.

[24] B.L. Tseng, C.Y. Lin, J.R. Smith, "Using MPEG-7 and MPEG-21 for Personalizing Video", *IEEE Multimedia*, vol. **11**, no. 1, pp. 84–87, January-March 2004.

[25] ISO/IEC 15938-3:2002(E), "Information technology – Multimedia Content Description Interface – Part 3: Visual", 2002.

[26] Y. Huh, D.S. Park, "Illumination environment description for adaptation of visual contents", ISO/IEC JTC1/SC29/WG11 M8341, Fairfax, USA, May 2001.

[27] B.S. Manjunath, P. Salembier, T. Sikora, P. Salembier, "Introduction to MPEG 7: Multimedia Content Description Language"", John Wiley & Sons, 1st edition, New York, June, 2002.

[28] H. Nishikawa, Y. Isu, S. Sekiguchi, K. Asai, "Description for Metadata Adaptation Hint", ISO/IEC JTC1/SC29/WG11 M8324, Fairfax, May 2002.

[29] H. Nishikawa, S. Sekiguchi, Y. Moriya, J. Yokosato, F. Ogawa, Y. Kato, K. Asai, "Metadata Centric Content Distribution Based on MPEG-21 Digital Item Adaptation", *Proc. Asia-Pacific Symposium on Information and Telecommunication Technologies (APSITT)*, Noumea, New Caledonia, November 2003.

[30] N. Adami, M. Corvaglia, R. Leonardi, "Comparing the quality of multiple descriptions of multimedia documents", *Proc. MMSP 2002*, Virgin Island, USA, December 2002.

[31] N. Adami, S. Benini, R. Leonardi, H. Nishikawa, "Report of CE on Metadata Adaptation – Integration", ISO/IEC JTC1/SC29/WG11 M9849, Trondheim, July 2003.

[32] I. Burnett, P. Ruskin, "Session Mobility for Digital Items", ISO/IEC JTC1/SC29/WG11 M8257, Fairfax, USA, May 2002.

8

Digital Item Adaptation – Coding Format Independence

Christian Timmerer, Sylvain Devillers and Anthony Vetro

8.1 INTRODUCTION

The increase in rich multimedia content is inevitable and has been convincingly demonstrated in the last decade, for example, owners of digital cameras already produce tons of images and videos in a relatively short time period. This especially pertains to the diversity of media formats and the variety of Quality of Service (QoS) requirements. In order to cope with this heterogeneity of multimedia content formats, networks and terminals, the development of an interoperable multimedia content adaptation framework has become a key issue.

To address this interoperability problem, the Moving Picture Experts Group (MPEG) has specified a set of normative description tools within the MPEG-21 Digital Item Adaptation (DIA) standard [1, 2] that can be used to support multimedia adaptation. Specifically, DIA enables the construction of interoperable adaptation engines that operate independent of the device and coding format [3]. Device independence is guaranteed through a unified description of the environment in which the content is consumed or through which it is accessed, namely, the Usage Environment Description (UED) as described in Chapter 7. Coding-format independence is accomplished by the *Adaptation Quality of Service (AQoS)* and *(generic) Bitstream Syntax Description ((g)BSD)* tools, as well as the *Universal Constraints Description (UCD)* and *BSDLink* tools. These tools are discussed in detail in this chapter.

The remainder of this chapter is organized as follows. In Section 8.2, we provide details about binary resource adaptation based on high-level syntax descriptions of the bitstream. Section 8.3 provides details about relating constraints, adaptation operations and resource quality within the so-called Adaptation QoS tool. The big picture of the BSD-based adaptation approach is described in Section 8.4, and Section 8.5 discusses the

The MPEG-21 Book   Ian S Burnett, Fernando Pereira, Rik Van de Walle, Rob Koenen
© 2006 John Wiley & Sons, Ltd

UCD tool, which provides a means for further constraining the usage environment as well as the usage of Digital Items. Finally, Section 8.6 concludes the chapter with some final remarks.

8.2 BINARY RESOURCE ADAPTATION BASED ON BITSTREAM SYNTAX DESCRIPTIONS

The key advantage of new scalable multimedia formats is that data are organized in such a way that by retrieving a part of the bitstream it is possible to render a lower quality version of the content in terms of quality, size, frame rate or other characteristics. Unlike conventional transcoding, the editing required on a scalable bitstream is limited to simple operations such as data truncation. In this way, the content is authored only once, and a variety of versions may be obtained on demand from a single bitstream without requiring prohibitive computing burden from the adapting device. However, the large variety of competing or complementary coding formats, each featuring its own data structure, hampers the deployment of this strategy since a device providing several contents coded in different standards needs as many dedicated software modules as the offered formats to manipulate them.

To solve this limitation and leverage the use of scalable multimedia formats, MPEG-21 DIA defines a generic framework based on XML for adapting the content. This framework is generic in the sense it can be applied to any binary multimedia format and can be used in any device involved in the production, exchange or consumption of the content, including servers, clients or any intermediate nodes such as network gateways or proxies[1]. In a nutshell, XML is used for describing the syntax of a bitstream. The resulting document, called *Bitstream Syntax Description (BS Description, BSD)* is then transformed, for example with an Extensible Stylesheet Language Transformations (XSLT) style sheet [4], and used by a generic processor to generate the required adapted bitstream [5]. The general high-level architecture is depicted in Figure 8.1.

The first step in this architecture comprises the BSD generation, which takes the bitstream as input and generates its corresponding BSD as described in Sections 8.2.2.3 or 8.2.3.3. The bitstream and its BSD are subject to actual adaptation. The BSD is transformed according to the information provided by the usage environment. Subsequently, this transformed BSD is used to generate the adapted bitstream within the bitstream generation process as described in Sections 8.2.2.2 and 8.2.3.2. In practice, however, the last two steps, that is, BSD transformation and bitstream generation, could be combined for efficiency. Note that Figure 8.1 is a simplified view of the architecture and that other elements – namely the Bitstream Syntax Schema introduced further below – intervene in the adaptation process, as in Figure 8.7 and Figure 8.8.

In the following section, we describe the Bitstream Syntax Description Language (BSDL) that is specified by DIA to describe the high-level structure of a bitstream with XML. Additionally, the key components of the generic Bitstream Syntax Schema are described.

---

[1] It should be noted though that this tool does not address text-based content such as HTML or SMIL for which specific adaptation strategies have been proposed in the literature.

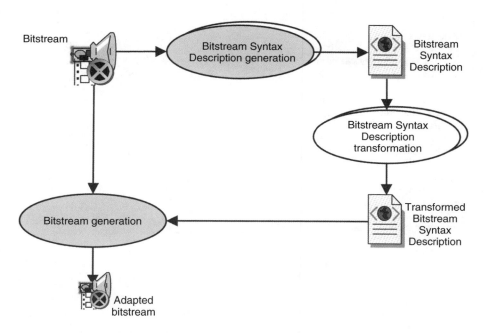

Figure 8.1 BSD-based adaptation architecture; adapted from [1]

8.2.1 DESCRIBING THE SYNTAX OF BINARY RESOURCES

A binary media resource consists of a structured sequence of binary symbols that is usually specific to a particular coding format; this sequence of binary symbols is referred to hereafter as a bitstream. In DIA, XML [6] is used to describe the high-level structure of a bitstream; the resulting XML document is called a *Bitstream Syntax Description*. This description is not meant to replace the original binary data, but acts as an additional layer, similar to metadata. In most cases, it will not describe the bitstream on a bit-per-bit basis, but rather addresses its high-level structure, for example, how the bitstream is organized in layers or packets of data. Furthermore, the bitstream description is itself scalable, which means it may describe the bitstream at different syntactic layers, for example, finer or coarser levels of detail, depending on the application. The bitstream description itself must also be adaptable to properly reflect bitstream adaptations. Lastly, unlike many metadata schemes such as MPEG-7, it initially does not deal with the semantics of the bitstream, that is, does not provide metadata information regarding the original object, image, audio or video it represents, but only considers it as a sequence of binary symbols. Example 8.1 shows an example of the BS Description of a JPEG2000 [7] bitstream.

In short, a JPEG2000 bitstream consists of a main header, which is itself made of several 'marker segments', followed by one or several 'tile(s)' containing the payload, that is, the packets of actual image data resulting from the encoding method. The marker segments give a number of parameters about the image such as its dimension, and about the compression method, such as the characteristics of the wavelet transformation used. In the JPEG2000 specification, these markers are referred to as mnemonics. For example, the SIZ marker provides the information relating to the image size. In

```
<Codestream>
  <MainHeader>
    <SOC>
      <Marker>FF4F</Marker>
    </SOC>
    <SIZ>
      <Marker>FF51</Marker>
      <LMarker>47</LMarker>
      <Rsiz>0</Rsiz>
      <Xsiz>640</Xsiz>
      <Ysiz>480</Ysiz>
      <!-- and so on -->
    </SIZ>
    <!-- and so on -->
  </MainHeader>
  <Tile>
    <Bitstream>
      <Packet>
        <SOP>
          <Marker>FF91</Marker>
          <LMarker>4</LMarker>
          <Nsop>0</Nsop>
        </SOP>
        <PacketData>155 242</PacketData>
      </Packet>
      <!-- and so on -->
    </Bitstream>
  </Tile>
</Codestream>
```

Example 8.1   Example of a Bitstream Syntax Description of a JPEG2000 bitstream.

our example, the XML elements describing the marker segments or their parameters are named after these mnemonics, but the naming used in the document is in no way mandated by MPEG-21.

Since the XML syntax does not allow the inclusion of binary data, the XML elements provide the values of successive bitstream symbols in hexadecimal (as for the `Marker` element) or decimal format (as for the `Xsiz` element). It is also possible to include longer segments of binary data by encoding them in Base 64. When the length of the data segment would make the description too verbose, a new specific format is defined and used to point to the segment in the original bitstream, as for the `PacketData` element. In this case, the two integer values respectively indicate the offset and the length in bytes of the data segment in the original bitstream. In this case, the actual data is not directly included in the BS Description, as for the other parameters, but pointed to by the description. The original bitstream will then be required for generating the adapted bitstream from its BS Description as will be seen below. In this respect, this XML document is a description and not a representation format.

It is therefore important to note that the content of a BS Description is hybrid in the sense it may contain actual data directly included in the XML elements on the one hand and references to data still in the described bitstream on the other hand. It is theoretically possible to directly embed all the binary symbols of the bitstream in the BS Description, in which case the original bitstream will not be required for further processing, but the resulting description would consequently become oversized. Conversely, a BS Description may only contain pointers and no explicit values. The choice between one and the other type of data depends purely on the application. Typically, the BS Description will detail the bitstream parameters in a readable format when they are required for the transformation as explained in the following text.

It may also be noted that the use of XML inherently implies a tree structure to the BS Description. This hierarchical organization, along with the naming of the XML tags and the choice of the types used (pointer or embedded value), is not mandated by MPEG-21 DIA and should be made according to the specific requirements of the application.

8.2.1.1 BSD Transformation

Once such a description is obtained, it is then possible to define a transformation of the XML document corresponding to the editing that should be applied to the original bitstream to obtain an adapted version. For example, with a JPEG2000 image, it is possible to obtain a smaller resolution image by removing a number of packets and modifying some parameters in the header. Similarly, by removing or modifying the corresponding XML elements in the BSD, one obtains a new, transformed BSD from which the adapted bitstream may be generated. For this, a processor will first parse the new description and then reconstruct the new bitstream by successively appending the parameters embedded in the description and copying the relevant data segments from the original bitstream when a pointer is found. This generation process is elaborated on further in the following text.

For editing the BS Description, MPEG-21 DIA does not mandate any specific XML transformation language. The well-known and widely used XSLT language [4] is a suggested solution. The only limitation is that XSLT requires one to load the full input document in memory, which may become an issue for very large descriptions or in streaming environments, in which case only a part of the description is available. In these cases, one will consider the promising developments of STX [8], an open-source project specifying a new XML transformation language applicable to streamed documents. The different processing models for XSLT and STX are illustrated in Figure 8.2 and Figure 8.3, respectively.

XSLT requires loading the whole tree representation of the XML document into memory before starting the actual transformation process, whereas STX gets by with the so-called events signaling the start or end of XML elements.

8.2.1.2 Bitstream Generation

Bitstream generation refers to the process of generating the adapted bitstream from the transformed description. To overcome the limitation mentioned earlier, this generation mechanism has to be format independent, that is, a single generic processor is required for this.

Going back to the example above, we see that the BS Description provides part of the information required to generate the bitstream, specifically the lexical value of the bitstream parameters. To use the BS Description to generate the bitstream, one needs to know how to binary-encode the value. For example, we need to specify that the `LMarker`

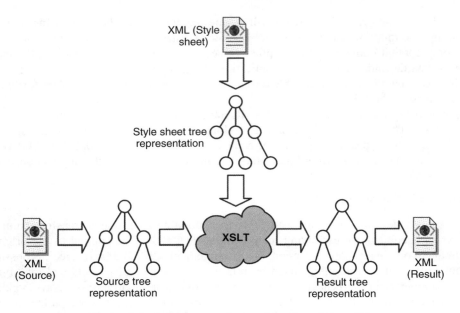

Figure 8.2   XSLT processing model; adapted from [9]

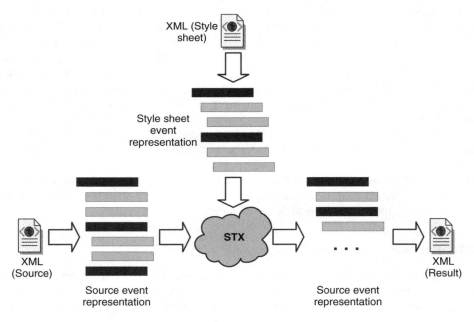

Figure 8.3   STX processing model; adapted from [10]

parameter should be encoded on a big-endian, two byte integer. Since this information is common to all the BSDs describing a JPEG2000 bitstream, it is possible to convey it in a model common to this family of descriptions, or using XML terminology, in a schema [11–13].

The BSD framework uses the information conveyed by the schema to generate a bitstream from its description. For this, a new language named *Bitstream Syntax Description Language (BSDL)* that is built on top of XML Schema is specified by MPEG-21 DIA. Schemas written in this language are called *Bitstream Syntax Schemas (BS Schemas)* and convey the information required by a generic processor to generate a bitstream from its description and vice versa. It may be noted that the BS Schema does not intervene in the transformation of the BS Description.

Figure 8.4 shows a BS Description, its BS Schema, and how the information of the BS Description can be found in the corresponding bitstream. The dotted line indicates the data type that is used for the respective BSD element during the bitstream generation process, for example, the value of the `Marker` element is encoded in hexadecimal and the value of the `Xsiz` element is encoded in unsigned integer.

However, it is beyond the scope of MPEG-21 DIA to standardize a set of BS Schemas for particular coding formats such as JPEG2000 or MPEG-4. Even for a same coding format, each application may have specific requirements on the level of detail of the description. The schema is therefore not unique for a given coding format. Therefore, MPEG-21 standardizes a *generic Bitstream Syntax Schema (gBS Schema)*, which is applicable to any format. As any other application-specific or coding format–specific schema, it is written in BSDL. It is meant to be generic by providing an abstract view of any bitstream by viewing it as a sequence of data segments called `gBSDUnit` and `Parameter` elements. The use of the gBS Schema then allows reducing the complexity of the adaptation by enabling optimized hardware implementation and hence allows the deployment of the BSD-based adaptations in constrained devices and environments in which computing resources are limited.

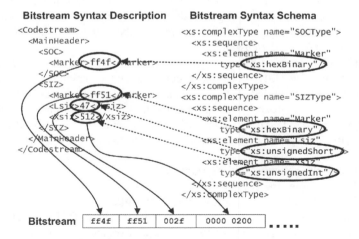

Figure 8.4 BSD, BS Schema and corresponding bitstream

8.2.2 BITSTREAM SYNTAX DESCRIPTION LANGUAGE

8.2.2.1 Overview: XML Schema, BSDL, gBS Schema

This section explains the main principles of BSDL. In particular, it explains how BSDL relates to XML Schema with respect to functional and syntactical aspects and describes the two 'levels' of the language, namely, BSDL-1 and BSDL-2.

BSDL and XML Schema

The functionality of a schema written in XML Schema is to express a set of constraints on the structure and data types of a set of XML documents. This set of constraints is then checked by a so-called validator, which returns a validation report. No other processing is implied beyond the validation of the instance. A BS Schema, on the other hand, is used by a BSDL processor to parse a bitstream and generate its description and vice versa. It is important to understand that, even though BSDL is syntactically built on top of XML Schema, the functionality of the two languages differs significantly. However, since BSDL makes use of the information conveyed by the schema, it is a requirement that a BS Description is valid with respect to its BS Schema. Conceptually, a BS Schema is first used as a regular schema to validate the description, and then to generate the bitstream from its description.

BSDL-1 and BSDL-2

BSDL is built by introducing two successive sets of extensions and restrictions over XML Schema. The first set is required for the bitstream generation process and is referred to as BSDL-1. The second set is required for the BSD generation process and is referred to as BSDL-2, where BSDL-1 is a subset of BSDL-2.

BSDL-1 defines a set of built-in data types and attributes, which carry specific semantics in the context of bitstream generation. Their syntax is specified in a schema named *Schema for BSDL-1 Extensions*. Any BS Schema can make use of these data types and attributes in addition to the regular XML Schema data types. When this is the case, the BS Schema then needs to import the Schema for BSDL-1 Extensions in order to be valid in the XML Schema meaning. BSDL-2 additionally introduces a set of new language extensions in the form of attributes and schema components, which carry specific semantics in the context of bitstream parsing and description generation. These language constructs are added to XML Schema as application-specific annotations, and their syntax is specified in a schema named *Schema for BSDL-2 Extensions*.

It is important to note the difference between the two schemas mentioned above. The extensions introduced by BSDL-2 define specific language mechanisms such as conditional statements and are added to the schema as application-specific information. The XML Schema validation process thus ignores them, and the BS Schema is not required to import the Schema for BSDL-2 Extensions. This schema is merely used as an internal tool for specifying the syntax of the language extensions; in this respect, it is similar to a schema for schema [12].

In the following, the schemas will be referred to as BSDL-1.xsd and BSDL-2.xsd, respectively. Furthermore, the prefixes bs1: and bs2: will be used as a convention to indicate which schema and namespace an extension belongs to.

The following gives two examples of BSDL extensions. First, BSDL-1 defines a built-in data type named bs1:byteRange to indicate a data segment in the original bitstream. From the XML Schema point of view, it is considered as a mere list of two integers, and will be validated as such, with no further semantics implied. On the other hand, this data type carries a specific semantics for BSDL-1. Whenever an element with this data type is encountered in the BS Description, the BSDL processor will copy the indicated byte range from the original bitstream into the output bitstream. As a second example, BSDL-2 introduces a new attribute named bs2:nOccurs specifying the number of occurrences of a particle. Unlike the XML Schema attribute xsd:maxOccurs, its value may be a variable specified by an XPath expression [14]. It characterizes the element declaration in the BS Schema similarly to xsd:maxOccurs, but is ignored by XML Schema since it belongs to an unknown namespace. On the other hand, while parsing the bitstream to generate its description, this attribute will indicate the number of times the given element should be read from the bitstream.

BSDL Parsers
In addition to the BSDL language itself, MPEG-21 DIA specifies the normative processes generating the bitstream from its description and vice versa. The processors implementing this process are referred to as *BSDtoBin (Bitstream Syntax Description to Binary)* and *BintoBSD (Binary to Bitstream Syntax Description) parsers*. A *BSDL parser* is a generic term encompassing both. The BSDtoBin parser uses the (possibly transformed) BS Description and its BS Schema as input, and generates the bitstream. For this, the BS Schema is required to conform to the BSDL-1 specification, and hence to import BSDL-1.xsd if it uses the built-in BSDL-1 data types or attributes. Similarly, the BintoBSD parser takes the bitstream and the BS Schema as input, and generates its BS Description. For this, the BS Schema is required to conform to the BSDL-2 specification.

BSDL and gBS Schema
As seen above, MPEG-21 DIA normatively specifies a generic BS Schema, making use of the built-in data types defined in BSDL-1 and giving an abstract view of any bitstream. A description conforming to *gBS Schema*, called a *generic Bitstream Syntax Description (gBS Description, gBSD)*, can thus be processed by a dedicated and optimized parser named *gBSDtoBin (generic Bitstream Syntax Description to Binary)*, also specified in MPEG-21 DIA. Since this schema conforms to BSDL-1, a gBS Description may also be processed by the BSDtoBin parser as any BS Description conforming to a user-defined BS Schema. The gBSDtoBin parser may therefore be considered as a particular optimized profile of BSDtoBin. On the other hand, owing to its abstract nature, the gBS Schema does not conform to BSDL-2 and cannot be used to generate the gBS Description. The way a gBS Description is generated is beyond the scope of MPEG-21 DIA. However, some examples of how to generate gBS Descriptions are elaborated in Section 8.2.3.3. The relationship between BSDL-1 and gBS Schema is depicted in Figure 8.5. Both the gBS Schema and codec-specific BS Schemas import the Schema for BSDL-1 Extensions with the only difference that the former is normatively specified within DIA while the latter is not.

Limitations of BSDL
A limitation of BSDL exists that directly stems from the inherent verbosity of XML. Describing a bitstream at a very fine level, for example, on a bit-per-bit basis may result in a large document, which may become prohibitive when the BS Description is to be exchanged. It is therefore stressed that the main application of BSDL is to describe the bitstream structure at a high level of detail. It is also possible to reduce the document size by binarizing it, for example, with the MPEG-7 BiM [15] encoder.

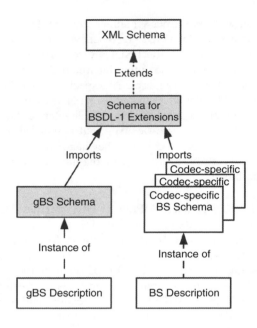

Figure 8.5   Schema hierarchy: relationship between BSDL-1 and gBS Schema; adapted from [1]

Furthermore, BSDL's purpose is not parsing a bitstream at any level of detail. In particular, the main part of a bitstream – qualified as the 'payload' – is usually the output of an encoding process in which the resulting sequence of bits can only be decoded via a specialized algorithm, but not by a BintoBSD parser. In this case, parsing of the bitstream with BSDL-2 relies on the availability of higher-level information about the structure of the bitstream, such as the use of start or end flags, or parameters giving the length of the payload, so that it is not required to actually decode the data in order to parse it. Therefore, BSDL cannot be used for testing the conformance of a bitstream.

Extensibility of BSDL
A BS Description principally contains the bitstream data or links to data segments that will be aggregated into the adapted bitstream. Additionally, it is possible to include extra information such as semantic metadata without interfering with this process. For this, data can be added to the BS Description as attributes, which are ignored by the BSDtoBin parser. Additionally, BSDL provides an escape mechanism (bsl:ignore attribute) that allows ignoring of a full XML fragment. For example, a BS Description may be enriched with MPEG-7 metadata describing the semantic content of a bitstream fragment identified by a fragment of this BS Description. Flags may also be added to the BS Description to facilitate the transformation process. It may be noted that this is in line with the philosophy of XML, where several layers of information can be provided by a single document, thanks to its extensible nature.

Multistep Adaptations
In some application scenarios, adaptation may happen in several successive steps, each step producing a further adapted version of the original bitstream. To avoid regenerating

a BS Description for each intermediary bitstream, it is possible to reuse the transformed BS Description that produced the adapted bitstream. For this, the address information contained in the transformed BS Description needs to be updated to correct the offsets of data segments. This process is nonnormative, but is briefly explained in the DIA specification for the gBSD case. A possible architecture for multistep adaptations is depicted in Figure 8.6. The left-hand side of the figure illustrates the dataflow as well as the involved adaptation modules. The right-hand side indicates the origin of the UEDs for the two adaptation steps. The server adapts the bitstream according to the usage environment that is available to this server. In general, this usage environment comprises a common set of characteristics of the subjacent terminals and networks of this server. The usage environment for the proxy is similarly gathered.

Each adaptation step performs the bitstream adaptation by means of the BSD-based adaptation framework according to the currently prevailing usage environment. It may be noted that in this architecture the BS Description transformation and bitstream generation is illustrated in a combined way. Additionally, the process for selecting the optimal parameter settings for the BS Description transformation is also part of this process, which is further described in Section 8.3.

8.2.2.2 Bitstream Generation: BSDL-1

As mentioned earlier, BSDL-1 is the set of extensions and restrictions over XML Schema required by the generic processor BSDtoBin to generate a bitstream from its

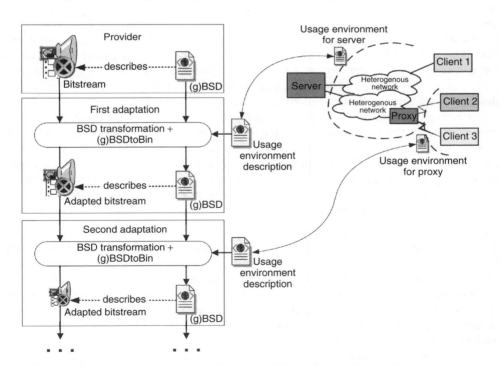

Figure 8.6 Multistep adaptation architecture with two (intermediary) adaptation nodes

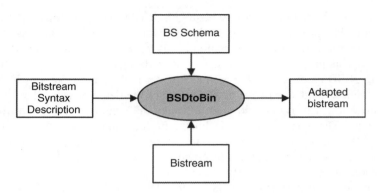

Figure 8.7   Bitstream generation using BSDtoBin; adapted from [1]

BS Description. This generation mechanism is illustrated in Figure 8.7. The BSDtoBin processor uses the information about types and structures conveyed by the BS Schema to parse the BS Description and generate the corresponding bitstream. It may be noted that the original bitstream also intervenes in the generation, as shown in Figure 8.7, when the BSD points to a segment of data.

Restrictions on Structural Aspects
A fundamental aspect of the BS Description is that only character data included in the XML elements participate in the bitstream generation. Conversely, all significant character data (i.e., excluding spaces, tabulations, carriage return and line feed characters) are binarized and appended to the output bitstream by the BSDtoBin parser unless the data is tagged with the `bs1:ignore` attribute. Also, attributes are ignored by the BSD-toBin process. The exceptions are attributes that carry specific semantics, such as the `xsi:type`. This property allows the adding of any kind of other information to the description without interfering with the bitstream generation process.

Furthermore, a type needs to be assigned to each element of the instance since the type is used by the BSDtoBin to specify the encoding scheme applied to the element content. Consequently, mixed content models are excluded from BSDL since in this case the character data inserted between the child elements of such an element have no type assigned by the schema. In other terms, XML elements should either contain character data (simple content elements) or a set of child elements (complex content elements) with no significant character data interspersed.

Extensions on Structural Aspects
The BSDL-1 structural extensions over XML Schema consist of two attributes, named `bs1:ignore` and `bs1:bitstreamURI`, that are used in the BS Description to carry specific semantics for the bitstream generation. These attributes are described in Table 8.1.

Note that since the `bs1:bitstreamURI` attribute applies to an element (and its descendant) and not to the full document, nothing prevents a single BS Description to describe several bitstreams: for example, a BS Description may contain a sequence of packets, each alternatively pointing to two input bitstreams. Generating the output bitstream with BSDtoBin is then similar to multiplexing the input bitstreams according to the information carried by the BS Description.

Table 8.1   BSDL-1 extensions on structural aspects

bs1:ignore attribute

The bs1:ignore attribute provides an escape mechanism that allows information to be added to
the description in such a way that it does not interfere with bitstream generation. For example,
the addition of semantic information could be achieved in this way.

bs1:bitstreamURI attribute

The bs1:bitstreamURI defines a property that applies to the characterized element and its
descendants. This property indicates the URI of the described bitstream, which will be used by
BSDtoBin to locate the original bitstream containing the segment of data to copy and append to
the output bitstream. It is possible to use relative URIs, in which case they should be
recursively resolved against the bitstream URI of the parent element, and so forth, up to the
document URI. The resolution mechanism is similar to that of the xml:base attribute [16].

Restrictions on Data Type Aspects

Restrictions on XML Schema Data Types

Since data types are used by BSDL to specify the encoding scheme applied to the character
data contained in the element, BSDL restricts the use of XML Schema built-in data types
to those for which a binary representation can be defined. For example, xsd:integer
represents the mathematical concept for an unbounded integer. However, no implicit
binary representation may be assigned to this type, and therefore it is excluded from
BSDL. On the other hand, xsd:int is derived from xsd:integer by restricting its
value space to the values that may be represented on four bytes (xsd:minInclusive
$= -2147483648$ and xsd:maxInclusive $= 2147483647$). BSDL thus imports this
type and assigns a binary representation of four bytes. The DIA specification lists the
built-in XML Schema data types allowed in BSDL.

Restrictions on xsd:anyType and xsd:anySimpleType

Another issue is related to the processing of elements with no specified type. XML
Schema allows an element to be declared in this way (with the use of xsd:anyType or
xsd:anySimpleType). However, BSDtoBin cannot process such an element unless a
type is specified in the instance with the xsi:type mechanism.

Restrictions on Derivation

XML Schema defines three types of simple type derivations, namely by list, restriction
and union. For a derivation by list (xsd:list), BSDtoBin will successively encode
and append the items of this list according to their type defined by the xsd:itemType
attribute of xsd:list. Derivation by restriction (xsd:restriction) is transparent to
BSDtoBin, except for the particular case of the xsd:maxExclusive facet, for which
a specific behavior is defined below. Lastly, the xsd:union mechanism allows the
declaration of two possible simple types for the same element. Unlike XML Schema,
BSDL then requires the instance to explicitly specify which type applies to the element
with the xsi:type attribute.

Extensions on Data Type Aspects

BSDL-1 introduces two built-in data types with specific semantics for the bitstream
generation, namely, bs1:byteRange and bs1:bitstreamSegment, as described
in Table 8.2.

Table 8.2   BSDL-1 extensions on data type aspects

`bs1:byteRange` data type

The `bs1:byteRange` data type indicates a byte range of the resource identified by the bitstream
  URI of the current element. Syntactically, it consists of a list of two nonnegative integers
  indicating the offset and the length of the data segment, respectively. Note that byte ranges are
  necessarily byte aligned, that is, it is not possible to indicate a segment starting on a non-byte
  aligned address.

`bs1:bitstreamSegment` data type

The `bs1:bitstreamSegment` data type targets the same functionality and is provided mainly
  for compatibility with the gBS Schema. With this built-in complex type, offset and length
  information are provided as attributes, which allows the adding of this information at several
  successive layers. Furthermore, it features a more complex addressing scheme, as explained in
  the section dedicated to the gBS Schema. Unlike `bs1:byteRange`, this data type cannot be
  used in BSDL-2 for the bitstream parsing process.

Use of Facets

The list of XML Schema data types supported by BSDL as seen above does not allow the
describing of binary symbols encoded on an arbitrary number of bits. Instead of defining
an additional list of built-in data types, which would be doomed to be too restricted,
BSDL specifies a mechanism for defining user-derived types with an arbitrary number of
bits by assigning a specific semantics to the `xsd:maxExclusive` facet.

In XML Schema, facets characterize a value space along independent axes or dimensions. Since BSDL does not consider the values of types but only their binary representations, facets are normally ignored by BSDL-1. As an exception, the `xsd:maxExclusive`
is used to indicate the number of bits with which an unsigned integer value should
be encoded when it applies to an unsigned integer built-in XML Schema data type
(`xsd:unsignedLong`, `xsd:unsignedInt`, `xsd:unsignedShort` or `xsd:
unsignedByte`) and when it directly constrains an XML Schema built-in data type. The
number of bits is then calculated as the logarithm in base 2 of the
`xsd:maxExclusive` value, rounded up to the next integer value. In all other cases,
it is ignored by BSDtoBin, as any other facet.

8.2.2.3 Bitstream Parsing and Description Generation: BSDL-2

BSDL-2 takes the BS Description approach a step further by building a new challenging
functionality. To achieve a fully generic framework, it is necessary for a format-unaware,
generic software module to be able to parse a bitstream based on the information provided
by its BS Schema and generate its description as depicted in Figure 8.8.

In this respect, BSDL comes close to other formal syntax description languages such
as Abstract Syntax Notation number One (ASN.1) [17] or Flavor [18], also known as
Syntactical Description Language (SDL) in the context of MPEG-4 [19].

To enable this new functionality, BSDL-2 defines an additional set of restrictions and
extensions on XML Schema, except for the structural aspects, for which no further restriction is introduced. The extensions are added to the BS Schema using the XML Schema
annotation mechanism and are therefore ignored by XML Schema and BSDL-1. A BS

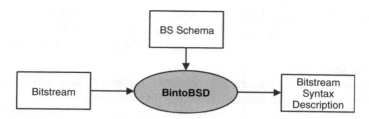

Figure 8.8 BS Description generation using BintoBSD [1]. Copyright ISO. Reproduced by permission

Schema conforming to BSDL-2 and used for parsing the bitstream and generating the description (i.e., BintoBSD) is therefore also conformant to BSDL-1 and can be used for BSDtoBin. The opposite is not necessarily true, a typical example being the gBS Schema, which cannot be used by BintoBSD.

Annotation Mechanisms of XML Schema

XML Schema provides two ways of adding application-specific annotations. First, all schema components can contain an `xsd:annotation` component, which itself can contain an `xsd:appinfo` component, intended as a placeholder for application-specific information. Secondly, all XML Schema components allow attributes with nonschema namespace, which leaves the possibility to add new attributes to some schema components by qualifying them with a specific namespace. BSDL-2 uses these features to define two kinds of language extensions to XML Schema:

- *BSDL components*: New schema components similar to XML Schema components; these are added to the schema via `xsd:annotation` and `xsd:appinfo`.
- *BSDL attributes*: New attributes characterizing an XML Schema component.

The two types of extensions are qualified with the BSDL-2 namespace and their syntax is specified in the Schema for BSDL-2 Extensions. Note that in XML Schema, application-specific information does not intervene in the XML validation, which means an XML Schema validator will not validate the content of the `xsd:appinfo` schema component against its schema. The same applies to attributes with nonschema namespace. It is therefore up to the BintoBSD parser to check that the attributes and schema components with BSDL-2 namespace follow the syntax specified in the Schema for BSDL-2 Extensions.

Use of Variables in BSDL-2

One of the key features of BSDL-2 is the use of variables, the value of which should be read from the bitstream. This is required, for example, when a bitstream contains a segment of data, the length of which is given by another field previously found in the bitstream. To parse this segment, the BintoBSD parser needs to retrieve this field and evaluate its value. It is therefore necessary to locate an upstream parameter (i.e., previously found in the bitstream) and evaluate its value.

While parsing a bitstream, the BintoBSD parser progressively instantiates the BS Description. As soon as a field is parsed, the corresponding XML element is instantiated and added to the description. A data field whose length or number of occurrences depend on a field found earlier in the bitstream can thus be read by locating and evaluating this parameter in the partially instantiated XML document. For this, BSDL-2 uses XPath expressions. When a BS Schema uses BSDL-2 extensions with XPath expressions, the BintoBSD parser should then evaluate them against the partially instantiated description.

Extensions on Structural Aspects

BSDL-2 defines a series of additional attributes for structural extensions, which are described in Table 8.3, Table 8.4, and Table 8.5 respectively.

Note that the values of xsd:minOccurs and xsd:maxOccurs should be compatible with the number of occurrences resulting from the evaluation of bs2:nOccurs. Typically, they should be set to respectively 0 and unbounded to handle all the possible cases.

Note that in both the cases, and for bs2:nOccurs, the value of xsd:minOccurs should handle the case the Boolean condition is evaluated as false and the particle is consequently not instantiated. In other terms, the value of xsd:minOccurs should be set to zero whenever a bs2:if or bs2:ifNext is used.

Restrictions on Data Type Aspects

Indefinite Length Data Types

Some of the XML Schema and BSDL-1 built-in data types, namely, xsd:normalized String, xsd:hexBinary, xsd:base64Binary and bs1:byteRange, have an

Table 8.3   BSDL-2 extensions on structural aspects – bs2:nOccurs

---

bs2:nOccurs attribute

XML Schema allows the declaring of the minimum and maximum number of occurrences of a
   particle (i.e., xsd:element, xsd:group, xsd:sequence, xsd:choice or xsd:all)
   with the attributes xsd:minOccurs and xsd:maxOccurs. Their value can only be declared
   as either a constant or as an unknown (unbounded). However, to handle the situation where a
   binary symbol or a segment of data should be read a variable number of times, and where this
   number is provided in a parameter provided upstream, BSDL-2 introduces a new attribute
   named bs2:nOccurs specifying the variable number of occurrences of a particle with an
   XPath expression. It is added to the particle declaration in the BS Schema similarly to
   xsd:maxOccurs, but is ignored by XML Schema and BSDL-1. On the other hand, while
   parsing the bitstream to generate its description, this attribute indicates the number of times the
   given element should be read from the bitstream.

Table 8.4   BSDL-2 extensions on structural aspects – bs2:if and bs2:ifNext

---

bs2:if and bs2:ifNext attributes

In addition to bs2:nOccurs, BSDL-2 introduces two attributes named bs2:if and
   bs2:ifNext specifying a condition on the (possibly multiple) occurrence of a particle. In the
   first case, the condition is based on data previously read and instantiated. The bs2:if attribute
   contains a Boolean XPath expression to be evaluated against the partially instantiated
   description. Conversely, the bs2:ifNext attribute expresses a condition on the value of the
   next bytes found in the bitstream. It contains either one or two hexadecimal strings; in the
   latter case, it expresses a range of allowed values. The BintoBSD parser then needs to buffer
   the relevant number of bytes and compare them to the content of bs2:ifNext. If the next
   bytes are equal to the indicated value (or contained in the indicated range when two strings are
   indicated), then the particle is instantiated.

Table 8.5 BSDL-2 extensions on structural aspects – `bs2:rootElement`

`bs2:rootElement` attribute

XML Schema allows the declaring of several global elements at the root of the schema, but does not specify which of these elements should be used as the document element, that is, the root element in the instance. Additional information is therefore required by the BintoBSD parser to know which element it should start to instantiate when several are declared globally in the BS Schema. This information may be provided by the application (e.g., by the command line) or by the BS Schema itself. In the latter case, the `bs2:rootElement` attribute is added to the `xsd:schema` component to indicate the qualified name of the root element.

indefinite length (Note that `bs1:bitstreamSegment` is not considered here since it cannot be used for BSDL-2). For BSDL-1, this is not a problem since the length is explicit in the lexical value provided in the instance. Conversely, BintoBSD requires this information to know how many bytes should be read for the element. For this, BintoBSD uses the information carried by a set of facets constraining the types. Such facets include XML Schema's `xsd:length`, along with new BSDL-2 facets introduced below, namely, `bs2:length`, `bs2:startCode` and `bs2:endCode`.

Derivation by List
Similarly, a list of data types (`xsd:list`) needs to have its length constrained by either `xsd:length` or `bs2:length` so that the BintoBSD parser knows how many symbols should be read for the instantiated element.

Derivation by Union
Lastly, a conditional statement needs to be added to the declaration of an `xsd:union` so that the BintoBSD determines which of the member types should be used for the instantiated type. This condition is provided by the `bs2:ifUnion` component.

Extensions on Data Type Aspects
BSDL-2 Facets
BSDL-2 introduces a set of new facets to specify constraints on BSDL and XML Schema data types with indefinite length. Since XML Schema does not allow a user to add his/her own facets, they are declared in the BSDL-2 namespace and added to the `xsd:restriction` component via the annotation mechanism, that is, the `xsd:annotation`/`xsd:appinfo` combination. The list of BSDL-2 facets consists of `bs2:length`, `bs2:startCode` and `bs2:endCode`. They are used to constrain `xsd:normalizedString`, `xsd:hexBinary`, `xsd:base64Binary` and `bs1:byteRange`. Note that if one of these data types is not constrained by either `xsd:length` or one of the BSDL-2 facets, then the bitstream is parsed until its end. These additional facets are described in Table 8.6.

Finally, the extensions on data type aspects include one additional XML component, which is described in Table 8.7.

8.2.3 GENERIC SYNTAX BITSTREAM DESCRIPTION

In contrast to the nonnormative BS Schema, which is specific to a given coding format, such as MPEG-4 or JPEG2000, the normative gBS Schema enables the use of BSDs in a generic and coding format–independent manner [20, 21]. Thus, this tool facilitates

Table 8.6 BSDL-2 extensions on data type aspects – facets

`bs2:length` facet

The `bs2:length` facet contains an XPath expression that should be evaluated against the partially instantiated description. Its value indicates the length in bytes of the considered data type. Its semantics is therefore similar to `xsd:length`, with the additional ability to specify a variable value.

`bs2:startCode` and `bs2:endCode` facets

The `bs2:startCode` and `bs2:endCode` facets define a flag indicating the end of a data segment. They contain one or two hexadecimal strings; in the latter case, the two strings define a range of allowed values. With such facets, the BintoBSD parser will read the bitstream until the read bytes are equal to the indicated flag (or contained in the indicated range in case two hexadecimal strings are provided). For `bs2:endCode`, the flag is understood as signaling the end of the current segment and will therefore be part of the data segment read by the BintoBSD parser. Conversely, `bs2:startCode` signals the start of the next segment and the flag itself will be excluded from the current segment.

The flags are defined with a hexadecimal value, and are consequently restricted to have a length in bits equal to a multiple of eight. Defining a range of values instead of a single one allows the simulating of a flag with an arbitrary number of bits. For example, using the range 0x08-0x0F comes to parse the bitstream until the five bit-long flag (four zero bits followed by a one bit) is found. On the other hand, the flag is always constrained to start on a byte-aligned address.

Table 8.7 BSDL-2 extensions on data type aspects – component

`bs2:ifUnion` component

As seen earlier, the use of `xsd:union` is allowed in BSDL-1 as long as the type is explicitly stated in the instance document with the `xsi:type` attribute. On the other hand, a deterministic decision mechanism is required in BSDL-2; otherwise, the BintoBSD parser cannot decide which type to instantiate. For this, it is possible to assign conditional statements to the member types of an `xsd:union` with the `bs2:ifUnion` schema component, which is not detailed here.

the processing of binary media resources without the need for coding format–specific schemas. The descriptions based on the gBS Schema are referred to as *generic Bitstream Syntax Descriptions (gBS Descriptions, gBSDs)*.

Additionally, the gBS Schema provides the following functionalities:

- semantically meaningful marking of syntactical elements described by the use of a 'marker' handle;

- description of a bitstream in a hierarchical fashion that allows grouping of bitstream elements for efficient, hierarchical adaptations;

- flexible addressing scheme to support various application requirements and random accessing of the bitstream;

- enables distributed adaptations in terms of multistep adaptations as proposed in [22].

Note that these features are also possible with a codec-specific BS Schema, since both the cases are written using the same BSDL-1. A BS Schema may or may not utilize these features, depending on the specific requirements of the application. Conversely, the normative gBS Schema has been designed to optimally fit the majority of expected applications.

With this generic functionality, the gBS Schema targets enhanced complex binary resource adaptation. In particular, the so-called 'marking' of elements and the hierarchical structure of the description can greatly simplify sophisticated bitstream manipulations, for example, remove an SNR layer from a JPEG2000 bitstream when the progression order is resolution, and semantic-related adaptations, for example, remove violent scenes from a video sequence. Furthermore, the gBSDtoBin process (see Section 8.2.3.2), which generates the (adapted) bitstream from a (transformed) gBS Description, does not require an application or codec-specific schema, since all the necessary information to regenerate the bitstream is included in the gBS Description and the semantics of the gBS Schema elements. This is particularly desired when the adaptation takes place on remote, constrained devices, since it saves bandwidth, memory and computational resources.

8.2.3.1 Components of the gBS Schema

The gBS Schema specifies three complex types: *gBSDType, gBSDUnitType* and *ParameterType*. Additionally, several simple types are specified for addressing, marking and labeling functionality.

gBSDType: The gBSDType supersedes the abstract type specified within the DIA schema tools by using the xsi:type attribute from the XML Schema instance namespace [12] in order to provide unique root elements for all DIA descriptions. Additionally, it provides addressing information consolidated within the addressAttributes for the entire gBS Description. The address attributes comprise the address mode and address unit to be used as well as the base URI of the bitstream described with this gBS Description. The address mode could be absolute, consecutive and offset, whereas, the address unit differentiates between bit and byte addressing. The absolute address mode is useful when describing large segments of a bitstream that may be dropped during the adaptation process, for example, frames, shots or scenes of a movie. On the other hand, consecutive and offset mode facilitates fast access to small, contiguous segments of a bitstream that may be updated only possibly due to the removal of other segments. These modes are often used in conjunction with bit addressing instead of byte address because of its possibility to describe the bitstream on a fine granular basis.

The base URI is specified using the bitstreamURI attribute that is specified as part of BSDL-1 and is resolved similarly to the xml:base attribute as specified in [16].

Example 8.2 shows a fragment of a gBS Description demonstrating the use of the gBSDType. This fragment also shows the usage of the classification scheme alias to define an alias for the classification scheme terms used throughout this DIA description.

gBSDUnitType: The gBSDUnitType enables the creation of gBSDUnit elements, which represent segments of a bitstream, but does not include the actual values of that segment. Therefore, it is used only for bitstream segments that might be removed during adaptation (e.g., due to frame dropping) or for carrying further gBSDUnit or Parameter

```
<?xml version="1.0" encoding="UTF-8"?>
<dia:DIA>
   <dia:DescriptionMetadata>
      <dia:ClassificationSchemeAlias alias="M4V"
      href="urn:mpeg:mpeg4:video:cs:syntacticalLabels"/>
   </dia:DescriptionMetadata>
   <dia:Description xsi:type="gBSDType" addressUnit="byte"
      addressMode="Absolute" bs1:bitstreamURI="akiyo.cmp">
   <!-- gBS Description comes here ... -->
   </dia:Description>
</dia:DIA>
```

Example 8.2   gBS Description example (gBSDType).

elements, which then results in a hierarchical structure for efficient bitstream manipulations. Each `gBSDUnit` carries several possible attributes:

- *start and length* – attributes conveying addressing information depending on the address mode.
- *syntacticalLabel* – attribute for including coding format–specific information identified via classification scheme terms.
- *marker* – attribute that provides a handle for application-specific (e.g., semantic) information to be used for performing complex bitstream adaptations in an efficient way.

Furthermore, it is also possible to include attributes specifying address information normally carried in the `dia:Description` element within a `gBSDUnit` element. This means that each `gBSDUnit` may also contain its own address mode and address unit. Additionally, the `bitstreamURI` could be also included within a `gBSDUnit` element, which is resolved relatively to its base URI and allows the locating of parts of the bitstream on different devices within a distributed multimedia environment.

Example 8.3 shows a fragment of a gBS Description describing an MPEG-4 coded movie at two levels of detail. The first level describes scenes/shots of the movie and marks them with application-specific violence levels, which can then be used for semantic adaptations. The second level describes the bitstream at a video object plane (VOP) level using the syntactical labels in order to determine the VOP type, for example, I-, P-, or B-VOP. The syntactical labels can be used for reducing the frame rate of the movie, thus saving bandwidth in a constrained heterogeneous terminal and network environment. Finally, each scene/shot is stored at a different location indicated by its bitstream URI. The bitstream URI of the first scene/shot (with marker 'violence-6') is relatively resolved against the base URI of the entire gBS Description as defined within the dia:Description element. The bitstream URI of the second scene/shot (with marker 'violence-4') is again an absolute URI but from a different server location.

ParameterType: The `ParameterType` allows the describing of a syntactical element of the bitstream, in particular, the value of a bitstream parameter that might need to be changed during the adaptation process. Therefore, the `ParameterType` provides the actual numerical value and its data type of the bitstream segment. The data type is required for correctly encoding the syntactical element in the bitstream generation process. Instead of defining all possible data types within the gBS Schema, they are specified by using

```
<dia:DIA>
    <dia:DescriptionMetadata>
        <dia:ClassificationSchemeAlias alias="M4V"
        href="urn:mpeg:mpeg4:video:cs:syntacticalLabels"/>
    </dia:DescriptionMetadata>
    <dia:Description xsi:type="gBSDType" addressUnit="byte"
        addressMode="Absolute"
        bs1:bitstreamURI="rtsp://mediasrv.somewhere.at/lotr2.cmp">
        <gBSDUnit syntacticalLabel=":MV4:VO" start="0" length="4"/>
        <gBSDUnit syntacticalLabel=":MV4:VOL" start="4" length="14"/>
        <gBSDUnit start="18" length="60038" marker="violence-6"
        bs1:bitstreamURI="lotr2_1.cmp">
            <gBSDUnit syntacticalLabel=":MV4:I_VOP" start="18"
                length="12270"/>
            <gBSDUnit syntacticalLabel=":MV4:P_VOP" start="12288"
                length="7589"/>
            <gBSDUnit syntacticalLabel=":MV4:B_VOP" start="19877"
                length="3218"/>
            <gBSDUnit syntacticalLabel=":MV4:B_VOP" start="23095"
                length="3371"/>
            <gBSDUnit syntacticalLabel=":MV4:B_VOP" start="26466"
                length="3511"/>
            <gBSDUnit syntacticalLabel=":MV4:P_VOP" start="29977"
                length="30079"/>
        </gBSDUnit>
        <gBSDUnit start="60056" length="1739135" marker="violence-4"
        bs1:bitstreamURI="rtsp://mediasrv.somewhere-else.at/lotr2_2.cmp">
            <gBSDUnit syntacticalLabel=":MV4:B_VOP" start="60056"
                length="12015"/>
            <gBSDUnit syntacticalLabel=":MV4:B_VOP" start="72071"
                length="10353"/>
            <gBSDUnit syntacticalLabel=":MV4:P_VOP" start="82424"
                length="15695"/>
            <!--... and so on ...-->
        </gBSDUnit>
        <!--... and so on ...-->
    </dia:Description>
</dia:DIA>
```

Example 8.3   gBS Description example (gBSDUnitType).

the xsi:type attribute in the instance, which is a common technique in the XML community. Nevertheless, frequently used basic data types are specified within an additional XML schema document. Similar to the gBSDUnitType, the ParameterType may also contain its own address information (address mode, address unit and bitstream URI).

Example 8.4 shows the fragment of a gBS Description that represents parts of the JPEG2000 main header (SIZ) where resolution and color information is stored. If the resolution of the image is reduced during the adaptation process, the Parameter elements ":J2K:XSiz", ":J2K:YSiz", ":J2K:XTsiz", and ":J2K:YTsiz" need to be updated accordingly. On the other hand, if the number of color components needs to be reduced, the ":J2K:Csiz" parameter requires updating and the ":J2K:Comp_siz" parameters must be removed by means of their marker values. In this example, the marker values "C0", "C1", and "C2" represent the YUV components respectively.

```
<gBSDUnit syntacticalLabel=":J2K:SIZ" start="2" length="49">
    <gBSDUnit addressMode="Consecutive" length="49">
        <gBSDUnit length="2"/>
        <Parameter name=":J2K:LMarker" length="2">
            <Value xsi:type="xsd:unsignedShort">47</Value>
        </Parameter>
        <Parameter name=":J2K:Rsiz" length="2">
            <Value xsi:type="xsd:unsignedShort">0</Value>
        </Parameter>
        <Parameter name=":J2K:Xsiz" length="4">
            <Value xsi:type="xsd:unsignedInt">640</Value>
        </Parameter>
        <Parameter name=":J2K:Ysiz" length="4">
            <Value xsi:type="xsd:unsignedInt">480</Value>
        </Parameter>
        <Parameter name=":J2K:XOsiz" length="4">
            <Value xsi:type="xsd:unsignedInt">0</Value>
        </Parameter>
        <Parameter name=":J2K:YOsiz" length="4">
            <Value xsi:type="xsd:unsignedInt">0</Value>
        </Parameter>
        <Parameter name=":J2K:XTsiz" length="4">
            <Value xsi:type="xsd:unsignedInt">640</Value>
        </Parameter>
        <Parameter name=":J2K:YTsiz" length="4">
            <Value xsi:type="xsd:unsignedInt">480</Value>
        </Parameter>
        <Parameter name=":J2K:XTOsiz" length="4">
            <Value xsi:type="xsd:unsignedInt">0</Value>
        </Parameter>
        <Parameter name=":J2K:YTOsiz" length="4">
            <Value xsi:type="xsd:unsignedInt">0</Value>
        </Parameter>
        <Parameter name=":J2K:Csiz" length="2">
            <Value xsi:type="xsd:unsignedShort">3</Value>
        </Parameter>
        <gBSDUnit syntacticalLabel=":J2K:Comp_siz" length="3"
            marker="C0"/>
        <gBSDUnit syntacticalLabel=":J2K:Comp_siz" length="3"
            marker="C1"/>
        <gBSDUnit syntacticalLabel=":J2K:Comp_siz" length="3"
            marker="C2"/>
    </gBSDUnit>
</gBSDUnit>
```

Example 8.4   gBS Description example (ParameterType).

8.2.3.2 Bitstream Generation Using gBS Descriptions

The behavior of the gBSDtoBin parser is normatively specified in DIA. This process generates the bitstream by using the information carried in a gBS Description, which has been possibly transformed.

The gBSDtoBin parser hierarchically parses the gBS Description and constructs the corresponding bitstream. For that purpose, it uses the addressing information from dia:

Description element or attributes of the gBSDUnit or Parameter elements for locating the bitstream segments to be copied from the source to the result bitstream. Additionally, the Parameter element contains the actual value and the required information for correctly encoding the syntactical elements into the resulting bitstream.

Assuming that a receiving terminal is not capable of decoding B-frames or owing to network conditions B-frames have to be dropped. The gBS Description that is transformed from that in Example 8.3 is shown in Example 8.5.

The gBSDtoBin parser, which uses the gBS Description from Example 8.5 to generate the adapted resource, is illustrated in Figure 8.9. The gBSDtoBin parser traverses the transformed gBS Description and copies only those bitstream segments that are still described by the gBS Description from the source bitstream and appends them to the result bitstream. In this case, B-VOPs (indicated by the value of the syntactical-Label attribute, i.e., :MV4:B_VOP) are discarded because they are no longer described by the transformed gBS Description.

To enable further adaptation steps (which may be located on different devices), it is necessary to update the transformed gBS Description with respect to the adapted bitstream.

```
<dia:DIA>
    <dia:DescriptionMetadata>
        <dia:ClassificationSchemeAlias alias="M4V"
        href="urn:mpeg:mpeg4:video:cs:syntacticalLabels"/>
    </dia:DescriptionMetadata>
    <dia:Description xsi:type="gBSDType" addressUnit="byte"
        addressMode="Absolute"
        bs1:bitstreamURI="rtsp://mediasrv.somewhere.at/lotr2.cmp">
        <gBSDUnit syntacticalLabel=":MV4:VO" start="0" length="4"/>
        <gBSDUnit syntacticalLabel=":MV4:VOL" start="4" length="14"/>
        <gBSDUnit start="18" length="60038" marker="violence-6"
        bs1:bitstreamURI="lotr2_1.cmp">
            <gBSDUnit syntacticalLabel=":MV4:I_VOP" start="18"
                length="12270"/>
            <gBSDUnit syntacticalLabel=":MV4:P_VOP" start="12288"
                length="7589"/>
            <!-- Removed gBSDUnit with syntacticalLabel=":MV4:B_VOP" -->
            <!-- Removed gBSDUnit with syntacticalLabel=":MV4:B_VOP" -->
            <!-- Removed gBSDUnit with syntacticalLabel=":MV4:B_VOP" -->
            <gBSDUnit syntacticalLabel=":MV4:P_VOP" start="29977"
            length="30079"/>
        </gBSDUnit>
        <gBSDUnit start="60056" length="1739135" marker="violence-4"
        bs1:bitstreamURI="rtsp://mediasrv.somewhere-else.at/lotr2_2.cmp">
            <!-- Removed gBSDUnit with syntacticalLabel=":MV4:B_VOP" -->
                <!-- Removed gBSDUnit with syntacticalLabel=":MV4:B_VOP" -->
                <gBSDUnit syntacticalLabel=":MV4:P_VOP" start="82424"
                length="15695"/>
            <!--... and so on ...-->
        </gBSDUnit>
        <!--... and so on ...-->
    </dia:Description>
</dia:DIA>
```

Example 8.5   Transformed gBS Description.

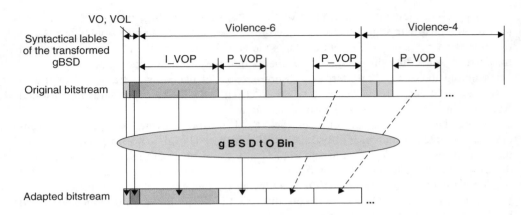

Figure 8.9   gBSDtoBin example removing B-frames by means of a transformed gBSD

This affects, in most cases, the address attributes such as start and length only. This is, for example, required in a distributed adaptation scenario where a server drops B-frames because the end devices are not able to decode these frame types and a home gateway removes violent scenes according to the age of the consumer.

8.2.3.3 Generating gBS Descriptions

The gBS Description generation process can be compared with the encoding of audiovisual resources and is therefore not normatively specified within DIA. This allows the design and development of various tools for different purposes.

One possibility would be to use the BintoBSD parser as described in Section 8.2.2.2 to generate a BS Description based on a coding format–specific BS Schema, which will be transformed into a gBS Description subsequently (Figure 8.10(b)). Traditional XML transformation tools or languages such as XSLT [4] could be used for this purpose. Generally, this kind of transformation could be separated into two steps.

1. The first step simply transforms one XML document (BS Description) based on one XML Schema (BS Schema) into a second XML document (gBS Description) based on another XML Schema (gBS Schema).
2. The second step comprises the introduction of additional hierarchical layers and appropriate marker values according to the need of an application. The required layers as well as the actual marker values could be obtained from an MPEG-7 description or any other additional metadata describing the resource. It should be noted that these steps could and should be combined for several reasons, such as efficiency and speed.

Please note that XSLT allows the combination of these two steps into one single style sheet.

Another possibility for generating gBS Descriptions is to directly engage in the encoding process of the resource and generate its gBS Description out of it (Figure 8.10(a)). The encoding process usually encompasses only the generation of a gBS Description

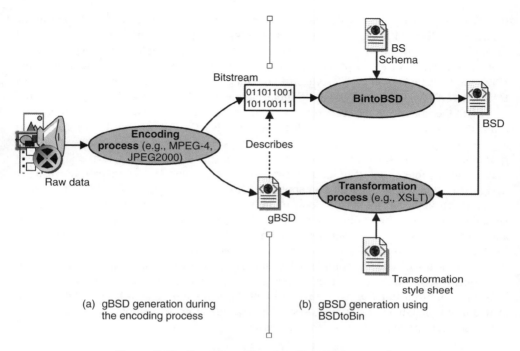

Bitstream

011011001
101100111

BS
Schema

BintoBSD

BSD

Encoding
process (e.g., MPEG-4,
JPEG2000)

Describes

Raw data

gBSD

Transformation
process (e.g., XSLT)

Transformation
style sheet

(a) gBSD generation during
     the encoding process

(b) gBSD generation using
     BSDtoBin

Figure 8.10   Possible architectures for gBSD generation

based on the syntactical elements of the resource. This would result in a low-level gBS
Description, which describes the resource with respect to only its structure and organization in headers, packets or frames. For the inclusion of additional layers and marker
values, further knowledge is required but this could be introduced similarly as described
earlier.

8.3 RELATING CONSTRAINTS, ADAPTATION OPERATIONS
AND RESOURCE QUALITY

In addition to the XML-based adaptation framework explained in the previous section,
DIA provides a set of tools for assisting an engine to take adaptation decisions based on
the input provided by the Digital Item author. For this, DIA defines a first tool named
AdaptationQoS (AqoS), declaring what types of adaptations may or should be applied to
the content in order to optimally fit a given external context [23]. Additionally, this tool
may be coupled to the so-called *Universal Constraints Description* (UCD) tool, briefly
introduced here and detailed in the next section, which describes a set of constraints
to be resolved. Both the tools are typically used with the BSD framework explained in
the previous section. In this case, a DIA tool named *BSDLink*, explained in Section 8.4,
provides a wrapper referencing the different elements involved in the adaptation (decision)
process.

   To introduce the AQoS tool, let us consider the following example: a colored image
with 640×480 pixels resolution has been encoded in a scalable way in the JPEG2000

format using four levels of decomposition, which means that it is possible to retrieve the resolutions 320×240, 160×120, 80×60 and 40×30 pixels in addition to the original resolution, in a colored or grayscale version, by performing simple editing operations.

Additionally, a BS Description is available along with the BS Schema[2] describing the JPEG2000 format and an XSLT style sheet defining the relevant transformation for the BS Description. This style sheet uses two input parameters, the first one defining the desired scaling of the image and the second one indicating the number of output color components (one for a grayscale version). Note that while the BS Description is always specific to a given image, the AQoS may be common to a set of images encoded with the same properties (e.g., the same resolution). Additionally, the BS Schema and the style sheet may be applicable to all JPEG2000 bitstreams descriptions. However, they are usually designed for the specific needs of an application in order to minimize the description size.

The functionality of the AQoS description is to declare which of the available versions should be chosen to optimally fit a given set of values characterizing the external context, and how to generate the relevant version with the provided BS Description and style sheet. Syntactically, the AQoS description consists of two main components: *Modules*, which provide a means to select an output value given one or several input values, and *IOPins*, which provide an identifier to these input and output values. In mathematics, a module is comparable to a function with several input variables represented by the IOPins. In a programming language, a module is similar to a function or method returning a scalar value, and IOPins may be considered as variables.

8.3.1 MODULES

The goal of AQoS is to select optimal parameter settings for media resource adaptation engines that satisfy constraints imposed by terminals and/or networks while maximizing QoS. In order to describe these relationships, three types of modules are specified by the standard, which allows an optimal choice for representing data according to the actual application requirements, for example, feasible set of adaptation operators, usage environment constraints, or parameters for adaptation engines.

The three types of modules, namely, Look-Up Table (LUT), Utility Function (UF), and Stack Function (SF), are illustrated in Figure 8.11 and the resulting representations are summarized in Table 8.8. Please note that the choice of representation depends mainly on the application requirements. In addition to the currently available types of modules, new representations can be added easily, if needed, because of the extensible design of the AQoS schema.

8.3.2 IOPINS

The second main component of AQoS, named *IOPin*, identifies a fixed or variable value. The uniqueness of its name is guaranteed by the use of the `xsd:ID` type of XML Schema, which constrains the name to be unique for the XML document. Additionally, it is possible to assign it a semantic label, for example, by using a term for a classification scheme, such as the ones defined in the DIA specification. However, to make use of this label in an unambiguous way, the different peers involved in the application need

---

[2] It should be noted that in case the BS Description is a gBS Description no BS Schema is needed.

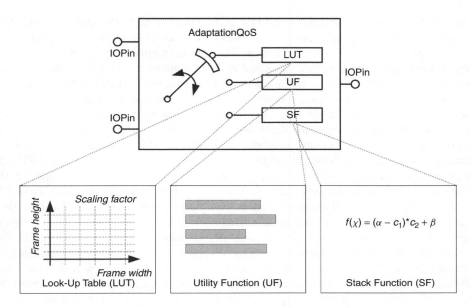

Figure 8.11 AdaptationQoS modules and IOPins

Table 8.8 Available AdaptationQoS modules

Module	Representation	Description
Look-Up Table (LUT)	M × N matrix, where the content may be defined as string, integer, floating-point values or NA value if a given combination of input coordinates is not valid.	LUTs are suitable for dense and discrete data and generally applicable for most adaptation requirements, but may incur additional overhead when data is sparse. The matrix coordinates themselves are bound to a set of discrete values declared in a so-called Axis and identified by an IOPin.
Utility Function (UF)	N-dimensional vector, where elements of vector may be defined as string, integer, floating-point values, or NA values.	UFs are suitable for sparse and discrete data because they are more efficient than look-up tables when many matrix coefficients are assigned an NA value.
Stack Function (SF)	Expression using the Reverse Polish Notation (RPN) [24], where an expression is written as a sequence or stack of operators and arguments.	SFs are suitable for functional and continuous data, that is, when a continuous approximation of the actual mapping by an equation is desirable. For example, `(a+b) * (c-d)` is written as the sequence "`a b + c d - *`". An XML syntax for expressing the arguments and operators is defined in the DIA specification.

to share the same understanding of this label. The value of an IOPin may be continuous or discrete. Additionally, it is possible to bind it to a set of discrete values declared in an Axis appended to the IOPin declaration. This last possibility may be useful when the IOPin represents the value to be determined in an optimization function, which is typical when using AdaptationQoS in conjunction with a Universal Constraint Description. In this case, the Axis then sets an additional constraint on the IOPin value.

An IOPin may be used as an input/output parameter or as an internal variable by the AQoS descriptor. In particular, an IOPin may be used to identify a value retrieved from the external context, such as the network bandwidth, in which case it acts as an input parameter. Alternatively, the input parameter of a style sheet declared by the BSDLink description may be provided by an output IOPin of the AQoS. In yet another case, the IOPin may be used as an internal variable that is not referenced anywhere else. Note that the same IOPin may also be used to retrieve an external parameter and also provide a value to the BSDLink, in which case it is used both as input and output. Thus, different modules can be connected through IOPins as depicted in Figure 8.12. In particular, this example illustrates the view-dependent texture adaptation as discussed in [25], which aims to reduce the number of wavelet levels and bitplanes to meet a given processing time constraint, while minimizing the perceived quality loss. Therefore, the different modules describe the relationship between angle (α), distance (d) and quality (Q), the relationship between levels (l), bitplanes (b) and quality (Q), and the relationship between levels (l), bitplanes (b) and processing time (T).

The input is determined by the viewing distance and angle as well as by the processing time constraint, whereas the outputs are the resulting bitplanes and levels that meet these constraints. One can easily see that the quality IOPin is used for internal purpose only, whereas the bitplanes and levels IOPins are used for two purposes, that is, internal (connecting module B and C) and external (output of the whole AQoS description).

However, when an IOPin is used as an input parameter, it is typically filled with a value retrieved from the external context. In this case, the IOPin contains a `GetValue` element expressing the assignment operation. The argument of this `GetValue` element may be of various types.

Suppose that the external value is provided in an XML document with a known structure, such as a Usage Environment Description (UED). In this case, the desired value can be identified with a URI enclosing a fragment identifier pointing to the relevant element. The URI specification [26] defines a fragment identifier appended to the main

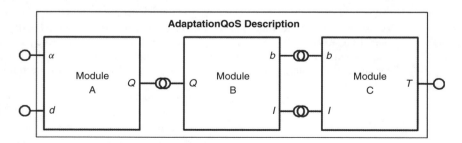

Figure 8.12   Connection of AdaptationQoS modules through IOPins

part identifying the resource, but does not specify a syntax for it, since this is supposed to be the property of the type of the indicated resource. For this purpose, the XPointer framework [27], which specifies a format to identify fragments of XML documents, is used in DIA as the normative way to select a value in an XML description, such as a UED. The XPointer framework is currently limited to locating XML elements, which may be restrictive since relevant data are sometimes included as attributes in UED. To solve this, an alternative syntax named *xpointer() scheme* [28] increases the expressiveness of the XPointer framework by using XPath expressions; however, this specification is still at W3C Working Draft stage. Once this specification reaches the W3C Recommendation stage, it could be used for the design and implementation of interoperable MPEG-21–based applications.

In other cases where UEDs are not available, a normative interface is required to communicate with legacy systems providing descriptions of the external context, but in a third format. For example, a device may be able to tell what display capabilities it offers, but without the ability to provide this information in the MPEG-21 DIA format. To exchange this information, the different peers then need to share the same understanding of what 'display size' means. In other words, they need to use a normative semantic reference. For this, DIA defines a list of normative semantic references by means of a classification scheme, which includes normative terms for the network bandwidth, the horizontal and vertical resolution of a display, and so on. When relevant, the definition of these terms points to the corresponding UED description tool.

It is important to note the difference between these two ways for retrieving a value from the external context. To use an analogy with object-oriented languages, locating an actual value in a given instance as in the first case is similar to pointing to an object, whereas using a semantic reference is similar to referencing a class. In the first case, the value is available in a known format, and the mechanism to retrieve this value (a URI resolver) is agnostic about the actual semantics of the value. In the other case, the value needs to be instantiated, that is, it is up to the application to express the request in a format understood by the targeted device.

8.3.3 ADAPTATIONQOS EXAMPLE

To further explain the use of the AQoS tool, consider the image example introduced in the introduction of Section 8.3, along with the AQoS description given in Example 8.6.

As in the introduction, an XSLT style sheet referenced by the BSDLink description, but not shown here, uses two input parameters to determine the transformation to be applied to the BS Description to obtain an optimally adapted image. The two input parameters reference the IOPins named *SCALE* and *NR_COLORS* of the AQoS description, which may be part of the same DID as shown in the example, but also available in another XML document.

To calculate these two values, the IOPins WIDTH, HEIGHT and COLOR_CAPABLE first need to be instantiated by the application. Let us consider that the targeted device has a grayscale display with a size of 300×400 pixels. To obtain the value of the SCALE IOPin, the WIDTH and HEIGHT values (300 and 400, respectively) are first rounded down to the closest discrete value declared in the corresponding axes (160 and 240, respectively)[3]. The indices of these values (the third and fourth coordinates, respectively)

---

[3] Note that the default rounding method is the floor() operation.

are then reported into the matrix; the corresponding coefficient then indicates a value of 4 for the SCALE IOPin. Similarly, the value of COLOR_CAPABLE IOPin yields 'false', which corresponds to the first element in the corresponding axis. This index gives the value 1 in the 1-dimension matrix declaring the output IOPin NR_COLORS. In conclusion, the AQoS description indicates that a 160×120 pixels, grayscale image optimally fitting the given external context can be obtained by assigning the following values to the respective IOPins: SCALE = 4 and NR_COLORS = 1.

```
<DIA>
  <Description xsi:type="AdaptationQoSType">
    <Module xsi:type="LookUpTableType">
      <Axis iOPinRef="WIDTH">
        <AxisValues xsi:type="IntegerVectorType">
          <Vector>40 80 160 320 640</Vector>
        </AxisValues>
      </Axis>
      <Axis iOPinRef="HEIGHT">
        <AxisValues xsi:type="IntegerVectorType">
          <Vector>30 60 120 240 480</Vector>
        </AxisValues>
      </Axis>
      <Content iOPinRef="SCALE">
        <ContentValues xsi:type="IntegerMatrixType" mpeg7:dim="5 5">
          <Matrix>
            16 16 16 16 16
            16  8  8  8  8
            16  8  4  4  4
            16  8  4  2  2
            16  8  4  2  1
          </Matrix>
        </ContentValues>
      </Content>
    </Module>
    <Module xsi:type="LookUpTableType">
      <Axis iOPinRef="COLOR_CAPABLE">
        <AxisValues xsi:type="BooleanVectorType">
          <Vector>false true</Vector>
        </AxisValues>
      </Axis>
      <Content iOPinRef="NR_COLORS">
        <ContentValues xsi:type="IntegerMatrixType" mpeg7:dim="2">
          <Matrix>
            1 3
          </Matrix>
        </ContentValues>
      </Content>
    </Module>
    <IOPin id="WIDTH">
      <!-- The term reference used below indicates
           the display width in pixels -->
      <GetValue xsi:type="SemanticalDataRefType"
        semantics="urn:mpeg:mpeg21:2003:01-DIA-AdaptationQoSCS-NS:6.5.9.1"/>
```

```
    </IOPin>
    <IOPin id="HEIGHT">
      <!-- The term reference used below indicates
           the display height in pixels -->
      <GetValue xsi:type="SemanticalDataRefType"
        semantics="urn:mpeg:mpeg21:2003:01-DIA-AdaptationQoSCS-NS:6.5.9.2"/>
    </IOPin>
    <IOPin id="COLOR_CAPABLE">
      <!-- The term reference used below indicates
           whether the display is color capable -->
      <GetValue xsi:type="SemanticalDataRefType"
        semantics="urn:mpeg:mpeg21:2003:01-DIA-AdaptationQoSCS-
          NS:6.5.9.26"/>
    </IOPin>
    <IOPin id="SCALE"/>
    <IOPin id="NR_COLORS"/>
  </Description>
</DIA>
```

Example 8.6   Example BSDLink and AdaptationQoS description.

8.3.4 USE OF ADAPTATIONQOS WITH UCD

The above example demonstrates the use of AQoS when external values are mapped into a LUT to calculate the output parameters. Another possibility is to resolve a set of constraints expressed by the Universal Constraint Description (UCD) tool detailed in Section 8.5. The term 'constraint' here is to be understood as a system of mathematical equations to be resolved, for example, that the image width should be smaller than the display width.

Using this new paradigm, the example above could be rephrased as follows: maximize the image width and height subject to the condition that they remain smaller than the display width and height. The first part of this statement (maximize the image width) is called an *optimization constraint*, whereas the second part (subject to) is called a *limit constraint*. The use of UCDs allows the defining of cost functions or heuristics, for example, to find an optimal compromise between different solutions. Furthermore, they can be exchanged as stand-alone descriptors, that is, independently from AQoS, hence enabling more elaborate application scenarios. On the other hand, this set of constraints may not necessarily be resolvable in an analytical way and may require costly processing, for example by trying all possible solutions (in case the IOPins have discrete values).

8.4 LINKING BSD-BASED DIA TOOLS

The required components for BSD-based media resource (bitstream) adaptation were elaborated and illustrated in the previous sections. In this section, we describe a tool that is used to link the various information assets that are input to an adaptation engine, such as the original high-quality media resource, its BSD, and the corresponding transformation instructions. To ease the referencing of these assets, a tool referred to as the *Bitstream Syntax Description Link* (BSDLink) tool has been specified. The BSDLink tool references the following information assets:

- The requested bitstream, that is, the resource in which the User is interested.

- The Bitstream Syntax Description (BSD), which describes the high-level structure of the bitstream as illustrated in Section 8.2. It could be either a coding format–specific BS Description or a generic BS Description (gBSD). Hence, in the following, a BS Description refers to both.
- The BS Description transformation instructions, for example, formulated in XSLT or STX, and an unlimited number of parameters, which can be used to control/steer the BSD-based media resource adaptation.
- The steering description, which provides the actual values for the parameters. Currently, only one kind of explicit steering description is defined within MPEG-21 DIA, that is, AQoS (cf. Section 8.3). However, owing to its extensible design, new types of steering descriptions and parameters can be added very easily.

8.4.1 BSDLINK INSIGHTS

The BSDLink tool provides the facilities to link steering descriptions and BS Descriptions in a flexible and extensible way. The BSDLink description is essentially a wrapper for the reference to these descriptions, such that the actual descriptions could be defined within one document or distributed over several documents at possible different locations. These references are explained in the following text.

8.4.1.1 Reference to the Bitstream and BS Description

The reference to the bitstream could be included as an attribute of the BS Description (`bs1:bitstreamURI`). Owing to the fact that this attribute is optional, the URI of the bitstream could be also provided by the `Resource` element within the DID or by the MPEG-21–enabled application. To provide maximum flexibility, a URI reference to the bitstream has been introduced in the BSDLink as well. Furthermore, reference to the BS Description of the bitstream is made through a URI as described in Section 8.2.

8.4.1.2 Reference to the BSD Transformation and its Parameters

The BSD Transformation contains instructions on how to manipulate the BS Description according to the parameters specified in the BSDLink description. The BS Description is an XML document and is consequentially transformed using standard XML transformation languages, such as XSLT or STX, which in general have standardized namespaces. Therefore, the type of manipulation instructions is identified via the namespace and the reference to the actual style sheet. XSLT specifies the syntax and semantics of a language allowing the transformation of XML documents into other XML documents.

In practice, XSLT has been widely adopted by the industry [29] and open-source community [30,31]. However, one major drawback is that the XML document and the XSLT style sheet need to be fully stored in memory in order to apply the transformation. This might be a burden when using large documents, for example, BS Description of a full movie, or performing the transformation in constrained environments, for example, within the network or wireless environments. STX seems to be the suitable candidate for replacing XSLT in this kind of usage scenario because it is based on SAX, Simple API for XML [32], which uses events instead of the XSLT's tree structure [9]. The BSDLink, however, allows the specification of multiple BSD transformations implementing the same functionality, which could be used to provide XSLT and STX versions of the same style sheet. The application will select the appropriate transformation as needed. Additionally, this could be used to reference style sheets on different devices.

The usage of parameters facilitates the implementation of generic style sheets, which can be reused for different types of applications. Currently, two types of parameters are specified within the BSDLink tool, namely, constant parameters and IOPin parameters. The former are constant values, which could be used in conjunction with a steering description based on MPEG-21 DID Choice/Selection. However, they could be also used for providing constant values only without any further implications. The latter, IOPin parameters, are references to IOPins that are specified as part of the AQoS tool. These IOPins provide an interface to parameter values contained within an AQoS description. Thus, in this case, the steering description is an AQoS description.

8.4.1.3 Reference to the Steering Description

The extensible linking mechanism allows the design of a rich variety of adaptation archi-tectures, for example, steered by the AQoS tool, UED tools, MPEG-7 tools, or steered via the Choice/Selection mechanism provided by MPEG-21 DID. In the follow-ing text, two types of steering descriptions are introduced in more detail, namely AQoS descriptions and MPEG-21 DID Choice/Selection.

AdaptationQoS description: The AdaptationQoS description provides means for adaptation decision taking as described in Section 8.3. Example 8.6 defines two IOPins (SCALE and NR_COLORS), which provide decisions based on the usage environment of the Digital Item. The actual values of these IOPins are used to steer the transformation of the BS Description, which is then used for generating the adapted bitstream.

MPEG-21 DID Choice / Selection : Part 2 of the MPEG-21 standard provides the means for configuring a Digital Item (Chapter 3 on DID). This configuration information could be used to adapt the media resource based on the references to the information assets stored in the BSDLink description according to the User's choice. A DID enabling this kind of adaptation is shown in Example 8.7. The component for which the first selection is conditional (NO_B_VOPS) contains a reference to the original media resource, whereas in the second case, the Resource element contains a reference to a BSDLink description that contains information for dropping all B-frames of the actual resource[4]. In particular, this BSDLink description (BSDLinkDropAllBVOPs.xml) contains at least one reference to a BSD transformation that removes all BS Description elements describing B-VOPs in the actual media resource. This transformed BS Description is then used to generate the adapted resource as described in Sections 8.2.2.2 and 8.2.3.2.

---

[4] The second edition of the MPEG-21 DID standard allows the direct embedding of XML data within the Resource element whereby the reference to the BSDLink description becomes obsolete.

```
<did:DIDL>
    <did:Item>
        <did:Choice choice_id="B_VOP_DROPPING" maxSelections="1">
            <did:Selection select_id="NO_B_VOPS">
                <did:Descriptor>
                    <did:Statement mimeType="text/plain">
                        <DIA>
                            <Reference uri="dia.xml#CAN_DECODE_BFRAMES"/>
                        </DIA>
                    </did:Statement>
                </did:Descriptor>
            </did:Selection>
            <did:Selection select_id="ALL_B_VOPS">
                <did:Descriptor>
                    <did:Statement mimeType="text/plain">
                        <DIA>
                            <Reference uri="dia.xml#CANNOT_DECODE_BVOPS"/>
                        </DIA>
                    </did:Statement>
                </did:Descriptor>
            </did:Selection>
        </did:Choice>
        <did:Component>
            <did:Condition require="NO_B_VOPS"/>
            <did:Resource mimeType="video/MP4V-ES"
    ref="http://www.somewhere.at/lotr2.cmp"/>
        </did:Component>
        <did:Component>
            <did:Condition require="ALL_B_VOPS"/>
            <did:Resource mimeType="text/xml"
    ref="http://www.somewhere-else.at/BSDLinkDropAllBVOPs.xml"/>
        </did:Component>
    </did:Item>
</did:DIDL>

<!-- DIA Description "dia.xml" -->
<DIA>
    <Description xsi:type="UsageEnvironmentType" id="CAN_DECODE_BVOPS">
        <UsageEnvironmentProperty xsi:type="TerminalsType">
            <Terminal>
                <TerminalCapability xsi:type="CodecCapabilitiesType">
                    <Decoding xsi:type="VideoCapabilitiesType">
                        <Format
  href="urn:mpeg:mpeg7:cs:VisualCodingFormatCS:2001:3.3.2">
                            <mpeg7:Name xml:lang="en">
                                MPEG-4 Visual Advanced Simple Profile @ Level 1
                            </mpeg7:Name>
                        </Format>
                    </Decoding>
                </TerminalCapability>
            </Terminal>
        </UsageEnvironmentProperty>
    </Description>
    <Description xsi:type="UsageEnvironmentType" id="CANNOT_DECODE_BVOPS">
```

```
<UsageEnvironmentProperty xsi:type="TerminalsType">
    <Terminal>
        <TerminalCapability xsi:type="CodecCapabilitiesType">
            <Decoding xsi:type="VideoCapabilitiesType">
                <Format
href="urn:mpeg:mpeg7:cs:VisualCodingFormatCS:2001:3.1.2">
                    <mpeg7:Name xml:lang="en">
                        MPEG-4 Visual Simple Profile @ Level 1
                    </mpeg7:Name>
                </Format>
            </Decoding>
        </TerminalCapability>
    </Terminal>
</UsageEnvironmentProperty>
    </Description>
</DIA>
```

Example 8.7   The User's choice/selection as steering description.

8.4.2 DIGITAL ITEM ADAPTATION USING THE BSDLINK

The BSDLink tool offers the design and implementation of a rich variety of DIA-based adaptations. In this section, the means by which this adaptation is achieved using the BSDLink tool is explained and an example implementation using AQoS as steering description is described.

Adaptation based on the information referenced by the BSDLink tool is depicted in Figure 8.13. The first input Digital Item contains the BSDLink description, which carries the information to generate a degraded version of the resource based on the usage environment. The second input Digital Item comprises information about the User's context, in particular, information about the usage environment of the Digital Item described by UEDs (cf. Chapter 7) and possible further constraints described by means of UCDs (cf. Section 8.5). The output Digital Item contains only the adapted resource and its corresponding BS Description. In practice, however, the entire BSDLink description could become part of the output Digital Item in order to enable the possibility of multistep adaptation distributed over several heterogeneous devices. In this case, not only the BS Description is updated according to the adapted resource, but the steering description might also be updated accordingly. When using AQoS as a steering description, such an update may include the removal of the outdated adaptation possibilities from the specific modules.

The processing of the Digital Items shown in Figure 8.13 is as follows:

1. The BSDLink is extracted from the first input Digital Item, which provides references to the information assets required to generate an adapted version taking into account the User's context as described within the second Digital Item comprising UEDs and/or UCDs.

2. The steering description, for example AdaptationQoS, forms the input for the Adaptation Decision-taking Engine (ADTE) together with the UED and/or UCD of the second input Digital Item. The ADTE provides adequate decisions to transform the BS Description according to the UED/UCD. If an AdaptationQoS description is used for the steering description, the decisions from the ADTE are represented in the form of the

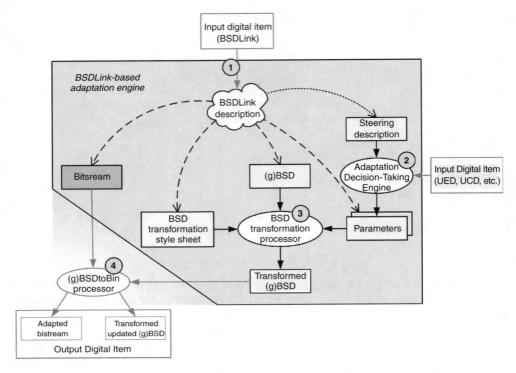

Figure 8.13 BSDLink-based adaptation

actual values of the variable parameters (IOPins) as defined in the BSDLink description. On the other hand, when the adaptation is steered by the User's `Choice/Selection`, the ADTE simply selects the appropriate BSDLink description for which the condition within the DID is evaluated to true. In this case, no explicit steering description is referenced through the BSDLink description. Furthermore, this BSDLink description contains constant parameter values, which provide input to the BSD transformation processor.

3. The BSD transformation processor transforms the BS Description according to the instructions within the BSD transformation style sheet taking into account the parameter values provided from the ADTE. The output of this processing step is a transformed BS Description, which is used to generate the adapted version of the bitstream that will satisfy the usage environment of the Digital Item.

4. The behavior of the process that generates a bitstream based on its BS Description is normatively specified within DIA and described in detail in Sections 8.2.2.2 and 8.2.3.2. The input to this process is the bitstream, which is referenced in the BSDLink, and the transformed BS Description, which is produced within the BSDLink-based adaptation process as described earlier. Basically, the output comprises the adapted bitstream only. However, to enable further adaptations using BS Descriptions, the transformed BS Descriptions become outdated and need to be updated according to the

adapted version of the bitstream. One possibility would be updating of the transformed BS Description during the bitstream generation.

8.4.2.1 Usage Example for the BSDLink

The previous sections presented a general overview about the BSDLink tool and how it can be used. In this section, a concrete implementation of the BSDLink tools is given. Therefore, a MPEG-4 Elementary Stream is adapted according to the decoding capabilities of the rendering device. The steering description is represented by an AQoS description, which provides the information if bidirectional encoded VOPs (also known as B-VOPs) should be removed according to the decoding capabilities. The following DIDs and DIA descriptions are used within this section:

- The first input Digital Item comprises a BSDLink description containing references to the following information assets:
  - The AdaptationQoS description as steering description
  - The reference to the actual bitstream
  - The gBS Description, which describes the syntax of the bitstream
  - The BSD transformation and parameters.
- The second input Digital Item describes the User's context in terms of decoding capabilities by means of the UED tools, that is, terminal capabilities.
- The transformed and updated gBS Description representing the adapted bitstream.

A UED describing the decoding capabilities of the receiving terminal is shown in Example 7.2 of Chapter 7. The decoding capabilities are specified by means of a reference to an MPEG-7 classification scheme term using its Universal Resource Name (URN). The term referenced by this URN specifies the MPEG-4 Simple Profile @ Level 1 format as specified in the MPEG-4 Visual standard. This UED provides the description of the usage environment of the requested Digital Item.

Example 8.8 shows the DID from the requested Digital Item that contains the BSDLink description as described above. There are two references to BSD transformations present that provide equivalent functionality but different implementations, that is, one BSD transformation is written using XSLT, whereas the other is written using STX. The type of transformation is indicated by their namespace within the `type` attribute. The actual parameter value for the 'removeBVOPs' parameter is retrieved from the AQoS description indicated by the `uri` attribute of the SteeringDescriptionRef element. In particular, the input/output pin with the id 'REMOVE_BVOPS' is asserted with the actual value during the adaptation decision-taking process.

The AQoS description, which acts as the steering description, is shown in Example 8.9. The input IOPin with the CODING_FORMAT id is characterized by means of a semantic data reference to a term as defined within the AQoS classification scheme. It represents the possible supported video decoding capabilities of the receiving terminal. The AQoS module, namely, a Look-Up Table, associated with this IOPin returns a Boolean value indicating if B-VOPs of the resource should be dropped or not with respect to the decoding capabilities of the receiving terminal. The IOPin with the REMOVE_BVOPS id provides this information.

```
<did:DIDL>
  <did:Item><did:Component><did:Descriptor><did:Statement
     mimeType="text/xml">
    <DIA>
     <Description xsi:type="BSDLinkType">
        <SteeringDescriptionRef uri="AQoS_AKIYO.xml"/>
        <BitstreamRef uri="rtsp://www.somewhere.at/AKIYO.cmp"/>
        <BSDRef uri="AKIYO_gBSD.xml"/>
        <BSDTransformationRef uri="mp4.xsl"
    type="http://www.w3.org/1999/XSL/Transform"/>
        <BSDTransformationRef uri="mp4.stx"
    type="http://stx.sourceforge.net/2002/ns"/>
        <Parameter xsi:type="IOPinRefType" name="removeBVOPs">
          <Value>REMOVE_BVOPS</Value>
        </Parameter>
      </Description>
    </DIA>
  </did:Statement></did:Descriptor></did:Component></did:Item>
</did:DIDL>
```

Example 8.8 BSDLink Description.

```
<DIA>
  <DescriptionMetadata>
    <ClassificationAlias alias="AQoS"
  href="urn:mpeg:mpeg21:2003:01-DIA-AdaptationQoSCS-NS"/>
  </DescriptionMetadata>
  <Description xsi:type="AdaptationQoSType">
    <Module xsi:type="LookUpTableType">
      <Axis iOPinRef="CODING_FORMAT">
        <AxisValues xsi:type="NMTokenVectorType">
          <Vector>urn:mpeg:mpeg7:cs:VisualCodingFormatCS:2001:3.1
             urn:mpeg:mpeg7:cs:VisualCodingFormatCS:2001:3.3</Vector>
        </AxisValues>
      </Axis>
      <Content iOPinRef="REMOVE_BVOPS">
        <ContentValues xsi:type="BooleanVectorType">
          <Vector>true false</Vector>
        </ContentValues>
      </Content>
    </Module>
    <IOPin id="CODING_FORMAT">
    <!-- Video Codec (decoding) Capabilities -->
      <GetValue xsi:type="SemanticalDataRefType"
        semantics=":AQS:6.5.4.6"/>
    </IOPin>
    <IOPin id="REMOVE_BVOPS"/>
  </Description>
</DIA>
```

Example 8.9 Steering Description (AdaptationQoS).

The fragment of a gBS Description as shown in Example 8.10 describes the syntax of the MPEG-4 Visual bitstream, for the Akiyo sequence. Therefore, the start and length of the Video Object (VO), Video Object Layer (VOL), and Video Object Planes (VOP) is represented in several `gBSDUnit` elements. The gBS Description of the entire bitstream looks very similar to the fragment shown in the following example, that is, it just contains more `gBSDUnit` elements.

The XSLT style sheet that transforms the gBS Description from Example 8.10 according to the parameter value retrieved from the ADTE is shown in Example 8.11. The parameter is represented by the `xsl:param` element and its current default value is set to true but can be superseded by the parameter value from the ADTE. Furthermore, the XSLT style sheet contains two templates. The processing of the XSLT style sheet is specified in [4]. The first template matches all nodes and attributes and copies them into the result tree. The second template matches `gBSDUnit` elements that have a `syntacticalLabel` attribute. If the parameter value is set to true and the `gBSDUnit` element represents a B-VOP, then this element is not copied from the source tree to the result tree, that is, it is removed from the gBS Description. Thereafter, the transformed gBS Description is used as input to the gBSDtoBin parser, which generates the adapted version of the bitstream as described in Section 8.2.3.2.

Additionally, the gBSDtoBin parser updates the transformed gBS Description as already described in Section 8.2.3.2. Finally, we note that in [33], an implementation of an adaptation engine using DIA tools around the BSDLink tool has been proposed.

```
<dia:DIA>
  <DescriptionMetadata>
    <ClassificationAlias alias="M4V"
  href="urn:mpeg:mpeg4:visual:cs:syntacticalLabels"/>
  </DescriptionMetadata>
  <dia:Description xsi:type="gBSDType" addressUnit="byte"
  addressMode="Absolute">
    <gBSDUnit syntacticalLabel=":M4V:VO" start="0" length="4"/>
    <gBSDUnit syntacticalLabel=":M4V:VOL" start="4" length="14"/>
    <gBSDUnit syntacticalLabel=":M4V:I_VOP" start="18" length="4641"/>
    <gBSDUnit syntacticalLabel=":M4V:P_VOP" start="4659" length="98"/>
    <gBSDUnit syntacticalLabel=":M4V:B_VOP" start="4757" length="16"/>
    <gBSDUnit syntacticalLabel=":M4V:B_VOP" start="4773" length="23"/>
    <gBSDUnit syntacticalLabel=":M4V:P_VOP" start="4796" length="178"/>
    <gBSDUnit syntacticalLabel=":M4V:B_VOP" start="4974" length="53"/>
    <gBSDUnit syntacticalLabel=":M4V:B_VOP" start="5027" length="39"/>
    <gBSDUnit syntacticalLabel=":M4V:P_VOP" start="5066" length="235"/>
    <gBSDUnit syntacticalLabel=":M4V:B_VOP" start="5301" length="66"/>
    <gBSDUnit syntacticalLabel=":M4V:B_VOP" start="5367" length="65"/>
    <gBSDUnit syntacticalLabel=":M4V:P_VOP" start="5432" length="274"/>
    <gBSDUnit syntacticalLabel=":M4V:B_VOP" start="5706" length="74"/>
    <gBSDUnit syntacticalLabel=":M4V:B_VOP" start="5780" length="68"/>
    <gBSDUnit syntacticalLabel=":M4V:I_VOP" start="5848"
    length="4670"/>
    <!--... and so on ...-->
  </dia:Description>
</dia:DIA>
```

Example 8.10 gBS Description fragment.

```
<xsl:stylesheet
  xmlns="urn:mpeg:mpeg21:2003:01-DIA-gBSD-NS"
  xmlns:gbsd="urn:mpeg:mpeg21:2003:01-DIA-gBSD-NS"
  xmlns:dia="urn:mpeg:mpeg21:2003:01-DIA-NS"
  mlns:xsl="http://www.w3.org/1999/XSL/Transform" version="1.0">
  <xsl:output method="xml" version="1.0" encoding="UTF-8"
    indent="yes"/>
  <xsl:param name="removeBVOPs" select="'true'"/>
  <!-- Match all: default template -->
  <xsl:template name="tplAll" match="@*| node()">
    <xsl:copy>
      <xsl:apply-templates select="@*| node()"/>
    </xsl:copy>
  </xsl:template>
  <!-- Match gBSDUnit: removes B_VOPs -->
  <xsl:template name="tplB_VOP"
   match="gbsd:gBSDUnit[@syntacticalLabel]">
    <xsl:choose>
      <xsl:when test="$removeBVOPs and @syntacticalLabel=':M4V:B_VOP'">
        <!-- Do nothing -->
      </xsl:when>
      <xsl:otherwise>
        <xsl:copy>
          <xsl:apply-templates select="node()| @*"/>
        </xsl:copy>
      </xsl:otherwise>
    </xsl:choose>
  </xsl:template>
</xsl:stylesheet>
```

Example 8.11   XSLT style sheet for B-VOP dropping.

8.5 UNIVERSAL CONSTRAINTS DESCRIPTION TOOL

To enable Users to further constrain the usage of a Digital Item, the Universal Constraints Description (UCD) tool has been specified. With this tool, it is possible to describe various types of constraints that impact the adaptation process [34]. End users or their application may, for example, constrain the resolution of the rendering device in order to satisfy the needs of the application itself or the display size. On the other hand, Digital Item providers may want to restrict the way that their media resources are adapted, for example, a minimum level of quality may be imposed. Besides specifying such constraints, several optimization criteria are also specified to guide the adaptation decision.

8.5.1 ADAPTATION DECISION-TAKING ARCHITECTURES USING MODES

Adaptation of a Digital Item using the UCD tool is based on decisions with regard to certain variables defined within the universal constraints, that is, an instance of the UCD tool. The actual values of these decisions could be related to the Digital Item itself or the usage environment of the Digital Item. In the former case, the possible values to which the constraints are applied could be obtained from appropriate MPEG-7 descriptions or the AQoS descriptions as described in Section 8.3. In the latter case, the UCD is used to further constrain UEDs (cf. Chapter 7). From this, it follows that an instance of the

UCD tool can travel both upstream from the consumer to an adaptation engine, which is referred to as push mode, and downstream from the content provider, which is referred to as pull mode. The hybrid mode combines the push and pull mode, that is, both UCDs are jointly used to provide adequate decisions for the adaptation process.

8.5.1.1 Push Mode

The decision-taking architecture where the UCD tool is used to further constrain the usage of the Digital Item is referred to as push mode. In this mode, the Digital Item provider would define the UCD together with the AQoS description, which can be used to describe trade-offs among different adaptation possibilities. A possible decision-taking architecture based on the push mode is depicted in Figure 8.14.

The AQoS description together with the UCD forms the input for the Adaptation Decision-Taking Engine (ADTE) at the provider side, whereas the consumer side provides the UEDs. The UCD could also reference data from the UED by means of the XPointer framework as a result of a prior negotiation. The output of the ADTE should be the optimal decision for the adaptation process under the given constraints. For example, frames of a movie may be dropped owing to the current network conditions; however, this operation should not make the resulting quality fall below a certain threshold as defined by the provider within the AQoS description.

8.5.1.2 Pull Mode

The pull mode covers the case where the Digital Item consumer wants to further constrain the usage environment of the Digital Item. Specifically, further constraints on User characteristics, terminal capabilities, network characteristics or natural environment characteristics may be imposed. The UCD may reference data contained in the UED; however, it may not originate from the same physical location. Furthermore, the UCD is capable of describing other constraints that are not covered by UED. This is achieved through an extensible design that allows semantic terms as defined by a classification scheme to be assigned to values that are part of the UCD expression. Figure 8.15 shows a possible decision-taking architecture, where the UCD is coming from the consumer side.

The primary advantage of this mode is the full flexibility of the DI consumer to pull the DI in a more controlled manner. For example, it is possible for one DI consumer to

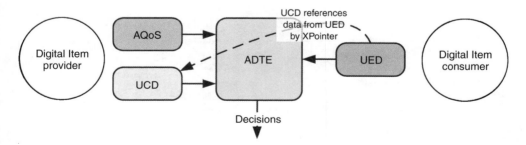

Figure 8.14   UCD-based push architecture for decision taking [1]. Copyright ISO. Reproduced by permission

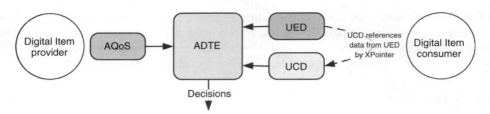

Figure 8.15 UCD-based pull architecture for decision taking [1]. Copyright ISO. Reproduced by permission

render a visual resource on their screen at a smaller resolution than that of the display, while another DI consumer displays the qualitatively best image with the given available bandwidth.

8.5.1.3 Hybrid Mode

In this hybrid mode, which is shown in Figure 8.16, the UCD may go downstream with the AQoS description (from the DI provider) and come upstream (from the DI consumer). The ADTE jointly considers both UCDs to provide an optimal decision. The hybrid mode preserves the advantages of the push and pull mode while eliminating their disadvantages. In practice, however, if both the provider and consumer side UCDs contain optimization constraints, then further knowledge may be needed by the ADTE to process these constraints (cf. Section 8.5.3).

8.5.2 *TYPES OF CONSTRAINTS*

In the previous subsections, several types of applications for the UCD tool have been discussed. The following subsections describe more details of the tool and show how decisions are made by means of concrete examples.

The UCD tool is based on establishing a mathematical abstraction where constraints are specified on variables representing resource and/or environment characteristics using values obtained either externally by using the XML XPath Language (XPath) [14] or by a direct specification of numeric constants. Additionally, when used in conjunction with the AQoS tool, these constraints are applied to IOPins in AQoS. Otherwise, they are applied to resource or usage environment characteristics indicated by their semantics.

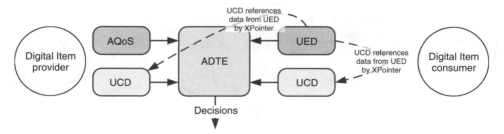

Figure 8.16 UCD-based hybrid architecture for decision taking [1]. Copyright ISO. Reproduced by permission

The adaptation constraints can be specified not only on the resource as a whole, but also differentiated with respect to individual units of the resource corresponding to logical partitions such as group-of-pictures (GOPs), region-of-interests (ROIs), tiles, frames, and so on. Such units are referred to as adaptation units.

8.5.2.1 Limitation Constraints

Constraints that restrict the solution space of a set of possible adaptation decisions are referred to as *limitation constraints*. Within MPEG-21 DIA, they are expressed by means of mathematical expressions in stack form using XML syntax. These mathematical expressions are processed similarly to the Reverse Polish Notation (RPN) as shown in Figure 8.17. It must be noted that limitation constraints have to be evaluated to either true or false.

An expression, as shown in Figure 8.17, is serialized in terms of arguments and operators into its XML form as specified within MPEG-21 DIA. Arguments are subsequently pushed on the stack until an operator appears. The operator will be applied on the arguments as specified in the standard, that is, unary operators pop one element from the stack, perform the calculations, and push the result back on the stack. Binary operators are processed in a similar way. After evaluating the expression, the final result remains on the stack.

8.5.2.2 Optimization Constraints

Limitation constraints restrict the solution space of possible decisions. In practice, however, this solution space could be still very huge and in most cases the users are more or

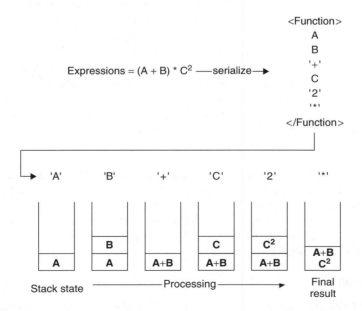

Figure 8.17 Base stack function processing [1]. Copyright ISO. Reproduced by permission

less interested in getting the qualitatively best resources. Therefore, one or more optimization constraints can be formulated within one UCD in order to provide optimal decisions for the adaptation of Digital Items. The next section provides a detailed overview on how decisions are made on the basis of the UCDs.

8.5.3 UCD-BASED DECISION TAKING

UCD-based decision taking is performed on variables with specific semantics that are related to resource or usage environment characteristics. In this context, the set of variables is denoted by the vector $I = \{i_0, i_1, \ldots, i_n\}, n \in N$. The purpose of the UCD is to specify an optimization problem involving these variables.

The UCD conveys numeric expressions $O_j(I), j = 0, 1, \ldots, n; n \in N$ called as *optimization constraints*, along with several Boolean expressions $L_k(I); k = 0, 1, \ldots, m; m \in N$ called as *limitation constraints*, which are used together to specify the following optimization problem involving I:

$$Maximize\ or\ Minimize\quad O_j(I), j = 0, 1, \ldots, n\quad subject\ to:$$

$$L_k(I) = true, k = 0, 1, \ldots, m$$

Let I^* represent a solution to the above problem that maximizes or minimizes expressions $O_j(I)$ subject to expressions $L_k(I)$ evaluating to `true`. The syntax of the UCD allows any number of optimization constraints to be specified.

- For $n = 0$, that is, no optimization constraints are specified, any solution I^* in the feasible solution space where the limitation constraints evaluate to `true` is acceptable.
- For $n = 1$, that is, a single optimization constraint is specified, a single objective optimization problem is defined, where a solution that maximizes or minimizes the only optimization metric within the feasible solution space is acceptable.
- For $n > 1$, that is, a multicriteria (also known a multiobjective) optimization problem is defined [35], any solution I^* in the Pareto optimal set included in the feasible region is acceptable. In multicriteria optimization literature, a set of points in the feasible region is said to be Pareto optimal if in moving from any point in the set to another in the feasible region, any improvement in one of the optimization metrics from its current value would cause at least one of the other optimization metrics to deteriorate from its current value. In other words, this is the set of best solutions that could be achieved without affecting at least one metric.

The above approach is applied on a single adaptation unit. However, if multiple adaptation units need to be considered, this approach is repeated for each adaptation unit. In some cases, the specification of dependencies on past decisions is desired to apply the decision of the first adaptation unit to all following adaptation units. Therefore, a special attribute within the SF is specified that indicates the number of subsequent adaptation units for which the same decision is valid.

8.5.3.1 Examples

The mathematical expressions are represented using the base SF as mentioned in Section 8.5.2.1.

The first example of a limitation constraint as shown in Example 8.12 describes the following mathematical expression: FrameWidth $\geq$ 0.75 $\times$ DisplayHorizontalResolution. This example has been extracted from a JPEG2000 use case where the user (or the application of the user) wants to restrict the space for rendering the image to 75 % of the horizontal resolution of the terminal. The limitation constraint using the vertical resolution of the terminal looks similar to this one.

Example 8.13 shows a case where the Digital Item provider constrains the possible adaptation of a JPEG2000 image with the help of an AQoS description. In particular, the frame width of the image should be greater or equal to the constant value 640. The possible values for the frame width are referenced through an IOPin within the AQoS description.

The optimization constraint as shown in Example 8.14 defines that the frame rate as described with an associated AQoS description should be maximized within the given solution space resulting from the limitation constraints.

Example 8.15 shows a complete UCD example for the JPEG2000 use case. This example is divided into constraints formulated by the DI consumer and the DI provider.

```
<LimitConstraint>
  <Argument xsi:type="SemanticalRefType"
     semantics="urn:mpeg:mpeg21:2003:01-DIA-MediaInformationCS-NS:17"/>
  <Argument xsi:type="ExternalIntegerDataRefType"
  uri="my_UED.xml#xmlns(dia=urn:mpeg:mpeg21:2003:01-DIA-
     NS)xpointer(//dia:DisplayCapability/dia:Resolution/@horizontal)"/>
  <Argument xsi:type="ConstantDataType">
    <Constant xsi:type="FloatType">
      <Value>0.75</Value>
    </Constant>
  </Argument>
  <!-- Multiply operation-->
  <Operation
  operator="urn:mpeg:mpeg21:2003:01-DIA-StackFunctionOperatorCS-NS:18"/>
  <!-- Bool IsLessThanOrEqualTo operation-->
  <Operation
  operator=" urn:mpeg:mpeg21:2003:01-DIA-StackFunctionOperatorCS-NS:38"/>
</LimitConstraint>
```

Example 8.12   Limitation constraints example.

```
<LimitConstraint>
  <Argument xsi:type="ExternalIOPinRefType" iOPinRef="#FRAME_WIDTH"/>
  <Argument xsi:type="ConstantDataType">
    <Constant xsi:type="FloatType">
      <Value>640</Value>
    </Constant>
  </Argument>
  <!-- Bool IsGreaterThanOrEqualTo operation-->
  <Operation
  operator="urn:mpeg:mpeg21:2003:01-DIA-StackFunctionOperatorCS-NS:39"/>
</LimitConstraint>
```

Example 8.13   Limitation constraints example.

```
<OptimizationConstraint optimize="maximize">
  <Argument xsi:type="ExternalIOPinRefType" iOPinRef="#FRAMERATE"/>
</OptimizationConstraint>
```

Example 8.14   Optimization constraints example.

Furthermore, it contains a possible AQoS description and the resulting decisions. The Digital Item consumer constraint provides limitation constraints that restrict the usage of the Digital Item to 75 % of the display's resolution. On the other hand, the Digital Item provider constraint limits a possible adaptation of the Digital Item to 20 % of the display's resolution. Additionally, the resolution should be maximized which is expressed by appropriate optimization constraints.

The AQoS description provides a scaling factor for given resolutions of a JPEG2000 resource within the Digital Item. Let us assume the following example use case: The Digital Item consumer wants to view the JPEG2000 resource on a device with a display resolution of 320×240 pixels. If we apply the DI consumer constraint, the resource needs to be adapted to a resolution of 240×180 pixels (75 %) or smaller. The DI provider constraints are applied in the same way, which results in a minimum possible resolution of 64×48 pixels (20 %). According to the AQoS description, only two possibilities remain in the feasible solution space: 160×120 and 80×60 pixels. Thereafter the optimization constraints are applied, that is, resolution should be maximized which results in a horizontal resolution of 160 pixels and a vertical resolution of 120 pixels, respectively. From this it follows that the scaling factor for the JPEG2000 resource is 2. This scaling factor provides the input for the resource adaptation engine, which adapts the actual resource according to the constraints specified by means of the UCD tool.

8.6 CONCLUDING REMARKS

In this chapter, we have described tools for creating high-level descriptions of the bitstream syntax to achieve format-independent adaptation, as well as tools that assist in making trade-offs between feasible adaptation operations and the constraints.

It should also be noted that while DIA is an essential part in moving closer to the MPEG-21 vision, there still exists some gaps in the overall communication between Users that will need to be addressed, including how descriptions are negotiated, transported and exchanged. Typically, other standardization forums handle protocol-like components such as this, that is, the Internet Engineering Task Force (IETF) in the networking space and the Digital Video Broadcasting project (DVB), the Association of Radio Industries & Businesses (ARIB), and the Advanced Television Systems Committee (ATSC) in the broadcasting space. It is expected that MPEG will need to collaborate with these bodies and forums on such endeavors.

Finally, in terms of research, there are a variety of interesting topics that are still quite open with regard to adaptation as we expand from content-based processing to context-based processing. For example, while the AQoS tools provide a good framework to enable decision taking, the optimal adaptation strategies in an environment-aware setting are not completely solved. Among other things, this involves being able to model the quality of the overall adaptation, including signal-level audiovisual adaptations, as well as semantic-level adaptations, and coming up with efficient algorithms to determine an optimal one.

```
<!-- Digital Item consumer constraint -->
<DIA>
 <Descirption xsi:type="UCDType">
  <AdaptationUnitConstraint><LimitationConstraint>
    <Argument xsi:type="SemanticalRefType" semantics="MediaInfo:17"/>
    <Argument xsi:type="SemanticalRefType" semantics="AQoS:6.5.9.1"/>
      <Constant xsi:type="FloatType"><Value>0.75</Value></Constant>
    </Argument>
    <!-- Multiply operation-->
    <Operation operator=":StackFunctionOperator:18"/>
    <!-- Bool IsLessThanOrEqualTo operation-->
    <Operation operator=":StackFunctionOperator:38"/>
  </LimitationConstraint>
  <LimitationConstraint><!-- same for frame height and vertical resolution
      respectively --></LimitationConstraint></AdaptationUnitConstraint>
 </Description>
</DIA>

<!-- Digital Item provider constraint -->
<DIA>
 <Descirption xsi:type="UCDType">
  <AdaptationUnitConstraint><LimitationConstraint>
   <Argument xsi:type="InternalIOPinRefType" iOPinRef="WIDTH"/>
   <!-- display's horizontal resolution -->
   <Argument xsi:type="SemanticalRefType" semantics=":AQoS:6.5.9.1"/>
   <Argument xsi:type="ConstantDataType">
    <Constant xsi:type="FloatType"><Value>0.20</Value></Constant>
   </Argument>
   <!-- Multiply operation -->
   <Operation operator=":SFO:18"/>
   <!-- Bool IsGreaterThanOrEqualTo operation -->
   <Operation operator=":SFO:39"/>
  </LimitationConstraint>
  <LimitationConstraint>
  <!-- same for frame height and vertical resolution respectively -->
  </LimitationConstraint></AdaptationUnitConstraint>
  <OptimizationConstraint optimize="maximize">
    <Argument xsi:type="SemanticalRefType" semantics="MediaInfo:17"/>
  </OptimizationConstraint>
  <OptimizationConstraint optimize="maximize">
    <Argument xsi:type="SemanticalRefType" semantics="MediaInfo:18"/>
  </OptimizationConstraint>
 </Descirption>
</DIA>

<!-- AdaptationQoS description -->
 cf. Example 8.6

<!-- Possible decisions within the given solution space -->
 SCALE="2"
```

Example 8.15   Complete UCD example.

There is also the notion of attempting to gauge the overall experiences of a User [36], which significantly complicates the problem. These issues represent challenging research topics in the domain of multimedia content adaptation for the coming years.

8.7 ACKNOWLEDGMENTS

This chapter summarizes the work and ideas of many people who have actively participated in the process of developing requirements, proposing technology and cooperating to help stabilize and mature the specification. Many thanks are due to all of those who have been involved and helped to bring this part of the MPEG-21 standard to completion.

REFERENCES

[1] ISO/IEC 21000-7:2004(E), Information Technology – Multimedia Framework – Part 7: Digital Item Adaptation, 2004.

[2] A. Vetro and C. Timmerer, "Digital Item Adaptation: Overview of Standardization and Research Activities", *IEEE Transactions on Multimedia*, vol. **7**, no. 3, June 2005, pp. 148–426.

[3] C. Timmerer and H. Hellwagner, "Interoperable Adaptive Multimedia Communication", *IEEE Multimedia Magazine*, vol. **12**, no. 1, January-March 2005, pp. 74–79.

[4] "XSL Transformations (XSLT) Version 1.0", World Wide Web Consortium (W3C) Recommendation, Available at http://www.w3.org/TR/xslt, November 1999.

[5] S. Devillers, C. Timmerer, J. Heuer, and H. Hellwagner, "Bitstream Syntax Description-Based Adaptation in Streaming and Constrained Environments", *IEEE Transactions on Multimedia*, vol. **7**, no. 3, June 2005, pp. 463–470.

[6] "Extensible Markup Language (XML) 1.0 (3rd Edition)", World Wide Web Consortium (W3C) Recommendation, Available at http://www.w3.org/TR/2004/REC-xml-20040204, February 2004.

[7] "JPEG2000 Part I", ITU-T Rec. T.800| ISO/IEC 15444-1:2000, 2000.

[8] P. Cimprich, "Streaming Transformations for XML (STX) Version 1.0", Working Draft, Available at http://stx.sourceforge.net/documents/, May 2003.

[9] M. Kay, "XSLT : Programmer's Reference", Wrox, 2nd edition, May 2001.

[10] O. Becker, "Streaming Transformations for XML-STX", *Proc. XMIDX 2003 – XML Technologies for Middleware*, Berlin, pp. 83–88, February 2003.

[11] "XML Schema Part 0: Primer", World Wide Web Consortium (W3C) Recommendation, Available at http://www.w3.org/TR/xmlschema-0, May 2001.

[12] "XML Schema Part 1: Structures", World Wide Web Consortium (W3C) Recommendation, Available at http://www.w3.org/TR/xmlschema-1, May 2001.

[13] "W3C, XML Schema Part 2: Datatypes", World Wide Web Consortium (W3C) Recommendation, Available at http://www.w3.org/TR/xmlschema-2, May 2001.

[14] "XML Path Language (XPath)", Version 1.0 World Wide Web Consortium (W3C) Recommendation, Available at http://www.w3.org/TR/xpath, November 1999.

[15] ISO/IEC 15938-1:2002(E), Information technology – Multimedia Content Description Interface – Part 1: Systems, 2002.

[16] "XML Base", World Wide Web Consortium (W3C) Recommendation, Available at http://www.w3.org/TR/xmlbase/, June 2001.

[17] "Abstract Syntax Notation One (ASN.1)", ITU Recommendation X.208, November 1988.

[18] A. Eleftheriadis, "Flavor: A Language for Media Representation", Proceedings *ACM Multimedia, '97 Conference*, Seattle, WA, November 1997, pp. 1–9.

[19] ISO/IEC 14496-1:2001(E) Information Technology – Coding of Audio-Visual Objects – Part 1: Systems, 2001.

[20] G. Panis, A. Hutter, J. Heuer, H. Hellwagner, H. Kosch, C. Timmerer, S. Devillers, and M. Amielh, "Bitstream Syntax Description: A Tool for Multimedia Resource Adaptation within MPEG-21", *EURASIP Signal Processing: Image Communication, Special Issue on Multimedia Adaptation*, vol **18**. No. 8, September 2003, pp. 721–747.

[21] C. Timmerer, G. Panis, H. Kosch, J. Heuer, H. Hellwagner, and A. Hutter, "Coding format independent multimedia content adaptation using XML", *Proc. SPIE Int'l Symp ITCom 2003 – Internet Multimedia Management Systems IV*, vol. **5242**, Orlando, USA, September 2003.

[22] D. Jannach, K. Leopold, H. Hellwagner and C. Timmerer, "A Knowledge Supported Approach for Multi-Step Media Adaptation", *Proc., 5th Int'l Workshop on Image Analysis for Multimedia Interactive Services (WIAMIS)*, Lisboa, Portugal, April 2004.

[23] D. Mukherjee, E. Delfosse, J.-G. Kim, and Y. Wang, "Optimal Adaptation Decision-Taking for Terminal and Network Quality of Service", *IEEE Transactions on Multimedia*, vol. **7**, no. 3, June 2005, pp. 454–462.

[24] C.L. Hamblin, "Computer Languages", *The Australian Journal of Science*, vol. **20**, 1957, pp. 135–139; reprinted in *The Australian Computer Journal*, vol. **17**, no. 4, November 1985, pp. 195–198.

[25] G. Lafruit, E. Delfosse, R. Osorio, W. van Raemdonck, V. Ferentinos, and J. Bormans, "View-Dependent, Scalable Texture Streaming in 3-D QoS With MPEG-4 Visual Texture Coding", *IEEE Transactions on Circuits and Systems for Video Technology*, vol. **14**, no. 7, August 2004, pp. 1021–1031.

[26] T. Berners-Lee, R. Fielding and L. Masinter, "Uniform Resource Identifiers (URI): Generic Syntax (RFC 2396)", Available at http://www.ietf.org/rfc/rfc2396.txt, August 1998.

[27] "XPointer Framework", World Wide Web Consortium (W3C) Recommendation, Available at http://www.w3.org/TR/xptr-framework, March 2003.

[28] "XPointer xpointer() Scheme", World Wide Web Consortium (W3C) Working Draft, Available at http://www.w3.org/TR/xptr-xpointer, December 2002,

[29] "Microsoft XML Core Services", Available at http://msdn.microsoft.com/xml, accessed 2005.

[30] "Xalan: XSL stylesheet processors in Java & C++", Available at http://xml.apache.org, accessed 2005.

[31] "SAXON: The XSLT and XQuery Processor", Available at http://saxon.sourceforge.net, accessed 2005.

[32] "Simple API for XML", Available at http://www.saxproject.org, accessed 2005.

[33] S. Hsiang, D. Mukherjee, and S. Liu, "Fully Format-Independent Adaptation Use Cases Using DIA SoFCD Tools", ISO/IEC JTC1/SC29/WG11 M10411, Waikoloa, December 2003.

[34] D. Mukherjee, G. Kuo, S. Liu and G. Beretta, "Proposal for a Universal Constraints Descriptor in DIA", ISO/IEC JTC1/SC29/WG11 M9772, Trondheim, July 2003.

[35] R.E. Steuer, "Multiple Criteria Optimization: Theory, Computation and Application", Krieger Publishing Company, August 1996.

[36] F. Pereira and I. Burnett, "Universal Multimedia Experiences for Tomorrow," *IEEE Signal Processing Magazine*, vol. **20**, March 2003, pp. 63–73.

9

Digital Item Processing

Frederik De Keukelaere and Gerrard Drury

9.1 INTRODUCTION

In Chapter 3, it was discussed how Digital Items (DIs) are declared using the Digital Item Declaration Language (DIDL) [1]. The DIDL is an Extensible Markup Language (XML)-based language [2] using the different constructs from the Digital Item Declaration (DID) model to create interoperable declarations of DIs. This chapter discusses how Digital Item Processing (DIP) [3, 4] can be used to further realize the MPEG-21 Multimedia framework [5].

To know the real reasoning behind the concept of DIP it is necessary to look at the basic thing that DIP acts upon: a DI. A DI is declared according to the semantics of the DID abstract model using the DIDL. One of the consequences of using DIDL is that the resulting DIDL documents are static. Static, because they only declare a DI and do not express how to actually process this declaration, and hence, the DI. Consider an example DID to gain an understanding of the different decisions that might be made when processing DIs declared in the DIDL.

In Example 9.1, a DIDL document is presented. This DID contains two different versions of a music track from a music album. The first version of the track is encoded in mp3 format [6], the second version of the track is encoded in the wma format [7]. The DID also includes a choice allowing the User of the DID to choose between both formats.

While this DID is fairly simple, there are already different ways the DIDL document can be processed by an MPEG-21 Peer, which is defined in [8] as any device or application that compliantly processes an MPEG-21 DI. Two possible ways of handling a DI, compliant to the semantics of the DID specification, are outlined below:

- An MPEG-21 Peer searches through the DID for the `Choices` (note that this formatting is used throughout this chapter for XML elements and attributes). The Peer detects the `Choice` and because it is capable of playing both the mp3 format and the wma format it randomly selects one of the two formats. Suppose now the Peer selected

The MPEG-21 Book   Ian S Burnett, Fernando Pereira, Rik Van de Walle, Rob Koenen
© 2006 John Wiley & Sons, Ltd

```
<?xml version="1.0" encoding="UTF-8"?>
<DIDL xmlns="urn:mpeg:mpeg21:2002:02-DIDL-NS">
  <Item>
    <Choice>
      <Descriptor>
        <Statement mimeType="text/plain">
          Please make a choice.
        </Statement>
      </Descriptor>
      <Selection select_id="MP3_FORMAT">
        <Descriptor>
          <Statement mimeType="text/plain">MP3 Format</Statement>
        </Descriptor>
      </Selection>
      <Selection select_id="WMA_FORMAT">
        <Descriptor>
          <Statement mimeType="text/plain">WMA Format</Statement>
        </Descriptor>
      </Selection>
    </Choice>
    <Component>
      <Condition require="MP3_FORMAT"/>
      <Resource mimeType="audio/mp3" ref="http://server/track01.mp3"/>
    </Component>
    <Component>
      <Condition require="WMA_FORMAT"/>
      <Resource mimeType="audio/wma" ref="http://server/track01.wma"/>
    </Component>
  </Item>
</DIDL>
```

Example 9.1   A static DI Declaration using the DID building blocks.

the wma format. After resolving the Choice, the MPEG-21 Peer searches for the available Components, that is, the second Component, and plays the Resource that is contained within the Component, that is, the Resource located at http://server/track01.wma.

- An MPEG-21 Peer searches through the DID for the Choices. The Peer detects the Choice and because it is capable of playing both the mp3 format and the wma format it selects both formats. After resolving the Choice, the MPEG-21 Peer searches for the available Components, that is, both Components, and plays, at the same time, both the Resources that are contained within the Components, that is, the Resource located at http://server/track01.mp3 and at http://server/track01.wma.

Although this is a simple example, it already illustrates two different ways of processing, each of which is valid with regards to the DID specification. In the end, the only thing the author of the DID wanted was a decent user interface that allowed the selection of an audio format; and a multimedia player to play the audio track in the selected format. From this example, it becomes clear that there is a need for a specification that enables the author of the DID to uniquely specify the way a DID should be processed. MPEG-21 DIP gives an answer to the following questions 'What happens when a DI arrives at an MPEG-21 Peer?', 'What is the author's suggested way of processing the DI?'

It should be noted that the term suggested has been used here to clarify the difference between enforcing a specific way of processing and describing a way of processing. MPEG-21 DIP describes how processing of a DI can be done. To achieve this, it standardizes the tools that allow a DID author to describe such processing. Enforcing that the described processing is actually performed is not mandatory according to the DIP specification. Enforcing should be done using other tools such as Rights Expressions and the Intellectual Property Management and Protection (IPMP) system of the MPEG-21 Peer.

Being able to express the suggested behavior of a DI allows the design of a DIP-enabled Peer that does not need to be aware of the content residing in the DI. Such an agnostic Peer does not need to know what is inside the DI, as long as it allows the User to interact with the DI via the supplied DIP information. For example, the Peer might receive a Movie DI. DIP allows the Peer to meaningfully interact with the Movie DI (e.g., play a teaser, display the movie poster, etc.), without even knowing what a Movie DI is. As long as the DI contains the necessary DIP information telling the Peer how the processing of such a DI could be done. Therefore, no specialized software for handling Movie DIs needs to be built.

To get a better understanding of the MPEG-21 DIP technology and its functionality, consider the following analogy. An Extensible HyperText Markup Language (XHTML) document [9] describing the data of a webpage is, without any additional information, a static page. Such a page looks quite simple and not very attractive to the end-user. In a second step to make the web page more appealing, interactivity could be added to the webpage using JavaScript. This could make the content of the page more dynamic. Finally to make the webpage really attractive to the end-user style sheets can be added using, for example, Cascading Style Sheets (CSS) [10].

The same principle can be applied to a DI. Starting from a DID describing the structure of the DI, it is possible to add interactivity to that DI using DIP. This allows the creation of dynamic DIDs. Finally, to make the presentation of the DI more attractive to an end-user, it is possible to use Synchronized Multimedia Integration Language (SMIL) [11] or MPEG-4 XMT-Ω [12] as a presentation language for the DI.

This chapter discusses how interactivity is added to DIs using the DIP technology. Section 9.2 starts by providing the terms and definitions used in DIP. Afterward, an architecture, positioning the different tools standardized by DIP in the bigger picture, is described in Section 9.3. Section 9.4 and Section 9.5 are all about including and using DIP information from a DIP engine perspective. This DIP engine will be the central part in DIP software implementations. Section 9.6 discusses how code providing the interactivity with the User can be written. Finally, 9.7 ends with a summarizing example containing all aspects of DIP discussed throughout this chapter.

9.2 TERMS AND DEFINITIONS

Before further explaining the concepts of DIP, an overview of the different terms and definitions that are normatively specified in ISO/IEC 21000-10 or other MPEG-21 Parts is given. Some of the terms and definitions that are explained here are explained in previous chapters. However, they are repeated here briefly to provide a common base for the discussion on DIP.

- A **DI** is a structured digital object, including a standard representation, identification and metadata.

Within the MPEG-21 Multimedia framework, a DI is the fundamental unit of transaction – every form of communication, every transmission of information is done through the exchange of such DIs.

- A **DID** is the declaration of the resources, metadata and their interrelationships of a DI, as specified in ISO/IEC 21000-2.
- The **DIDL** consists of an XML-based language, and the validation rules specified by ISO/IEC 21000-2, used for declaring DIs.
- A **DIDL document** is an XML document using the DIDL to declare DIs according to ISO/IEC 21000-2.
- A **DIDL element** is an XML element of the DIDL.

To be able to express DIs in a format that can be used by a computer, the DIDL was developed by MPEG. This language is based on XML, which can easily be interpreted by computers. The DIDL provides the constructs for creating a computer-interpretable interoperable DI. A DI is declared as a DID in a DIDL document using DIDL elements from ISO/IEC 21000-2. An extensive explanation of the DID specification can be found in Chapter 3.

- The **DIP** specification provides the tools that enable the authors of DIs to describe suggested processing for their DIs. Note that DIP is *not* a way to require a specific way of processing a DI. This should be enforced by using other parts of MPEG-21.

Again, it should be noted that the term suggested has been used here to emphasize the difference between enforcing a specific way of processing and describing a way of processing. MPEG-21 DIP describes how processing of a DI can be done. To achieve this, it standardizes the tools that allow a DID author to describe such processing. Enforcing that the described processing is actually performed is not mandatory according to the DIP specification. Enforcing should be done using other tools such as Rights Expressions and the IPMP system of the MPEG-21 Peer.

- A **DIP engine** is the component of an MPEG-21 DIP-enabled Peer that supports the DIP functionality. For example, generating an Object Map, executing the DIMs, and so on. The execution of DIMs includes calling of DIBOs, loading of DIXOs, execution of DIXOs and interpreting the DIML code.

Since the DIP engine is the central part of a DIP environment, Section 9.5.2, in which a walk-through for DIP is explained, fully discusses how the DIP engine performs its tasks and functionalities.

- A **Digital Item Method (DIM)** is a tool for expressing the suggested interaction of an MPEG-21 User (as defined in [8]) with a DI at the DID level. It consists of calls to Digital Item Base Operations, Digital Item eXtension Operations and code written in the Digital Item Method Language.
- The **Digital Item Method declaration (DIM declaration)** is the declaration of the Digital Item Method as being part of a particular DIDL document. Therefore, the DIM declaration is a piece of XML code in the DIDL documents.

- A **Digital Item Method Argument (DIM argument)** is an argument to a Digital Item Method as represented in the Digital Item Method Language. Therefore, a DIM argument is an object in the ECMAScript environment.

- The **Digital Item Method definition (DIM definition)** is the code written in the Digital Item Method Language that defines the Digital Item Method and that is either embedded inline with the Digital Item Method declaration or located separately and referenced from the Digital Item Method declaration. Therefore, the DIM definition is a textual string containing the ECMAScript implementation of the DIM.

- The **Digital Item Method Language (DIML)** is the language in which Digital Item Methods are expressed. It provides the syntax and structure for authoring a Digital Item Method using Digital Item Base Operations and Digital Item eXtension Operations.

Within MPEG-21 DIP, the Digital Item Method Language (DIML) is an extension of ECMAScript [13]. The ECMAScript language, which is also used in MPEG-4, was chosen as the language for DIML because there are existing implementations which are lightweight in terms of memory consumption and footprint, and it is available on many devices. It can be seen as a standardized version of JavaScript [14]. JavaScript is best known from its usage in web pages where it adds dynamic behavior to the standard HyperText Markup Language (HTML) [9] functionality. For example, JavaScript allows changing the color of a button when a mouse moves over the button. ECMAScript, which is in fact a lightweight scripting language, allows the creation of quite powerful scripts using the various constructs of the language. Besides the traditional loop constructs, for example, for and while, the ECMAScript language also supports more advanced features such as exception handling. This exception handling capability is an important aspect of the language in the context of DIP. Since a Digital Item Method (DIM) is expressed in the DIML, a DIM is an ECMAScript method. To avoid any confusion with the Java programming language [15], it should be noted that ECMAScript and JavaScript are scripting languages, while Java is a compiled language.

- A **Digital Item Base Operation (DIBO)** provides access to functionality implemented within the MPEG-21 environment. DIBOs are used for authoring DIMs.

The Digital Item Base Operations (DIBOs) that are available at an MPEG-21 Peer can be seen as a set of functionalities that can be used for authoring a DIM. The set of DIBOs is therefore a library of predefined functions. For example, they can be used to write a DIM that plays the third track of a music album.

- The **Digital Item Base Operation implementation (DIBO implementation)** is the manner in which a particular implementer of a DIBO chooses to implement the normative semantics of the DIBO.

Because the DIP specification only standardizes the semantics of a DIBO, it is possible that there will be several possibilities for implementing the DIBO semantics while still conforming to those semantics. For example, an implementation of the print DIBO might print a resource in Braille, whereas other implementations might print it regularly. Both implementations are valid according to the DIP semantics.

- A **Digital Item eXtension Operation (DIXO)** is an operation allowing extended functionality, that is, functionality that is not accessible using the normative set of DIBOs, to be invoked by a Digital Item Method.

- A **Digital Item eXtension Operation Language (DIXO Language)** is a programming language in which DIXOs are defined.

Since it is not possible to foresee every interaction with a DI, it will (most likely) not be possible to create a set of DIBOs that encompass all the possible interactions with a DI. For example, if a company wants to use its own proprietary operation that is not included in the set of DIBOs, and therefore wants to go beyond the standardized set, they would need some mechanism for extending the set of DIBOs. The Digital Item eXtension Operations (DIXOs) and the framework that is created to realize them provide an answer to that problem. It realizes the extension of the set of DIBOs in a standard manner.

- A **Java-Digital Item eXtension Operation (J-DIXO)** is an operation allowing extended functionality to be invoked by a Digital Item Method implemented in the Java programming language.

To allow interoperability at the level of the languages that are used when implementing DIXOs, MPEG has currently identified the Java programming language as one of the possible languages in which DIXOs could be implemented. Therefore, the DIP specification provides a set of bindings to the DIBOs, in the Java language, for use in the implementation of DIXOs. It should be noted here that the Java programming language is only one of the possible languages for expressing DIXOs. It is likely that in the future other programming languages (e.g., C++, C#, etc.) will follow.

Together with this set of normative bindings, there is an informative part of the DIP specification that illustrates how an execution model for Java-Digital Item eXtension Operation (J-DIXOs) can be constructed. In the rest of this chapter, it is referred to as the **J-DIXO execution environment**.

- A **Digital Item Declaration Object (DID Object)** is an object representation in the Digital Item Method Language of an element in the Digital Item Declaration Language associated with an Object Type.

To allow a DID element to be accessed in the ECMAScript environment, it is necessary to create an object representation of that DID element. Within DIP, the DID elements are accessible from the DIMs using the DIBOs. The object representations of the DID elements are handles to access them from within the DIMs.

- An **Object Type** is the type of the Digital Item Declaration Object specified by an `ObjectType` element in the *descriptor* (note that this formatting is used throughout this chapter for DID Model entities) of the associated DIDL element.

- An **Argument Type** is the type of the Digital Item Method Argument specified by an `Argument` element of the associated Digital item Method declaration.

- An **Object Map** maps the DIDL elements in the DIDL document to the DID Objects of the arguments of the Digital Item Methods using the associated Object Types and Argument Types.

The Object Type and the Object Map have been created to allow DIDL elements, represented as DID Objects, to act as arguments for the DIMs. They will be fully discussed in Sections 9.4.1 and 9.5.1, respectively.

9.3 ARCHITECTURE AND RELATIONSHIP WITH OTHER MPEG-21 PARTS

Within an MPEG-21 DIP-enabled Peer, the DIP engine can be seen as the central part of the Peer. The communication of DIP with the other parts (e.g., DID) can be regulated by the DIP engine. Figure 9.1 gives an overview of a possible architecture for an MPEG-21 Peer. At the bottom of Figure 9.1, different engines are presented. Each of those engines correspond to a certain part of MPEG-21 and present a processing unit that is capable of handling the information defined by those different parts of MPEG-21. Please note that the term engine is not formally defined within those parts, but it is used here to indicate the parts of a Peer responsible for handling the information. For example, a DID engine will most likely be a DID parser, similar to the parser that has been discussed in Chapter 3, with some additional features for communicating with the DIP engine.

The communication of the DIP engine with the other engines, and hence with the different parts of MPEG-21, is realized by executing DIMs in the DIP engine. The DIMs are constructed using the DIBOs and/or DIXOs. DIMs will be discussed in detail in Section 9.6.

In the current set of DIBOs, there are base operations that allow the manipulation of the DID (DID-related DIBOs), base operations related to Digital Rights Management (DRM) information (REL and RDD related DIBOs), base operations that allow adaptation of multimedia resources (DIA related DIBOs), and so on. Each of those DIBOs will communicate with the different engines that are responsible for implementing the tools that are standardized in that specific part of MPEG-21. Note that using DIP does not necessarily result in the creation of a secure and governed MPEG-21 Peer. DIP could be

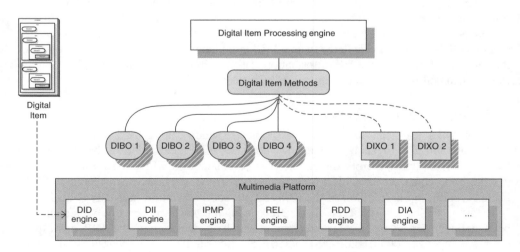

Figure 9.1   An architecture for MPEG-21 digital item processing

used to create such a Peer, but it is not required by the other MPEG-21 parts to use DIP for that purpose. More information about DIBOs will be given in Section 9.6.3.

If, for a certain DIM author, the base operations are not suitable for the task he or she wants to perform, new extensions to the base operations can be constructed using DIXOs. They can create a new form of communication with the different engines in the MPEG-21 DIP-enabled Peer or with other parts of the underlying multimedia platform of that Peer. More information about DIXOs will be given in Section 9.6.4.

From Figure 9.1, it becomes clear that the DIP engine plays a central role in the DIP architecture. As described earlier, it is responsible for communicating with the different engines for the other parts of the MPEG-21 specification. However, at the same time it is also responsible for communicating with the other parts of the underlying multimedia platform of the MPEG-21 Peer. The underlying multimedia platform of the MPEG-21 Peer can be seen as the software/hardware platform on which the MPEG-21 DIP-enabled Peer is created. It will typically contain multimedia codecs, network libraries, user interface libraries, and so on. Therefore, the multimedia platform includes everything that is specific for that DIP-enabled Peer. To ensure interoperability for DIP, the communication with the underlying multimedia platform is realized using the set of DIBOs and DIXOs that are available at the MPEG-21 Peer.

The DIP engine is also responsible for communicating with the User, defined in MPEG-21 as any entity that interacts in the MPEG-21 environment or makes use of DIs, including both human users and machine users. Communicating with the User, typically includes asking the User to select a DIM that needs to be executed; ask the User to select an argument for a certain DIM; ask the User configuration questions; and so on. A full walk-through of how a DIP engine can communicate with a User is given in Section 9.5.2.

9.4 INCLUDING DIGITAL ITEM PROCESSING INFORMATION IN A DIGITAL ITEM

The Digital Item is the fundamental unit of transaction within the MPEG-21 Multimedia framework. Every bit of information that is exchanged within the MPEG-21 Multimedia framework is transferred within a DI. Therefore, DIP information must also be carried within a DI. Since DIs are declared in an XML-based language, the DIP specification specifies an XML format for the DIP information. This section discusses how this information can be included in a DID such that it can be used by a DIP engine.

9.4.1 OBJECT TYPE DECLARATION

To allow DIDL elements to become DID Objects, which can be used as arguments for DIMs, the Object Map has been defined. Until now, only a definition of the Object Map has been given, and therefore the chapter has been silent on where the information that is required to construct the Object Map is actually stored and how it is declared in XML. In this section, the declaration of the Object Type and its location in a DID is discussed.

To associate an Object Type to a DIDL element, the ObjectType element has been defined. This ObjectType element, which is declared in the DIP namespace urn:mpeg:mpeg21:2005:01-DIP-NS contains a Uniform Resource Identifier (URI) [16] identifying the type that is declared using the ObjectType element.

```
<?xml version="1.0" encoding="UTF-8"?>
<dip:ObjectType xmlns:dip="urn:mpeg:mpeg21:2005:01-DIP-NS">
  urn:foo:type1
</dip:ObjectType>
```

Example 9.2   The declaration of an Object Type.

```
<?xml version="1.0" encoding="UTF-8"?>
<DIDL xmlns="urn:mpeg:mpeg21:2002:02-DIDL-NS"
      xmlns:dip="urn:mpeg:mpeg21:2005:01-DIP-NS">
  <Container>
    <Item>
      <Descriptor>
        <Descriptor>
          <Statement mimeType="text/xml">
            <dip:ObjectType>urn:foo:MoviePoster</dip:ObjectType>
          </Statement>
        </Descriptor>
        <Component>
          <Resource mimeType="image/jpeg" ref="poster01.jpg"/>
        </Component>
      </Descriptor>
      <Descriptor>
        <Statement mimeType="text/xml">
          <dip:ObjectType>urn:foo:Movie</dip:ObjectType>
        </Statement>
      </Descriptor>
      <Component>
        <Resource mimeType="video/mpeg" ref="movie01.mpg"/>
      </Component>
    </Item>
  </Container>
</DIDL>
```

Example 9.3   The declaration of an Object Type in a DIDL document.

In Example 9.2, a declaration of the urn:foo:type1 Object Type is presented. This Object Type can be associated with a DIDL element in order to allow it to be used as an argument for a DIM.

Example 9.3 illustrates how this Object Type declaration can be included in a Digital Item Declaration. The Digital Item Declaration represents a movie collection containing one movie, declared in the Item element. The Object Type of that Item is urn:foo:Movie. The Item contains a Descriptor with the poster of the movie. The Object Type of that poster is urn:foo:MoviePoster.

9.4.2 DIGITAL ITEM METHOD DECLARATION

Besides the association of the Object Types to the different DID elements, it is necessary to declare the DIMs in the DI. A DIM is declared within a Component element. It must be constructed such that

- the Component contains a DIM information *descriptor* represented by a DIP Method Info element in a Descriptor/Statement construction

- the Digital Item Method definition is referenced by or embedded in a Resource child of the Component

The DIM information *descriptor* contains the XML declaration of the information needed for executing a DIM. It is constructed using the MethodInfo element and is declared in the DIP namespace urn:mpeg:mpeg21:2005:01-DIP-NS. Its child element, the Argument element, contains the Argument Type declaration, that is, a URI. The Argument Type declaration is done in document order. Therefore, the first occurrence of the Argument element maps to the first argument of the DIM definition, the second Argument element to the second argument of the DIM definition, and so on. The MethodInfo element is a wrapper for the list of arguments that need to be selected before executing the DIM. If the DIM definition has no arguments, the MethodInfo element can still be present but it will contain zero child Argument elements.

Example 9.4 demonstrates how a MethodInfo descriptor for a Digital Item Method requiring two arguments is declared. The DIM declaration containing this MethodInfo *descriptor* requires an argument of the Argument Type urn:foo:type1 and one of the Argument Type urn:foo:type2.

Example 9.5 illustrates how a Digital Item Method can be declared in a Component element. The MethodInfo *descriptor* is included in the Descriptor element of the Component and the implementation of the method is embedded in the Resource.

As an alternative to the embedded implementation, it is also possible to declare the implementation of the Digital Item Method by reference to the file that actually stores the ECMAScript code. Example 9.6 illustrates how the ref attribute of the Resource

```
<?xml version="1.0" encoding="UTF-8"?>
<dip:MethodInfo xmlns:dip="urn:mpeg:mpeg21:2005:01-DIP-NS">
  <dip:Argument>urn:foo:type1</dip:Argument>
  <dip:Argument>urn:foo:type2</dip:Argument>
</dip:MethodInfo>
```

Example 9.4 The declaration of a MethodInfo element.

```
<?xml version="1.0" encoding="UTF-8"?>
<Component xmlns="urn:mpeg:mpeg21:2002:02-DIDL-NS">
  <Descriptor>
    <Statement mimeType="text/xml">
      <dip:MethodInfo xmlns:dip="urn:mpeg:mpeg21:2005:01-DIP-NS">
        <dip:Argument>urn:foo:type1</dip:Argument>
        <dip:Argument>urn:foo:type2</dip:Argument>
      </dip:MethodInfo>
    </Statement>
  </Descriptor>
  <Resource mimeType="application/mp21-method">
    function dim_01( arg1 , arg2)
    {
      /* implementation goes here */
    }
  </Resource>
</Component>
```

Example 9.5 A Digital Item Method Declaration with an embedded implementation.

```
<?xml version="1.0" encoding="UTF-8"?>
<Component xmlns="urn:mpeg:mpeg21:2002:02-DIDL-NS">
  <Descriptor>
    <Statement mimeType="text/xml">
      <dip:MethodInfo xmlns:dip="urn:mpeg:mpeg21:2005:01-DIP-NS">
        <dip:Argument>urn:foo:type1</dip:Argument>
        <dip:Argument>urn:foo:type2</dip:Argument>
      </dip:MethodInfo>
    </Statement>
  </Descriptor>
  <Resource mimeType="application/mp21-method"
            ref="implementation.dim"/>
</Component>
```

Example 9.6   A Digital Item Method Declaration with a referenced implementation.

element allows referring to the file containing the ECMAScript code, called *implementation.dim*.

Besides having `Argument` child elements, it is also possible for the `MethodInfo` element to have an `autoRun` attribute. This attribute is an optional boolean flag indicating that the DIM is designed to be run automatically. If someone wants to design a DIP-enabled Peer that runs DIMs without any User interaction to choose a DIM to run, this flag can be used to determine which DIM is designed to be run automatically. If there is more than one DIM with the `autoRun` flag turned on, the Peer can choose any DIM with that flag. Note that this does not imply that a DIM having the `autoRun` attribute with value true will *always* run automatically. Such constraints need to be enforced by external means (e.g., rights expressions, profiles, etc.).

Finally, it is also possible for the `MethodInfo` element to have a `profileCompliance` attribute. This attribute allows the DIM author to signal a profile or profiles to which the DIM conforms. If this attribute is present, each member in the list represents one such profile. Although there are no standardized profiles for DIP yet, the DIP specification already contains some example profiles for DIP.

When declaring DIMs in DIDs, it is necessary for a DIP-enabled Peer to be able to detect the DID elements in which a DIM is declared. To realize this, it would be possible for a DIP-enabled Peer to search through the DID for all `Components` containing a `MethodInfo` element and select those as DIMs that could possibly be executed. However, MPEG decided not to use this solution because that would make it impossible to differentiate between DIMs declared in a DID for storage, for example, for future shipment, and DIMs declared to be executable in that DID. Therefore, MPEG decided to define a `Label` element to flag the DIMs that can be executed. The value of the `Label` element is urn:mpeg:mpeg21:2005:01-DIP-NS:DIM for DIMs.

Example 9.7 demonstrates how the `Label` element can be added to the `Component` by means of a `Descriptor/Statement` construction. This example is ready to be recognized by an MPEG-21 DIP-enabled Peer as an executable DIM.

9.4.3 J-DIXO DECLARATION

To allow J-DIXOs to be used from within DIMs, it is necessary that the J-DIXOs can be located by the DIP engine. This allows the DIP engine to initialize the J-DIXO execution

```
<Component>
  <Descriptor>
    <Statement mimeType="text/xml">
      <dip:MethodInfo>
        <dip:Argument>urn:foo:MoviePoster</dip:Argument>
      </dip:MethodInfo>
    </Statement>
  </Descriptor>
  <Descriptor>
    <Statement mimeType="text/xml">
      <dip:Label>urn:mpeg:mpeg21:2005:01-DIP-NS:DIM</dip:Label>
    </Statement>
  </Descriptor>
  <Resource mimeType="application/mp21-method">
    function removePoster( arg1 )
    {
      /* implementation goes here */
    }
  </Resource>
</Component>
```

Example 9.7   A DIM declaration with a `Label` element.

environment and to call the J-DIXOs from within the DIMs. The declaration of the J-DIXO inside a DID is similar to the declaration of a DIM inside a DID. It is declared in a `Component` element and this element must be constructed such that

- the `Component` contains the J-DIXO classes *descriptor* in a `Descriptor/ Statement` construction
- the J-DIXO definition is contained within the `Resource` of the `Component`

The J-DIXO classes *descriptor* is the XML definition of the class(es) corresponding to the implementation of the J-DIXO(s). It is constructed using the `JDIXOClasses` and the `Class` elements. Both elements are declared in the DIP namespace urn:mpeg:mpeg21:2005: 01-DIP-NS. Each of the `Class` elements contain a string indicating the fully qualified Java class name of the J-DIXO class that is intended to be invoked as a J-DIXO.

Example 9.8 demonstrates how a `JDIXOClasses` *descriptor* for one J-DIXO is declared. The J-DIXO corresponding to this declaration has the fully qualified Java class name com.company.DIXOClass.

Besides the information that is contained within a J-DIXO classes *descriptor*, it is also necessary to include (a reference to) the actual implementation of the J-DIXO in the Digital Item Declaration. This is realized by declaring the `JDIXOClasses` *descriptor* in a `Component` element. This `Component` has one `Descriptor/Statement` containing

```
<?xml version="1.0" encoding="UTF-8"?>
<dip:JDIXOClasses xmlns:dip="urn:mpeg:mpeg21:2005:01-DIP-NS">
  <dip:Class>com.company.DIXOClass</dip:Class>
</dip:JDIXOClasses>
```

Example 9.8   The declaration of a `JDIXOClasses` descriptor.

```
<?xml version="1.0" encoding="UTF-8"?>
<Component xmlns="urn:mpeg:mpeg21:2002:02-DIDL-NS">
  <Descriptor>
    <Statement mimeType="text/xml">
      <dip:JDIXOClasses xmlns:dip="urn:mpeg:mpeg21:2005:01-DIP-NS">
        <dip:Class>com.company.DIXOClass</dip:Class>
      </dip:JDIXOClasses>
    </Statement>
  </Descriptor>
  <Resource mimeType="application/java" encoding="base64">
    /* Base64 encoded implementation goes here */
  </Resource>
</Component>
```

Example 9.9 A J-DIXO declaration with an embedded implementation.

```
<?xml version="1.0" encoding="UTF-8"?>
<Component xmlns="urn:mpeg:mpeg21:2002:02-DIDL-NS">
  <Descriptor>
    <Statement mimeType="text/xml">
      <dip:JDIXOClasses xmlns:dip="urn:mpeg:mpeg21:2005:01-DIP-NS">
        <dip:Class>com.company.DIXOClass</dip:Class>
      </dip:JDIXOClasses>
    </Statement>
  </Descriptor>
  <Resource mimeType="application/java" ref="DIXOClass.class"/>
</Component>
```

Example 9.10 A J-DIXO declaration with a referenced implementation.

the JDIXOClasses *descriptor* and one Resource containing the implementation of the J-DIXO. Similar to the inclusion of the DIM definition, it is possible to include the J-DIXO definition both as an embedded Resource (e.g., Example 9.9) and as a referenced Resource (e.g., Example 9.10).

Similar to DIMs, when declaring J-DIXOs in DIDs it is necessary for a DIP-enabled Peer to be able to detect the DID elements in which a J-DIXO is declared. For the same reasons as for the DIMs, J-DIXOs are flagged by a Label element with the value urn:mpeg:mpeg21:2005:01-DIP-NS:DIXO:Java.

To conclude this section on the inclusion of J-DIXO information in Digital Item Declarations, the PerformCustomAuthentification DIXO illustrates how a J-DIXO can be included in a Digital Item Declaration. Suppose there is a company using their own authentication system that is not standardized by any standardization body. Before the multimedia content provided by the company can be consumed, it is necessary to authenticate the MPEG-21 Peer as being a legitimate User of the multimedia content. Since the authentication mechanism that is used by the company is proprietary, and therefore not part of the authentication systems supported by MPEG-21, it is necessary to extend the set of DIBOs to include the proprietary authentication system. This can be done by creating the PerformCustomAuthentification DIXO. This DIXO can be included in a DID, as depicted in Example 9.11.

```
<?xml version="1.0" encoding="UTF-8"?>
<Component xmlns="urn:mpeg:mpeg21:2002:02-DIDL-NS">
  <Descriptor>
    <Statement mimeType="text/xml">
      <dip:JDIXOClasses xmlns:dip="urn:mpeg:mpeg21:2005:01-DIP-NS">
        <dip:Class>
          com.company.authentication.PerformCustomAuthentification
        </dip:Class>
      </dip:JDIXOClasses>
    </Statement>
  </Descriptor>
  <Descriptor>
    <Statement mimeType="text/xml">
      <dip:Label xmlns:dip="urn:mpeg:mpeg21:2005:01-DIP-NS">
        urn:mpeg:mpeg21:2005:01-DIP-NS:DIXO:Java
      </dip:Label>
    </Statement>
  </Descriptor>
  <Resource mimeType="application/java" encoding="base64">
    /* Base64 encoded implementation of
       PerformCustomAuthentification goes here */
  </Resource>
</Component>
```

Example 9.11   Including a PerformCustomAuthentification J-DIXO in a DID.

Whenever the authentication of the MPEG-21 Peer is necessary for consuming the multimedia content, the Digital Item Method implementation can include a call to the PerformCustomAuthentification J-DIXO. When this call to the PerformCustomAuthentification is found in the DIM, the J-DIXO is executed, the authentication is performed, and the multimedia content will be available for consumption.

9.5 USAGE OF DIGITAL ITEM PROCESSING INFORMATION

9.5.1 THE OBJECT MAP

Since the DI is the fundamental unit of transaction within the MPEG-21 Multimedia framework, it is required that the different elements of such a DI can be processed using the MPEG-21 DIP engine.

In MPEG-21 DIP, the logic of the processing program is expressed using the DIML. This language, which is an extension of ECMAScript, needs to be able to use the information contained within the DID. Since all of the variables within the ECMAScript environment are objects, it will be necessary to create an object representation of the DID elements, called *DID Objects*, in the ECMAScript environment.

For example, suppose that there exists a movie collection that is expressed using the DIDL. Such a DID can be expressed as a *container*, representing the movie collection, with several *items*, representing the different movies. In the different *items*, there might be a *descriptor*, representing the posters of the movies, and a *component* representing the actual movie data. This *component* contains a *descriptor* describing the technical information about the movie data, such as the used video and audio codec, the bit rates, and so on. A schematic representation of this DID is provided in Figure 9.2 at the left side.

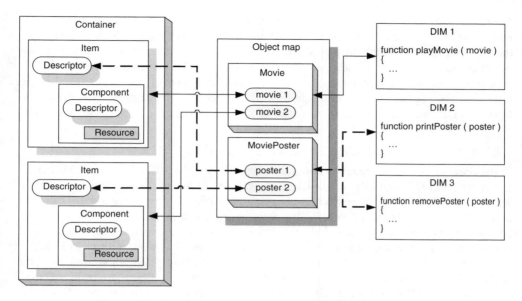

Figure 9.2 Linking DID elements and DIM arguments together

Suppose that there are three different DIMs: a playMovie DIM, a printPoster DIM, and a removePoster DIM. The playMovie DIM plays the movie that is passed to the DIM as an argument; the printPoster DIM prints the poster; and the removePoster DIM removes the poster from the DI. A schematic representation of these DIMs is provided in Figure 9.2 at the right side.

If the only information that is considered is the DID and the ECMAScript implementations, there is no way that the DIP engine can know what parts of the DID represent the posters, and what parts of the DID represent the movies. The Object Map solves this problem by linking the different elements of a DID to an Object Type, which can be used when selecting the arguments for a DIM. A schematic representation of the Object Map is provided in Figure 9.2 at the center.

The information that is stored within the Object Map will now be examined in closer detail. The Object Map contains a list of the different Object Types that occur in the DID. For each of those Object Types, it contains a list of DID Objects that are of the corresponding type. For example, for the MoviePoster Object Type, it contains two links to the first *descriptors* of the *items* that are contained within the DID.

At the same time, for each of the Object Types, it contains a link to the DIMs requiring an Argument Type with a value equal to that Object Type. For example, for the MoviePoster Object Type, it contains a link to the printPoster DIM and to the removePoster DIM.

On the basis of the information in the Object Map, possible arguments can be selected for DIMs. Or vice versa, possible DIMs can be selected for DID Objects. For example, suppose the User has selected the playMovie DIM and needs to know on which DID elements that DIM can be performed. To find this out, the DIP engine can be asked what DID elements are available as arguments for that DIM. The DIP engine uses the Object

Map to look up what DID elements correspond to the Object Type Movie, that is, the two *items* that are declared in the DID. Therefore, the DIP engine returns those two *items* as the DID elements on which the playMovie DIM can be performed. An alternative example in which possible DIMs are selected for a DID Object will be given in Section 9.5.3.

Note that the information needed to construct an Object Map is actually included in the DID. The same applies to the DIMs and the information about the arguments of the DIMs. How this DIP information can be included in DIDs has been discussed in Section 9.4.

9.5.2 A WALK THROUGH OF A DIP-ENABLED MPEG-21 PEER

In the previous sections, the relationship between DIP and the different parts of MPEG-21 in general, and DIP and DID in particular, have been discussed. This section discusses a walk-through that gives an answer to the question: 'What happens when a DID arrives at an MPEG-21 DIP-enabled Peer?' This walk-through is outlined in Figure 9.3 and can be used to implement a generic DIP enabled MPEG-21 Peer. It should be noted that it is *not* a normative part of the MPEG-21 DIP specification. The walk-through is included in this chapter because it could possibly be used as a starting point when implementing an MPEG-21 DIP-enabled Peer. Therefore, this walk-through is a bit more restricted than the possibilities allowed by the DIP specification. For example, whether the DIP engine asks the User to first select a DIM and afterwards a DID Object, or vice versa, is up to the implementation of the DIP engine. In this walk-through, the User selects a DIM and afterwards the DID Objects that are suitable as arguments for the DIM.

1. As a first step, a DID arrives at an MPEG-21 Peer. In this step, the Peer recognizes it received a DID, for example, by looking at the namespace in which the XML code is expressed, and transfers the control to the DID engine.

2. The DID engine opens the DID and searches through the DID to find DIP information. This information is detected by searching for DIP elements from the DIP namespace.

3. If the DID contains DIP information, the DID engine transfers the DIP information to the DIP engine.

4. The DIP engine builds a list of DIMs and creates the Object Map.

5. If there is a DIM with an `autoRun` attribute with the value true in the DID, this DIM is selected automatically. In that case, the walk-through continues in bullet 8.

6. The list of DIMs is presented to the MPEG-21 User. This list of DIMs can be seen as a menu of possible actions the MPEG-21 User can request from the DIP engine.

7. The MPEG-21 User selects a DIM and his/her selection is transferred to the DIP engine.

8. The DIP engine ensures the DIM implementation is loaded and available for execution in the ECMAScript environment. If there are any DIXOs used in the DIM, those DIXOs are loaded and initialized.

9. On the basis of the information present in the Object Map, the DIP engine generates a list of the arguments valid for the selected DIM.

10. The list of valid arguments is presented to the MPEG-21 User. This list can be seen as a menu of possible elements on which the selected DIM can be performed.

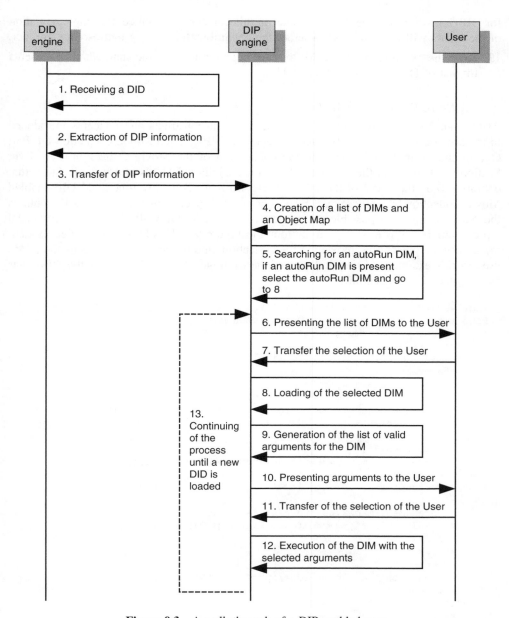

Figure 9.3   A walk-through of a DIP-enabled peer

11. The MPEG-21 User selects an argument and notifies the DIP engine of the selection.

12. The DIM is now executed by calling the DIBOs and DIXOs listed in the DIM implementation.

   Note: the execution of the DIBOs and DIXOs can trigger several rights checks. For example, when a play DIBO (note that this formatting is used throughout this chapter for DIBOs), which will be discussed later in this chapter, is executed, the implementation of the DIBO will need to check whether the conditions for playing the resource are met.

13. After the execution of the DIM, the DIP engine returns to the state where it presents the list of DIMs to the MPEG-21 User.

9.5.3 A REAL-WORLD EXAMPLE – PART ONE

At this point in the discussion of DIP, the different constructs and concepts introduced in the previous sections, can be used to declare a DIP-enabled Movie Collection DI. Part One of the example will focus on the declaration of the Movie Collection. Part Two, Section 9.7.1, extends this example with some additional DIMs and provide the actual definition (i.e., the code) of the DIMs. Part Two realizes a fully functional DIP-enabled Movie Collection DI using several aspects of the DIP specification. Example 9.12 contains the XML listing of a possible declaration of such a Movie Collection DI. A graphical representation of this example can be found in Figure 9.2. This DI contains three Items, the first two represent the movies in the collection, and the last Item contains the DIMs used in this example. In this first part of the example, the DIML code of the DIMs has been replaced by placeholders.

```xml
<?xml version="1.0" encoding="UTF-8"?>
<DIDL xmlns="urn:mpeg:mpeg21:2002:02-DIDL-NS"
      xmlns:dip="urn:mpeg:mpeg21:2005:01-DIP-NS">
  <Container>
    <Item>
      <Descriptor>
        <Descriptor>
          <Statement mimeType="text/xml">
            <dip:ObjectType>urn:foo:MoviePoster</dip:ObjectType>
          </Statement>
        </Descriptor>
        <Component>
          <Resource mimeType="image/jpeg" ref="poster01.jpg"/>
        </Component>
      </Descriptor>
      <Descriptor>
        <Statement mimeType="text/xml">
          <dip:ObjectType>urn:foo:Movie</dip:ObjectType>
        </Statement>
      </Descriptor>
      <Component>
        <Resource mimeType="video/mpeg" ref="movie01.mpg"/>
      </Component>
    </Item>
    <Item>
      <Descriptor>
        <Descriptor>
          <Statement mimeType="text/xml">
            <dip:ObjectType>urn:foo:MoviePoster</dip:ObjectType>
          </Statement>
        </Descriptor>
        <Component>
          <Resource mimeType="image/jpeg" ref="poster02.jpg"/>
```

```
      </Component>
    </Descriptor>
    <Descriptor>
      <Statement mimeType="text/xml">
        <dip:ObjectType>urn:foo:Movie</dip:ObjectType>
      </Statement>
    </Descriptor>
    <Component>
      <Resource mimeType="video/mpeg" ref="movie02.mpg"/>
    </Component>
  </Item>
  <Item>
    <Component>
      <Descriptor>
        <Statement mimeType="text/xml">
          <dip:MethodInfo>
            <dip:Argument>urn:foo:Movie</dip:Argument>
          </dip:MethodInfo>
        </Statement>
      </Descriptor>
      <Descriptor>
        <Statement mimeType="text/xml">
          <dip:Label>
            urn:mpeg:mpeg21:2005:01-DIP-NS:DIM
          </dip:Label>
        </Statement>
      </Descriptor>
      <Resource mimeType="application/mp21-method">
        function playMovie( arg1 )
        {
          /* implementation goes here */
        }
      </Resource>
    </Component>
    <Component>
      <Descriptor>
        <Statement mimeType="text/xml">
          <dip:MethodInfo>
            <dip:Argument>urn:foo:MoviePoster</dip:Argument>
          </dip:MethodInfo>
        </Statement>
      </Descriptor>
      <Descriptor>
        <Statement mimeType="text/xml">
          <dip:Label>
            urn:mpeg:mpeg21:2005:01-DIP-NS:DIM
          </dip:Label>
        </Statement>
      </Descriptor>
      <Resource mimeType="application/mp21-method">
        function printPoster( arg1 )
        {
          /* implementation goes here */
        }
```

```
    </Resource>
  </Component>
  <Component>
    <Descriptor>
      <Statement mimeType="text/xml">
        <dip:MethodInfo>
          <dip:Argument>urn:foo:MoviePoster</dip:Argument>
        </dip:MethodInfo>
      </Statement>
    </Descriptor>
    <Descriptor>
      <Statement mimeType="text/xml">
        <dip:Label>
         urn:mpeg:mpeg21:2005:01-DIP-NS:DIM
        </dip:Label>
      </Statement>
    </Descriptor>
    <Resource mimeType="application/mp21-method">
      function removePoster( arg1 )
      {
        /* implementation goes here */
      }
    </Resource>
  </Component>
 </Item>
</Container>
</DIDL>
```

Example 9.12   A movie collection DID including DIP information – Part One.

Now let us look at the walk-through from Section 9.5.2, which can be used to process this Movie Collection DI. When this DI arrives at the Peer, the DIP information is extracted and passed to the DIP engine. For the Movie Collection DI, the DIP engine will detect the three different DIMs: playMovie, printPoster and removePoster each of them requiring one argument of the Argument Type urn:foo:Movie, urn:foo:MoviePoster, and urn:foo:MoviePoster, respectively. Two different Object Types are detected: urn:foo:Movie and urn:foo:MoviePoster.

On the basis of this information, the Object Map will be constructed. The information contained within the Object Map is presented in Figure 9.2. It links the urn:foo:MoviePoster Object Type to the two Descriptors and to the Argument Type of the printPoster and removePoster DIMs. Similar to the urn:foo:MoviePoster Object Type, it links the urn:foo:Movie Object Type to the two Components and to the Argument Type of the playMovie DIM.

Since there were no autoRun DIMs detected, the three DIMs are presented to the User and the User selects a DIM. Suppose the removePoster DIM has been selected. To actually execute the removePoster DIM, it is required to have an argument of the Argument Type urn:foo:MoviePoster. On the basis of the Object Map information, the list of valid arguments is generated, and the two Descriptors containing the JPEG images are returned as valid arguments for the DIM. The two Descriptors are presented to the User. After the User has selected one of the Descriptors, the DIM is executed. When the execution of the DIM has finished, the list of available DIMs is presented to the User and the process restarts.

Alternatively, the DIP engine could have asked the User to select a DID Object first. Now, suppose a User has selected one of the `Descriptors` containing a poster and he wants to know what he can do with that Descriptor. To find this out, he asks the DIP engine what DIMs are available for that `Descriptor`. The DIP engine then uses the Object Map and looks up the Object Type urn:foo:MoviePoster. Two DIMs require an argument of a matching `Argument Type` urn:foo:MoviePoster, the printPoster DIM and the removePoster DIM. Therefore, the DIP engine returns the printPoster DIM and removePoster DIM as the two possible actions that can be performed on the selected `Descriptor`. Whether the DIP engine asks the User to first select a DIM and afterwards a DID Object, or vice versa, is up to the implementation of the DIP engine.

9.6 DIGITAL ITEM METHODS

DIMs are a key component in DIP. They allow an author of a DI to include their suggestions on how to interact with a DI (in this case an author refers to any entity that is able to create or modify a DI). Thus DIMs allow the DI author to include a dynamic interactive component in the otherwise static DID.

This section will often make use of Programmer's Notes and Implementer's Notes. The former gives information to the programmer using the discussed concepts in the implementation of the DIM definition, that is, in the DIML code. The latter gives information to the implementer of the DIP specification, for example, for implementing a DIP engine with the necessary DIBO implementations, and so on.

9.6.1 DIGITAL ITEM METHOD COMPONENTS

As explained in the introduction to this chapter, DIDL itself does not specify any standard way for a DI author to provide suggested interactions of a User with the DI. DIMs add this capability to the MPEG-21 Multimedia framework and this is in fact the major objective of the DIP specification.

An MPEG-21 Peer is any device or application that compliantly processes an MPEG-21 DI [8]. Hence, it is expected to be able to process the building blocks of a DI as represented by a DID. Since these building blocks are common to any DID, it can be reasonably expected that any Peer will provide a base level of functionality for processing the elements of a DID (as expressed in DIDL [1] and described in Chapter 3). DIMs can be seen as providing the DI author with a standard interface to access this functionality of the Peer for the purpose of providing suggested interactions with the DI at the level of the DID.

Note that DIMs provide the capabilities for an author to express suggested interactions with the building blocks of the DI as defined in part 2 of MPEG-21. For example, a DIM can be used to evaluate a DID *condition*, and to request a *resource* to be played. However, a DIM is not intended to provide the capabilities within the DIM definition itself for user control of the playing *resource*, or for processing of the *resource* itself (such as transcoding). This functionality is left to the Peer to provide in a manner that is appropriate to the Peer (such processing might occur in the implementation of a DIBO or DIXO called from the DIM). Also, it does not address any interactivity inherent within a *resource* itself, for example, if the *resource* is an interactive QuickTime™ VR presentation.

As an example, a DI that is a digital music album might contain *items* that represent tracks of the music album, with each track *item* containing an mp3 *resource* for an audio *component* of the track. The DI might also include DIMs to play a track or to add a new track. The play track DIM is authored to access the mp3 *resource* of a selected track *item* and requests the *resource* to be played. The add track DIM is authored to create and insert a well-structured new track *item* in the proper place within the digital music album DI. When a DIP-enabled Peer processes the DI, it can provide the User with a list of accessible DIMs, and the User is able to choose a DIM as appropriate.

For a DI author to be able to express their suggested interactions of a User with a DI, MPEG-21 DIP specifies the DIML for defining DIMs. The DIML provides access to the base functionality of an MPEG-21 Peer by a standard set of DIBOs. In addition, extended functionality can be made available to DIMs by the use of DIXOs. Those concepts will be discussed extensively in the next sections.

In Figure 9.4, an overview of the relationship between DIMs, DIBOs, DIXOs and the underlying multimedia platform can be found.

- At the top level, the Digital Item Methods, defined by the DIM definitions can be found. From within the DIM definition, it is possible to call the different DIBOs, other DIMs (using the runDIM DIBO), and DIXOs (using a specialized DIBO for calling DIXOs defined in a given DIXO language, e.g., runJDIXO).

- The DIBO implementations can call modules provided by the multimedia platform. The runDIM DIBO calls an identified DIM and can be used to call a DIM from another DIM. Specialized DIBOs for calling DIXOs defined in a given language, for example, runJDIXO, initiate the execution of DIXOs and are used to call DIXOs from a DIM.

- Below the DIBOs, the DIXOs are presented. DIXO definitions may call DIBOs via the standardized bindings to the DIBOs in the DIXO language as well as other modules provided by the multimedia platform. A DIXO is called from a DIM by a specialized DIBO for calling DIXOs defined in the DIXO language, for example, runJDIXO.

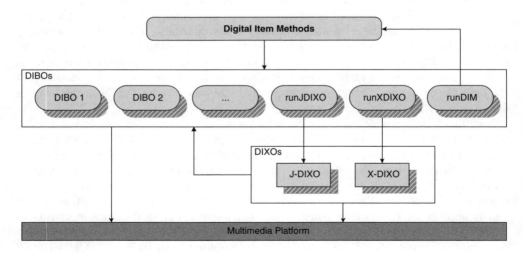

Figure 9.4 Digital item method components

- The multimedia platform refers to the environment in which the DIBOs are implemented and the DIMs are executed. Calling the platform is realized by APIs not specified by DIP. This could include modules providing functionality related to other parts of MPEG-21.

9.6.2 DIGITAL ITEM METHOD LANGUAGE

The DIML specifies the normative language for defining DIMs and from which the DIBOs are able to be called. It consists of three parts:

- Core language (discussed in Section 9.6.2.1)
- DIML global values and objects (discussed in Section 9.6.2.2)
- DIML object types (discussed in Section 9.6.2.3)

9.6.2.1 Core Language

The DIML is based on ECMAScript 3rd edition [9]. This provides a standardized core language specification for DIML, including specification of the following features of DIML:

- Lexical conventions and syntactical structure
- Flow control structures (i.e., if/then, while, etc.)
- Primitive data types (String, Boolean and Integer)
- Composite data types (Objects and arrays)
- Standard arithmetic, logical and bitwise operators
- Exception handling
- Error definitions
- Support for regular expressions.

DIML extends ECMAScript by defining additional object types and properties required for processing a DI. An object type in ECMAScript should not be confused with an Object Type in DIP. An object type in ECMAScript defines the structure of an ECMAScript object. It can be roughly compared to a header file in the C programming language. Objects in ECMAScript can have properties. Those properties can be:

- value properties, for example, `object.integervalue`
- object properties, for example, `object.otherObject`
- function properties, for example, `object.method(args)`. Note that the function properties are called DIBO properties if the specified function is a DIBO.

In ECMAScript terminology, the 'host environment' is provided by the DIP engine and additional DIML specific objects are 'host objects'. Hence, an MPEG-21 Peer that supports DIMs must provide the execution environment for DIMs. This includes a DIML (ECMAScript) interpreter and implementation of the additional DIML object types, global properties and functions (including the DIBOs).

9.6.2.2 DIML Global Properties

DIML extends the standard ECMAScript global object with the addition to the global object of the following DIML specific properties. Examples of how these value properties and object properties can be used will be given later in Section 9.6.3, where the DIBOs using them are discussed.

MSG_INFO, MSG_WARNING, MSG_ERROR, and MSG_PLAIN

MSG_INFO, MSG_WARNING, MSG_ERROR, and MSG_PLAIN are value properties of the global object. They are integer valued message type codes indicating messages are respectively of a general informational nature (MSG_INFO), providing a warning (MSG_WARN-ING), indicating some error condition has occurred (MSG_ERROR), or generic in nature (MSG_PLAIN).

Programmer's note: This value can be passed as the messageType argument to the alert DIBO (which will be discussed later in this chapter).

Implementer's note: An implementation of the alert DIBO might use this value passed as the messageType argument to modify the presentation of the message, for example to display an appropriate icon in a GUI, or to use an appropriate prefix in a log file.

diddDocument

The didDocument is a global object property. It represents a Document Object Model (DOM) Document of the current DID instance document. This is the DID containing (directly in the DID or by inclusion) the DIM declaration that is currently executing. In the case in which a series of DIMs are in the execution stack, for example, via one or more calls to the runDIM DIBO (explained later in this chapter), it is the DID containing the DIM declaration of the first DIM on the execution stack.

The didDocument is to the DIP environment what the global document object is to client side JavaScript in web browsers.

DID

This global object property is an object that contains DIBOs providing DID related functionality (see Section 9.6.3.2).

DII

This global object property is an object that contains DIBOs providing Digital Item Identification (DII) related functionality (see Section 9.6.3.3).

DIA

This global object property is an object that contains DIBOs providing Digital Item Adaptation (DIA) related functionality (see Section 9.6.3.4).

DIP

This global object property is an object that contains DIBOs providing DIP related functionality (see Section 9.6.3.5).

REL

This global object property is an object that contains DIBOs providing REL related functionality (see Section 9.6.3.6).

Other MPEG-21 Parts

The list of global object properties is restricted to those MPEG-21 Parts that were finalized when DIP was developed. It is possible that in the future other MPEG-21 Parts will be added to this list by means of amendments to the DIP specification.

9.6.2.3 DIML Object Types

DIML extends the set of native ECMAScript objects with the additional DIML specific objects defined below. Examples of how these object types can be used will be given later in Section 9.6.3, where the DIBOs using them are discussed.

DIPError

This DIML object inherits properties and functions from the native ECMAScript `Error` object. A `DIPError` exception is thrown when a DIP-related run-time error occurs during execution of a DIM. A `DIPError` exception can be caught using the `try - catch` statements (defined by ECMAScript). More specific DIP-related exceptions that inherit from `DIPError` may also be thrown. Note that standard ECMAScript `Error` exceptions may also be thrown during execution of a DIM.

Value Properties
General_exception

The `GENERAL_EXCEPTION` value property is an integer value of 1. It indicates that a general DIP error, not covered by any other error code below, has occurred.

Invalid_param

The `INVALID_PARAM` value property is an integer value of 2. It indicates a parameter passed to a DIBO is invalid. For example, this `DIPError` exception can occur when a string parameter is passed to a DIBO requiring an integer parameter.

Invalid_permission

The `INVALID_PERMISSION` value property is an integer value of 3. It indicates that an attempt was made to execute an operation for which required permission in the host environment is unavailable. For example, if the host environment does not allow execution of resources in DI, a `DIPError` returning this value will be generated if the `execute` DIBO (which will be explained later in this chapter) is called.

Not_found

The `NOT_FOUND` value property is an integer value of 4. It indicates something required to complete an operation was not found. The missing entity is dependent on the operation that was attempted.

Adaption_failed

The `ADAPTION_FAILED` value property is an integer value of 5. It indicates an error occurred during an attempt to adapt a resource.

Playback_failed

The `PLAYBACK_FAILED` value property is an integer value of 6. It indicates an error occurred during an attempt to play.

Execute_failed

The `EXECUTE_FAILED` value property is an integer value of 7. It indicates an error occurred during an attempt to execute.

Print_failed

The PRINT_FAILED value property is an integer value of 8. It indicates an error occurred during an attempt to print.

DIBO Properties

getDIPErrorCode()

This DIBO returns an integer value indicating the specific error that caused the exception represented by this error object to be thrown. The value returned may be one of the specified values above, or some other value specified by other parts of MPEG-21.

ObjectMap

This DIML object represents the Object Map and provides the DIM author with access to the information contained in the Object Map. For information on the Object Map, see Section 9.5.1.

DIBO Properties

getArgumentList(index)

This DIBO returns an array of string values representing an Argument Type list.

As described in Section 9.4.2, if a DIM requires arguments, the corresponding Argument Types are listed in the Method Info *descriptor* and represented in XML by the Argument child elements of the MethodInfo element. Each DIM that requires arguments will have such an Argument Type list and in the Object Map these themselves form a list of Argument Type lists. In the Object Map, the Argument Type lists are listed in document order. Also if there is an Argument Type list whose Argument Types and their order are identical to another Argument Type list, only one such Argument Type list containing such a permutation of Argument Types will be in the Object Map (however a link to each DIM requiring such an Argument Type list is maintained in the Object Map). The getArgumentList DIBO returns one of these Argument Type lists and the index parameter is a number value that indicates the zero-based index of the Argument Type list in the Object Map.

This DIBO generates a DIPError exception with error code INVALID_PARAM if the index parameter is:

- not a number value; or
- not a valid index (a valid index is a number ranging from zero to one less than the value returned from getArgumentListCount()).

getArgumentListCount()

This DIBO returns a number value and is the number of Argument Type lists with Argument Types in a specific order that are defined in the Object Map. This corresponds to the number of unique permutations for all lists of DIP Argument children of all MethodInfo *descriptors* in the DIDL document. (See *getArgumentList* for further information).

getMethodCount(argumentList)

This DIBO returns a number value and is the number of DIMs that are declared to accept as parameters the Argument Types listed in argumentList (such an Argument Type list could be retrieved by calling getArgumentList()). This corresponds to the number of DIM declarations in the DIDL document that have a DIP MethodInfo *descriptor* with zero or more child Argument elements such that the number and values of those Argument elements match the number and names of the specified Argument Types.

The `argumentList` parameter is an array of string values that indicates the Argument Types. A zero length array passed as the `argumentList` parameter indicates DIMs declared to accept no Arguments (this includes DIM declarations that have a `Method-Info` *descriptor* with zero child `Argument` elements, and those that do not have a `MethodInfo` *descriptor*).

This DIBO generates a `DIPError` exception with error code `INVALID_PARAM` if the `argumentList` parameter is not an array of string values.

getMethodsWithArgs(argumentList)

This DIBO returns an array of `Element` objects corresponding to the DIDL `Component` elements representing the DIM declaration for DIMs that accept as parameters the Argument Types listed in `argumentList`. This corresponds to the list of DIDL `Component` elements, in document order, in the DIDL document that have a DIP `MethodInfo` descriptor with zero or more child `Argument` elements such that the number and values of those `Argument` elements match the number and names of the specified Argument Types.

The `argumentList` parameter is an array of string values that indicates the Argument Types. A zero length array passed as the `argumentList` parameter indicates DIMs declared to accept no arguments.

This DIBO generates a `DIPError` exception with error code `INVALID_PARAM` if the `argumentList` parameter is not an array of string values.

getMethodWithArgs(argumentList, index)

This DIBO returns an `Element` object corresponding to the DIDL `Component` element representing the DIM declaration for a DIM that accepts as parameters the Argument Types listed in `argumentList`.

The `argumentList` parameter is an array of string values that indicates the Argument Types. A zero length array passed as the `argumentList` parameter indicates DIMs that are declared to accept no Arguments.

The `index` parameter is a number value that indicates the zero-based index of the DIM in the list of DIMs that accept as parameters the Argument Types listed in `argumentList`. The order of the DIMs in the list of DIMs corresponds to the document order of the DIM declarations in the DIDL document that have a DIP `MethodInfo` *descriptor* with zero or more child `Argument` elements such that the number and values of those `Argument` elements match the number and names of the specified Argument Types.

This DIBO generates a `DIPError` exception with error code `INVALID_PARAM` if:

- the `argumentList` parameter is not an array of string values;
- the `index` parameter is not a number value; or
- the `index` parameter is not a valid index (a valid index is a number ranging from zero to one less than the value returned from `getMethodCount(argumentList)`).

getObjectOfType(typeName, index)

This DIBO returns an `Element` object of a given Object Type.

The `typeName` parameter is a string value that indicates the Object Type name.

The `index` parameter is a number value that indicates the zero-based index of the Object in the list of Objects of the given Object Type. The order of the Objects in the list of Objects corresponds to the document order of the elements in the DIDL document that have a child DIP `ObjectType` *descriptor* with a value matching the specified Object Type.

This DIBO generates a `DIPError` exception with error code `INVALID_PARAM` if:

- the `typeName` parameter is not a string value;
- the `index` parameter is not a number value; or
- the `index` parameter is not a valid index (a valid index is a number ranging from zero to one less than the value returned from `getObjectsOfTypeCount(typeName)`).

getObjectsOfType(typeName)
This DIBO returns an array of `Element` objects of a given Object Type. This corresponds to the list of elements, in document order, in the DIDL document that have a child DIP `ObjectType` *descriptor* with value matching the specified Object Type. If there are no objects of the given Object Type, a zero length array shall be returned.

The `typeName` parameter is a string value that indicates the Object Type name.

This DIBO generates a `DIPError` exception with error code `INVALID_PARAM` if the `typeName` parameter is not a string value.

getObjectsOfTypeCount(typeName)
This DIBO returns a number value and is the number of Objects that are defined in the Object Map to be of a given Object Type. This corresponds to the number of elements in the DIDL document that have a child DIP `ObjectType` descriptor with value matching the specified Object Type.

The `typeName` parameter is a String that indicates the Object Type name.

This DIBO generates a `DIPError` exception with error code `INVALID_PARAM` if the `typeName` parameter is not a string value.

getObjectTypeCount()
This DIBO returns a number value and is the number of Object Types that are defined in the Object Map. This corresponds to the number of unique values for all DIP `Object-Type` *descriptors* in the DIDL document.

getObjectTypeName(index)
This DIBO returns a string value representing the Object Type name.

The `index` parameter is a number value that indicates the zero-based index of the Object Type in the Object Map. The order of the Object Types in the Object Map corresponds to the document order of the DIP `ObjectType` *descriptors* in the DIDL document and with subsequent instances of duplicate values removed from the list.

This DIBO generates a `DIPError` exception with error code `INVALID_PARAM` if the `index` parameter is:

- not a number value; or
- not a valid index for an Object Type in the Object Map (a valid index is a number ranging from zero to one less than the value returned from `getObjectTypeCount()`).

PlayStatus
This DIML object represents the status of a DID Object that has been acted upon by a call to the `play` DIBO (which will be discussed later in this chapter). It provides the DIM author with access to basic state information of the DID Object.

The status transitions of the `PlayStatus` object are dependent on whether media associated with the playing Object is time-based media (that is, changes over time) or

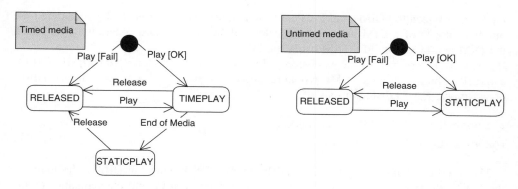

Figure 9.5 PlayStatus status transitions

static media (does not change over time). The status may also transition as a result of a subsequent call to the `release` DIBO passing the `PlayStatus` object returned from a previous call to the `play` DIBO. Figure 9.5 presents a diagram showing the status transitions.

Value Properties
Released
The `RELEASED` value property is an integer value of 0. This value indicates the associated *resource* is not currently playing. Note that a *resource* can be in the `RELEASED` state because an error prevented a request to play the *resource* from successfully completing, or as a result of an explicit request to release the *resource*.

A `RELEASED` *resource* has no other preserved state information. Playing a `RELEASED` *resource* will start with relevant state information in an initial state. For example, for a *resource* with time-based media, playing a `RELEASED` resource will commence from the media start time.

Staticplay
The `STATICPLAY` value property is an integer value of 1. This value indicates the associated *resource* is currently playing. Time-based state information related to playing the *resource*, if relevant, is paused for a `STATICPLAY` resource.

Timeplay
The `TIMEPLAY` value property is an integer value of 2. This value indicates the associated *resource* is currently playing. Time-based state information related to playing the *resource*, if relevant, is advancing for a `TIMEPLAY` resource. Note that only time-based media can have a status of `TIMEPLAY`.

DIBO Properties
getStatus()
This DIBO returns an integer indicating the status of the resource associated with the `PlayStatus` object. The returned value is one of the value properties specified above for the `PlayStatus` prototype.

9.6.3 DIGITAL ITEM BASE OPERATIONS
The DIBOs provide the DIM author with the standard interface to access the functionality of the MPEG-21 Peer for processing a DI by a DIM. That is, the DIBOs provide the

interface between the DIMs and the Peer functionality. The DIBOs can be considered the building blocks of a DIM and are similar to a library of standard functions from which the DIM author can build a DIM using the DIML.

The DIBOs form part of the extension of ECMAScript that makes up the DIML. DIBOs can include both additional DIML global functions as well as functions of additional DIML object types. Each DIBO consists of:

- a normatively defined interface, and
- normatively defined semantics.

The interface defines how to call a DIBO from a DIM using the DIML (the name of the DIBO, the parameters of the DIBO, the return type, if any), and the semantics define the functionality of the Peer that is called upon by the DIBO.

The semantics of the DIBOs are applicable in general to processing of a DI in a wide range of applications and with a wide range of resources. That is, the semantics of the DIBOs attempt to make as few assumptions as possible about the application of a DI or nature of the resources and metadata that it contains. For example, the play DIBO requests that the Peer plays a *component*, or *descriptor*, but the DIBO semantics make as few assumptions as possible about the content of the DID entity, or how the entity is to be presented as appropriate for the application of the DI or environment in which it is to be presented.

It is the Peer that provides the functionality of the DIBOs, and hence the Peer is able to provide the requested functionality in a manner that it sees as appropriate. For example, a Peer that is an application running on a desktop computer might implement the play DIBO to present an mp3 *resource* with a 'player' interface that provides substantial control (pause, fast forward, reverse, set current media time, etc.) over the playing of the mp3 *resource*. For a PDF *resource*, it might pass the presentation of the resource to a third party PDF-viewing application. For an mp3 *resource*, an MPEG-21 Peer on a mobile phone may simply commence playing of the audio data, and allow only a 'stop' control; for a PDF *resource* it may display a message indicating that type of media is not supported. Note this example is intended to show only how different MPEG-21 Peers might implement the functionality of a given DIBO in different manners. A well-constructed DI might be able to exclude access to unsupported resources on a given Peer. In each of the cases, the semantics of the play DIBO are still implemented, that is, 'play this *resource*', however, the Peer has been able to provide the functionality in a manner appropriate to its context.

In general, DIBOs fall in to one of the following categories:

- Operations that access and manipulate the DID at the DIDL level. For example, adding child nodes, modifying element attributes, and so on. The interface and semantics for these DIBOs are defined by the DOM Level 3 Core Specification [18], and, in particular, the ECMAScript bindings.
- Operations that load and serialize DIDL documents. The interface and semantics for these DIBOs are defined by the DOM Level 3 Load and Save Specification [19], and, in particular, the ECMAScript bindings.
- Operations related to particular parts of MPEG-21. The interface and semantics for these DIBOs are defined in the DIP specification and are described below.

DIBOs that are functions of DIML object types were described earlier along with the object type in Section 9.6.2.3. The following sections describe how one can use them.

Implementer's note: When implementing a DIBO, it is not required that the DIBO semantics be completely implemented within the body of the DIBO implementation. It is expected that DIBO implementations would utilize typical modular programming methodologies, for example, making calls to libraries of modules that implement functionality utilized not only by DIBOs but also in general processing of a DI according to various parts of MPEG-21. Referring back to the DIP-enabled Peer architecture in Section 9.3, then the DIBO implementations would be expected to make calls to functions implemented in the engines related to other parts of MPEG-21, as well as calls to functions provided by the underlying multimedia platform. In regards to the DIBO semantics, the requirement from the DIM author's perspective is that the complete DIBO semantics are implemented by the author's call to the DIBO. The actual mechanics of implementing the DIBO semantics is left to the DIBO implementer.

9.6.3.1 DIDL Document Access, Manipulation, Loading and Saving

DIDL is an XML-based declaration language for authoring DIDs that declare DIs. DIML is an ECMAScript-based scripting language for manipulating DIs. Hence, there is a requirement in DIML for DIBOs that provide the capabilities to access and manipulate the DIDL elements of a DIDL document.

The DOM interface specified by W3C is an interface designed for this purpose. It is an interface that is widely known and used, and is well suited to accessing and manipulating XML documents. The W3C DOM specifications also include ECMAScript bindings.

Hence, DIML extends standard ECMAScript by also including the ECMAScript bindings for the following W3C DOM specifications.

DOM Level 3 Core [18]	This provides the basic operations and objects for accessing and manipulating DIDL documents and DIDL elements.
DOM Level 3 Load and Save [19]	This provides the basic operations and objects for loading and saving DIDL documents.

The serialization-related operations defined by DOM Load and Save are also related to actions defined in part 6 of MPEG-21, the Rights Data Dictionary (RDD). As such, there might be rights (for example, expressed using part 5 of MPEG-21, REL) associated with the entities on which these operations act. In this case, the DIP engine will be required to check for permissions to exercise such rights.

Using the DOM Level 3 Core and DOM Level 3 Load and Save, it is possible to implement the removePoster DIM declared in Example 9.12. Example 9.13 contains this implementation. The argument `arg1` that will be passed to this DIM will be one of the `Descriptors` containing the posters. In the first step, the parent of the `Descriptor` is assigned to the variable `parent`. From this `parent` variable, the `removeChild` function is asking the DOM3 implementation to remove the `Descriptor` containing

```
function removePoster ( arg1 )
{
  var parent=arg1.parentNode;
  parent.removeChild(arg1);
  var registry=DOMImplementationRegistry.getDOMImplementation("LS");
  var domWriter = registry.createLSSerializer();
  var modifiedDID=domWriter.writeToString(didDocument);
}
```

Example 9.13   Implementation of the removePoster DIM.

the poster. In the next three lines, the DOM Level 3 Load and Save are used to write the result to a string variable called `modifiedDID`.

9.6.3.2 DID Related DIBOs

These DIBOs provide access to functionality that is related to part 2 of MPEG-21, DID. DID is described in detail in Chapter 3 of this book. The functionality accessed by these DIBOs relate to the semantics of DIDL elements. More general functions for accessing and manipulating DIDL elements are defined by the DOM API, as described in Section 9.6.3.1. These DID-related DIBOs are function properties of the global `DID` object.

configureChoice

Syntax

`configureChoice(choice)`

Semantics
A DIM author calls this DIBO to request the User to configure a specified *choice*. The *choice* is configured by making decisions about the *predicates* embodied by the child *selections* of the *choice*. If a *selection* is chosen by the User, its *predicate* becomes true; if it is rejected, its *predicate* becomes false; if it is unresolved, its *predicate* becomes undecided.

Implementer's note: The manner of configuring the *choice* is implementation dependent. For example, a DIBO implemented as part of a desktop computer software application might present a GUI allowing a human user to choose the *selections* of the *choice*. A DIBO implemented as part of a nonterminating network node might implement some form of automated configuration process.

Arguments
The `choice` argument is a DOM `Element` object that corresponds to the DIDL `Choice` element representing the *choice* to be configured.

Return Value
The `configureChoice` DIBO returns a boolean value. If the returned value is true, this indicates the *choice* configuration was modified by the User. If the returned value is false, this indicates the *choice* configuration was not modified by the User.

Exceptions
If the `choice` argument is not a DOM `Element` object representing a DIDL `Choice` element, a `DIPError` will be thrown with its DIP error code set to `INVALID_PARAMETER`.

setSelection

Syntax

```
setSelection(selection, state)
```

Semantics

A DIM author calls this DIBO to set the state of a specified *selection*. The state of a *selection* is set by making a decision about the *predicate* embodied by the *selection*. In the case of the `setSelection` DIBO, this decision is being made by the DIM author when they call the DIBO. The decision is indicated by the value of the `state` parameter, as described below.

Arguments

The `selection` argument is a DOM `Element` object that corresponds to the DIDL `Selection` element representing the *selection* whose state is to be set.

The `state` argument is a string value representing the decision being made about the state of the *predicate* embodied by the *selection*. Its value can be `true`, `false`, or `undecided`.

Exceptions

If the `selection` argument is not a DOM `Element` object representing a DIDL `Selection` element, a `DIPError` will be thrown with its DIP error code set to `INVALID_PARAMETER`. If the value of the `state` argument is not one of `true`, `false`, or `undecided`, a `DIPError` will be thrown with its DIP error code set to `INVALID_PARAMETER`.

areConditionsSatisfied

Syntax

```
areConditionsSatisfied(element)
```

Semantics

A DIM author calls this DIBO to test whether the *conditions* associated with a specified DID entity are satisfied or not. *Conditions*, in association with *selections* and *choices*, enable the run-time configuration of a DI by allowing the DID author to express the conditional inclusion of parts of the DI. A *condition* specifies the required state (either true or false) of one or more *predicates* for the *condition* to be satisfied. If there is more than one *condition*, then it is sufficient for one or more of the *conditions* to be satisfied to conditionally include the associated DID entity. For detailed information on *choices, selections, conditions* and their representation in DIDL, see Chapter 3 of this book.

If the specified DID entity is itself a *condition*, then the *condition* is tested by checking the required states of the *predicates* in the *condition* with their states (as the result of configuring *choices* or setting the state of *selections*). If the *condition* is satisfied, that is, those predicates required to be true are true, and those predicates required to be false are false, then a boolean value of true is returned. Otherwise, a boolean value of false is returned. Note that a value of false is returned when the *condition* is not satisfied, and also when the *condition* is unresolved because in that case one or more *predicates* is undecided.

If the specified DID entity is not a *condition*, then the *condition* children of the DID entity are tested. If the DID entity has no *condition* children, its inclusion is unconditional and a boolean value of true is returned. If there are one or more *condition* children, then each *condition* is tested as described in the preceding paragraph. If at least one *condition* is satisfied, then a boolean value of true is returned. Otherwise, if no *conditions* are satisfied (or cannot be resolved), then a boolean value of false is returned.

Arguments
The `element` argument is a DOM `Element` object that corresponds to the DIDL element representing the DID entity whose *conditions* are to be tested.

Return Value
The `areConditionsSatisfied` DIBO returns a boolean value as described above.

Exceptions
If the `element` argument is not a DOM `Element` object representing a DIDL element, a `DIPError` will be thrown with its DIP error code set to `INVALID_PARAMETER`.

Example
In Example 9.14, there is a DID containing a *choice* and a DIM for configuring that *choice*. In the DIM the `selection` variable is assigned the `Selection` element containing the WMA_FORMAT *predicate* on its `select_id`. Afterwards, the *predicate* of this *selection* is assigned the value true using the `setSelection` DIBO. Then `choice` variable is assigned the `Choice` element based on its `choice_id`. At this point, the predicate MP3_FORMAT has the value undecided and the predicate WMA_FORMAT has the value true.

The fourth line of the DIM definition demonstrates the use of the `configureChoice` DIBO. This DIBO could possibly pop up a GUI for further configuring the *choice*. Finally it is checked if the `Component` with id `component_01` is available after configuring the `choice`. The value of the `available` variable will be dependent on the configurations made by the `configureChoice` DIBO. If the User changed the value of the *predicate* MP3_FORMAT to true, the `Component` will be available.

9.6.3.3 DII Related DIBOs

These DIBOs provide access to functionality that is related to part 3 of MPEG-21, DII. DII is described in Chapter 3 of this book. These DII-related DIBOs are function properties of the global `DII` object.

getElementsByIdentifier
Syntax

```
getElementsByIdentifier(sourceDID, identifier)
```

Semantics
A DIM author calls this DIBO to retrieve DIDL elements that are identified using a DII Identifier. A DI or parts of the DI can be identified by utilizing a *descriptor* that contains a DII `Identifier` element. For further information on DII, see Chapter 3 of this book.

The DIDL elements that will be retrieved are those that contain a `Descriptor` child element that itself contains a child DII `Identifier` element whose value matches the value specified by the `identifier` argument.

```xml
<?xml version="1.0" encoding="UTF-8"?>
<DIDL xmlns="urn:mpeg:mpeg21:2002:02-DIDL-NS"
      xmlns:dip="urn:mpeg:mpeg21:2005:01-DIP-NS">
  <Item>
    <Choice choice_id="choice_01">
      <Descriptor>
        <Statement mimeType="text/plain">
          What format do you want?
        </Statement>
      </Descriptor>
      <Selection select_id="MP3_FORMAT">
        <Descriptor>
          <Statement mimeType="text/plain">MP3 Format</Statement>
        </Descriptor>
      </Selection>
      <Selection select_id="WMA_FORMAT">
        <Descriptor>
          <Statement mimeType="text/plain">WMA Format</Statement>
        </Descriptor>
      </Selection>
    </Choice>
    <Component id="component_01">
      <Condition require="MP3_FORMAT"/>
      <Resource mimeType="audio/mpeg" ref="http://server/track01.mp3"/>
    </Component>
    <Component>
      <Descriptor>
        <Statement mimeType="text/xml">
          <dip:Label>urn:mpeg:mpeg21:2005:01-DIP-NS:DIM</dip:Label>
        </Statement>
      </Descriptor>
      <Resource mimeType="application/mp21-method">
        function test_DID_DIBOs()
        {
          var selection = didDocument.getElementById( "WMA_FORMAT" );
          DID.setSelection(selection,"true");
          var choice = didDocument.getElementById( "choice_01" );
          DID.configureChoice( choice );
          var component=didDocument.getElementById( "component_01" );
          var available=DID.areConditionsSatisfied(component);
        }
      </Resource>
    </Component>
  </Item>
</DIDL>
```

Example 9.14 Calling DID related DIBOs.

Arguments

The sourceDID argument is a DOM Document object that represents the DID document instance from which the required DIDL elements are to be retrieved.

The identifier argument is a string value that specifies the value of the DII Identifier to be used to identify the DIDL elements to be retrieved.

Return Value

The `getElementsByIdentifier` DIBO returns an array of DOM `Element` objects, each representing any DIDL element in the `sourceDID` that is identified by the criteria specified by the `identifier` argument. If no DIDL elements match the criteria, an empty array will be returned.

Exceptions

If the `didDocument` argument is not a DOM `Document` object representing a DID document instance or if the `identifier` parameter is not a string value, a `DIPError` will be thrown with its DIP error code set to `INVALID_PARAMETER`.

getElementsByRelatedIdentifier

Syntax

```
getElementsByRelatedIdentifier(sourceDID, relatedIdentifier)
```

Semantics

A DIM author calls this DIBO to retrieve DIDL elements that are identified using DII related identifiers. Information related to a DI or parts of the DI can be identified by utilizing a *descriptor* that contains a DII `RelatedIdentifier` element. For further information on DII, see Chapter 3 of this book.

The DIDL elements that will be retrieved are those that contain a `Descriptor` child element that itself contains a child DII `RelatedIdentifier` element whose value matches the value specified by the `relatedIdentifier` argument.

Arguments

The `sourceDID` argument is a DOM `Document` object that represents the DID document instance from which the required DIDL elements are to be retrieved.

The `relatedIdentifier` argument is a string value that specifies the value of the DII `RelatedIdentifier` to be used to identify the DIDL elements.

Return Value

The `getElementsByRelatedIdentifier` DIBO returns an array of DOM `Element` objects, each representing any DIDL element in the `sourceDID` that is identified by the criteria specified in the `relatedIdentifier` argument. If no DIDL elements match the criteria, an empty array will be returned.

Exceptions

If the `sourceDID` argument is not a DOM `Document` object representing a DID document instance or if the `relatedIdentifier` parameter is not a string value a `DIPError` will be thrown with its DIP error code set to `INVALID_PARAMETER`.

getElementsByType

Syntax

```
getElementsByType(sourceDID, type)
```

Semantics

A DIM author calls this DIBO to retrieve DIDL elements that are identified using DII Type. A DI can be typed by utilizing a *descriptor* that contains a DII `Type` element. For further information on DII, see Chapter 3 of this book.

The DIDL elements that will be retrieved are those that contain a `Descriptor` child element that itself contains a child DII `Type` element whose value matches the value specified by the `type` argument.

Arguments
The `sourceDID` argument is a DOM `Document` object that represents the DID document instance from which the required DIDL elements are to be retrieved.

The `type` argument is a string value that specifies the value of the DII `Type` to be used to identify the DIDL elements to be retrieved.

Return Value
The `getElementsByType` DIBO returns an array of DOM `Element` objects, each representing any DIDL element in the `sourceDID` that is identified by the criteria specified by the `type` argument. If no DIDL elements match the criteria, an empty array will be returned.

Exceptions
If the `didDocument` argument is null or is not a DOM `Document` object representing a DID document instance, a `DIPError` will be thrown with its DIP error code set to `INVALID_PARAMETER`.

Example
Example 9.15 shows the use of the DII related DIBOs to retrieve DIDL elements that are identified using the tools specified by DII. The variable `byIdentifier` will be assigned an array containing the first `Item` element of the DIDL document. The `byRelatedIdentifier` and `byType` variables will contain the same value after executing the DIM. This illustrates how the first `Item` can be accessed using the three different DII DIBOs.

9.6.3.4 DIA Related DIBOs

These DIBOs provides access to functionality that are related to part 7 of MPEG-21, DIA. DIA is described in Chapters 7 and 8 of this book. The `adapt` DIBO presented in the following is also related to actions defined in part 6 of MPEG-21, the RDD. As such, there might be rights (for example, expressed using part 5 of MPEG-21, REL) associated with the entities on which these DIBOs act. In this case, DIBO implementations will be required to check for permissions to exercise such rights. These DIA-related DIBOs are function properties of the global `DIA` object.

Adapt
Syntax

```
adapt(element, metadata)
```

Semantics
A DIM author calls this DIBO to explicitly request that an attempt to adapt a specified *component* or *descriptor* be done. Note that the verb attempt is used because it is possible that the adaptation might fail if, for example, the Peer does not have any adaptation capabilities or does not have sufficient rights to perform adaptations, and so on. Since DIA provides an extensive set of tools to support DID and resource adaptation, DIA information that might be used in the adaptation could be implicitly included (by

```
<?xml version="1.0" encoding="UTF-8"?>
<DIDL xmlns="urn:mpeg:mpeg21:2002:02-DIDL-NS"
      xmlns:dii="urn:mpeg:mpeg21:2002:01-DII-NS"
      xmlns:dip="urn:mpeg:mpeg21:2005:01-DIP-NS">
  <Item>
    <Descriptor>
      <Statement mimeType="text/xml">
        <dii:Identifier>urn:dii:example:identifier</dii:Identifier>
      </Statement>
    </Descriptor>
    <Descriptor>
      <Statement mimeType="text/xml">
        <dii:RelatedIdentifier>
          urn:dii:example:relatedidentifier
        </dii:RelatedIdentifier>
      </Statement>
    </Descriptor>
    <Descriptor>
      <Statement mimeType="text/xml">
        <dii:Type>urn:dii:example:type</dii:Type>
      </Statement>
    </Descriptor>
    <Component>
      <Descriptor>
        <Statement mimeType="text/xml">
          <dip:Label>urn:mpeg:mpeg21:2005:01-DIP-NS:DIM</dip:Label>
        </Statement>
      </Descriptor>
      <Resource mimeType="application/mp21-method">

function test_DII_DIBOs() {
   var byIdentifier = DII.getElementsByIdentifier(didDocument,
                                "urn:dii:example:identifier");
   var byRelatedIdentifier = DII.getElementsByRelatedIdentifier(
                       didDocument, "urn:dii:example:relatedidentifier");
   var byType = DII.getElementsByType(didDocument,
                                "urn:dii:example:type");

}
      </Resource>
    </Component>
  </Item>
</DIDL>
```

Example 9.15   Calling DII related DIBOs.

the DI author) in *descriptors* of the specified DIDL element, or explicitly passed by the DIM author in the metadata parameter. DIA is described in Chapters 7 and 8 of this book.

This DIBO should be considered as providing the DIM author with the capability to explicitly request at a point of their choosing in a DIM that an adaptation of a specified DIDL element be attempted. It does not exclude the possibility that the MPEG-21 Peer might attempt an adaptation independently of any explicit call to the adapt DIBO by the DIM author. For example, a Peer might choose to also check for and attempt an adaptation of a DIDL element passed to the play DIBO. If the DIDL element passed to the play DIBO is the result of an explicit call to the adapt DIBO by the DIM author, then no further adaptation might be required.

The adaptation will be attempted by the underlying MPEG-21 Peer. The adaptation should be done utilizing any available and applicable information. For example, the adapt DIBO implementation might first check if the DIM author has provided some information in the metadata argument. If this information is supplied and the information is understood, then the adaptation could be attempted using that information. If the DIM author does not supply any information in the metadata argument, the DIBO implementation could check for *descriptors* of the specified element containing information it understands to be used for adaptation.

Implementer's note: An implementation of the adapt DIBO is not required to completely implement the adaptation within the body of the DIBO itself. For example, referring back to the DIP-enabled Peer architecture in Section 9.3, the adapt DIBO implementation requests the DIA engine to do any actual adaptation. In some architectures, for example, mobile devices, the DIA engine might even be implemented remotely, such as on an adaptation server.

Implementer's note: It is recommended that an implementer of the adapt DIBO considers support for the following tools defined by DIA (see Chapters 7 and 8):

- Usage Environment Descriptions
- Bitstream Syntax Description-based adaptation
- Peer and network quality of service
- Usage Constraints Descriptions
- DIA Configuration

Arguments
The element argument is a DOM Element object that represents a DIDL Component or Descriptor element to be adapted.

The metadata argument is either the null value or an array of DOM Element objects that represent additional information that can be considered when attempting the adaptation. For example, these could be DOM Element objects representing DID *descriptors* that specify DIA information relevant to the desired adaptation.

Return Value
The adapt DIBO returns a DOM Element object representing the adapted DIDL element. The qualified name of the DIDL element returned will be equivalent to that of the DIDL element passed as the element argument. This DOM Element will contain the adapted *resource* or *statement* either as an embedded *resource* or *statement* or as a referenced *resource* or *statement*.

If an adaptation was not attempted, then a null value will be returned. An adaptation might not be attempted, for example, if the underlying MPEG-21 Peer does not support adaptation of the specified DIDL element on the basis of the available information.

Exceptions
If the element argument is not a DOM Element object representing a DIDL Component, or Descriptor, a DIPError will be thrown with its DIP error code set to INVALID_PARAMETER.

If the `metadata` argument is not an array of DOM `Element` objects or null, a `DIPError` will be thrown with its DIP error code set to `INVALID_PARAMETER`.

If an adaptation is attempted, but an error occurs preventing the completion of the adaptation, a `DIPError` will be thrown with its DIP error code set to `ADAPTATION_FAILED`.

Example
Example 9.16 illustrates how a DIA Usage Environment Description can be used to give information to the DIA adaptation engine. The actual adaptation is triggered by the call to the `adapt` DIBO. After the execution of the test_DIA_DIBOs DIM, the variable `adaptedComponent` contains an adapted version of component_01 suited for a QCIF resolution (luminance with 176 by 144 samples).

9.6.3.5 DIP-related DIBOs

These DIBOs provide access to functionality that is related to part 10 of MPEG-21, DIP. DIP is described in this chapter. The `play`, `print`, and `execute` DIBOs are also related to actions defined in part 6 of MPEG-21, the RDD. As such, there might be rights (for example, expressed using part 5 of MPEG-21, REL) associated with the entities on which these DIBOs act. In this case, DIBO implementations will be required to check for permissions to exercise such rights. These DIP-related DIBOs are function properties of the global `DIP` object.

Play
Syntax

```
play(element, async)
```

Semantics
A DIM author calls this DIBO to request that an attempt to play a specified *component* or *descriptor* is done. The DIM author can request that the entity be played synchronously or asynchronously.

In RDD terms, playing an entity is the act of rendering the entity into a transient and directly perceivable representation. The implementation of the `play` DIBO must implement these semantics. However, the manner of doing so is an implementation choice. For example, one DIBO implementer might choose to implement the playing of mp3 media internally, while another DIBO implementer might choose to call a third party application to do so. Two DIBO implementers that both choose to implement playing of mp3 media internally might choose to provide different user interfaces and different levels of control over playback. For example, a mobile device might provide ability only to stop the playback, while a software product for a desktop computer environment might provide more extensive controls such as pause, fast forward, and so on.

While the presentation of the entity played is an implementation choice, it might utilize available information associated with the entity. For example, *descriptors* associated with the entity might contain information that the `play` DIBO implementation could use in the presentation of the entity. If the entity played is a *component*, then the media type of the associated *resource* can be obtained from the DIDL representation via the `mimeType` attribute of the `Resource` element. If the entity played is a *descriptor* containing a *component*, the media type of the associated *resource* can be obtained similarly. If the entity

```xml
<?xml version="1.0" encoding="UTF-8"?>
<DIDL xmlns="urn:mpeg:mpeg21:2002:02-DIDL-NS"
      xmlns:dip="urn:mpeg:mpeg21:2005:01-DIP-NS"
      xmlns:dia="urn:mpeg:mpeg21:2003:01-DIA-NS"
      xmlns:mpeg7="urn:mpeg:mpeg7:schema:2001">
  <Item>
    <Descriptor id="descriptor_01">
      <Statement mimeType="text/xml">
        <dia:DIA>
          <dia:Description xsi:type="dia:UsageEnvironmentType">
            <dia:UsageEnvironmentProperty xsi:type="dia:TerminalsType">
              <dia:Terminal>
                <dia:TerminalCapability xsi:type="dia:DisplaysType">
                  <dia:Display id="d0">
                    <dia:DisplayCapability
                            xsi:type="dia:DisplayCapabilityType">
                      <dia:Mode>
                        <dia:Resolution horizontal="176"
                                        vertical="144"/>
                      </dia:Mode>
                    </dia:DisplayCapability>
                  </dia:Display>
                </dia:TerminalCapability>
              </dia:Terminal>
            </dia:UsageEnvironmentProperty>
          </dia:Description>
        </dia:DIA>
      </Statement>
    </Descriptor>
    <Component id="component_01">
      <Resource mimeType="video/mpeg" ref="movie01.mpg"/>
    </Component>
    <Component>
      <Descriptor>
        <Statement mimeType="text/xml">
          <dip:Label>urn:mpeg:mpeg21:2005:01-DIP-NS:DIM</dip:Label>
        </Statement>
      </Descriptor>
      <Resource mimeType="application/mp21-method">
        function test_DIA_DIBOs()
        {
          var metadata = didDocument.getElementById( "descriptor_01" );
          var component = didDocument.getElementById( "component_01" );
          var adaptedComponent = DIA.adapt(component,metadata);
        }
      </Resource>
    </Component>
  </Item>
</DIDL>
```

Example 9.16 Calling a DIA related DIBO.

played is a *descriptor* containing a *statement*, then the media type of the *statement* can be obtained from the DIDL representation via the `mimeType` attribute of the `Statement` element.

Implementer's note: An implementation of the `play` DIBO is not required to completely implement the playing of the entity within the body of the DIBO itself. For example, referring back to the DIP-enabled Peer architecture in Section 9.3, the `play` DIBO implementation might call on functionality provided by the underlying multimedia platform to actually play the entity.

Arguments

The `element` argument is a DOM `Element` object that represents either a DIDL `Component` or `Descriptor` element to be played.

The `async` argument is a boolean value. If the value is true, then the entity is played asynchronously, in which case control is returned immediately to the DIM once the playing of the entity is initiated. If the value is false, then the entity is played synchronously, in which case control is returned to the DIM only after playing is complete. More explicitly, if the entity is played synchronously, then control is not returned to the DIM until the status code of the `PlayStatus` object associated with this call to the `play` DIBO transitions to `STATICPLAY` or `RELEASED`.

Return Value

The `play` DIBO returns a `PlayStatus` object (as described in Section 9.6.2.3). The returned `PlayStatus` object can be thought of as encapsulating the status of the specified `element` in association with this particular call to the `play` DIBO. As well as its use in providing the DIM author the status of the playing entity, it can also be utilized as an argument to the `release` DIBO. The `release` DIBO will be explained later in this chapter.

Exceptions

If the `element` argument is not a DOM `Element` object representing a `Component`, or a `Descriptor`, a `DIPError` will be thrown with its DIP error code set to `INVALID_PARAMETER`.

If the `async` parameter is not a boolean value, a `DIPError` will be thrown with its DIP error code set to `INVALID_PARAMETER`.

If playback is attempted, but an error occurs preventing the completion of the playback, a `DIPError` will be thrown with its DIP error code set to `PLAYBACK_FAILED`.

Release

Syntax

```
release(playStatus)
```

Semantics

A DIM author calls this DIBO to request that playing of an entity be stopped and that any associated state information be released.

Arguments

The `playStatus` argument is a `PlayStatus` object (as described in Section 9.6.2.3). This is a `PlayStatus` object returned from a previous call to the `play` DIBO. After calling the release DIBO, the status code of the PlayStatus object will transition to RELEASED. If the status was already RELEASED, then the release DIBO does nothing.

Exceptions

If the playStatus parameter is not a PlayStatus object, a DIPError will be thrown with its DIP error code set to INVALID_PARAMETER.

Print

Syntax

```
print(element)
```

Semantics

A DIM author calls this DIBO to request that an attempt to print a specified *component*, or *descriptor* be done.

In RDD terms, printing an entity is the act of rendering the entity into a fixed physical and directly perceivable representation. The manner of doing so appropriate to the entity being printed is an implementation choice.

While the presentation of the entity printed is an implementation choice, it might utilize available information associated with the entity. Such information is available to the print DIBO implementer in a similar manner as for the play DIBO.

Implementer's note An implementation of the print DIBO is not required to completely implement the printing of the entity within the body of the DIBO itself. For example, referring back to the DIP-enabled Peer architecture in Section 9.3, the print DIBO implementation might call on functionality provided by the underlying multimedia platform to actually print the entity.

Arguments

The element argument is a DOM Element object that represents either a DIDL Component or a Descriptor element to be played.

Return Value

The print DIBO returns a boolean value. If true, then the specified entity was able to be printed ok, otherwise if the specified entity was not printed, false is returned.

Exceptions

If the element argument is not a DOM Element object representing a DIDL Component, or Descriptor, a DIPError will be thrown with its DIP error code set to INVALID_PARAMETER.

If printing is attempted, but an error occurs preventing the completion of the printing, a DIPError will be thrown with its DIP error code set to PRINT_FAILED.

Execute

Syntax

```
execute(element)
```

Semantics

A DIM author calls this DIBO to request that an attempt to execute a specified *component*, or *descriptor*, be done.

In RDD terms, execute refers to the primitive computing process of executing a digital resource. The manner of doing so appropriate to the entity being executed is an implementation choice.

While the execution of the specified entity is an implementation choice, it might utilize available information associated with the entity. Such information is available to the execute DIBO implementer in a similar manner as for the play DIBO.

Programmer's note: It should be noted that including executable *resources* in a DI and calling the execute DIBO for those *resources* could make your DI less portable if the executable resource is not able to be executed across different platforms. This lack of portability is at the level of the executable resource, not at the DIM level, as the execute DIBO will still be recognized and executed on a DIP-enabled Peer; however, it might only be able to execute the *resource* (and return true) on one platform, and not be able to execute the resource on other platforms (thus returning false).

Implementer's note: An implementation of the execute DIBO is not required to completely implement the executing of the entity within the body of the DIBO itself. For example, referring back to the DIP-enabled Peer architecture in Section 9.3, the execute DIBO implementation might call on functionality provided by the underlying multimedia platform to actually execute the entity.

Implementer's note: It is the responsibility of the DIBO implementation and DIP engine to ensure security related issues are addressed when executing executable resources. For example providing a warning to the User, checking of certificates, if present, and so on.

Arguments
The element argument is a DOM Element object that represents either a DIDL Component or a Descriptor element to be executed.

Return Value
The execute DIBO returns a boolean value. If true, then the specified entity was able to be executed ok, otherwise if the specified entity was not executed, false is returned. An entity might not be executed, for example, if the digital resource is not recognized as an executable resource by the MPEG-21 Peer.

Exceptions
If the element argument is not a DOM Element object representing a DIDL Component, or Descriptor, a DIPError will be thrown with its DIP error code set to INVALID_PARAMETER.

If more than one supported executable digital resource is associated with the specified entity, a DIPError will be thrown with its DIP error code set to INVALID_PARAMETER.

If execution is attempted, but an error occurs preventing the completion of the execution, a DIPError will be thrown with its DIP error code set to EXECUTE_FAILED.

getObjectMap

Syntax

```
getObjectMap(document)
```

Semantics
This DIBO allows the retrieval of an Object Map from a DIDL document.

Arguments
The document argument is a DOM Document object that represents a DIDL document with the Object Type and DIM declarations information needed to construct the Object Map (see Section 9.5.1). The Object Map is represented in DIML by an ObjectMap object (see 9.6.2.3).

Return Value
An ObjectMap object representing the Object Map of the DIDL document.

Exceptions
If the document argument is not a DOM Document object representing a DIDL document, a DIPError will be thrown with its DIP error code set to INVALID_PARAMETER.

getValues
Syntax

```
getValues(dataTypes, requestMessages)
```

Semantics
A DIM author calls this DIBO to request the User to supply values for boolean, string or number data. Optional messages for each requested data value may also be provided by the DIM author. The manner by which the User is requested to supply the data values is a DIBO implementation choice. For example, a data entry GUI might be presented to a human user to allow the human to enter the data values. In another implementation that might be part of an automated test environment, the values might be automatically supplied from a configuration file and the request messages, if any, used to generate log messages (note that in this case the automated test environment is the User).

Arguments
The dataTypes argument is an array of one or more string values. Each entry in the array must contain a string value equal to one of the values Boolean, String, or Number. These entries indicate the primitive data type for each data value that the User will be requested to supply.

The requestMessages argument may be null, or an array. If the requestMessages argument is an array, it must be an array of equal size to the array passed as the dataTypes argument. Each entry in the array must be either null or a string value. If a string value is provided, then the value may be used by the getValues DIBO implementation as a request message for the data value whose type is at the corresponding index in the dataTypes array.

Return Value
The getValues DIBO returns null or an array. If no data values were supplied by the User, then null is returned. Otherwise an array whose size is equal to the size of the array passed as the dataTypes argument is returned. Each entry in the array will be null or a value whose primitive data type will be as specified by the corresponding entry in the dataTypes array.

Exceptions
If the dataTypes argument contains an array entry whose value is not one of Boolean, String, or Number, a DIPError will be thrown with its DIP error code set to INVALID_PARAMETER.

If the `requestMessages` argument is not null and is not an array of the same size as `dataTypes` with each array member being a string value or null, a `DIPError` will be thrown with its DIP error code set to `INVALID_PARAMETER`.

getObjects
Syntax

```
getObjects(objectTypes, requestMessages)
```

Semantics
A DIM author calls this DIBO to request the User to choose from the current DID instance document available DID Objects of specified Object Types. Optional messages for each requested DID Object Type may also be provided by the DIM author. The manner by which the User is requested to choose the DID Objects is a DIBO implementation choice. For example, a GUI might be presented to a human user to allow the human to choose the DID Objects. In another implementation, that might be part of an automated test environment; the DID Objects might be automatically chosen according to a configuration file and the request messages, if any, used to generate log messages (note that in this case the automated test environment is the User).

Programmer's note: It should be noted that using the `getObjects` DIBO can prevent the DIP engine from determining what DID Objects will be acted upon by executing a DIM. For this reason, it is strongly recommended that, whenever possible, DID Objects to be acted upon in a DIM always be passed as DIM arguments (and hence they will be described by DIP `Argument` child elements of a DIP `MethodInfo` element).

Implementer's note: The DID Objects for a particular specified Object Type from which the User may choose are those Objects in the Object Map that map to the specified Object Type. That is, those (available) DIDL elements that have a child *descriptor* containing a DIP `ObjectType` element with value matching the specified Object Type.

Arguments
The `objectTypes` argument is an array of one or more string values. Each entry in the array must contain a string representing an Object Type. These entries indicate the Object Type for each DID Object that the User will be requested to choose.

The `requestMessages` argument may be null or an array of string values. If the `requestMessages` argument is an array, it must be an array of equal size to the array passed as the `objectTypes` argument. Each entry in the array must be either null or a string value. If a string value is provided, then the value may be used by the `getObjects` DIBO implementation as a request message for the DID Object whose Object Type is at the corresponding index in the `objectTypes` array.

Return Value
The `getObjects` DIBO returns null or an array. If no Objects were chosen by the User then null is returned. Otherwise, an array whose size is equal to the size of the size of the array passed as the `objectTypes` argument is returned. Each entry in the array will be null or a DOM `Element` representing a DIDL element that maps to an Object whose Object Type will be as specified by the corresponding entry in the `objectTypes` array.

Exceptions

If the `objectTypes` argument is not an array of string values, a `DIPError` will be thrown with its DIP error code set to `INVALID_PARAMETER`.

If the `requestMessages` argument is not null and is not an array of the same size as `dataTypes` with each array member being a string value or null, a `DIPError` will be thrown with its DIP error code set to `INVALID_PARAMETER`.

getExternalData

Syntax

```
getExternalData(mimeTypes, requestMessages)
```

Semantics

A DIM author calls this DIBO to request the User to choose resources located external to the DI. Optional messages for each requested resource may also be provided by the DIM author. The manner by which the User is requested to choose the external resources is a DIBO implementation choice.

Arguments

The `mimeTypes` argument is an array with one element for each of the resources to be chosen. Each entry in this array is itself an array of one or more string values. Each string value specifies an allowed Multipurpose Internet Mail Extensions (MIME) type for the corresponding resource in the form `type/subtype`, where the values for `type` and `subtype` are as for MIME media types [20, 21]. This argument thus indicates the allowable MIME types for each resource that the User will be requested to choose.

The `requestMessages` argument may be null or an array. If the `requestMessages` argument is an array, it must be an array of equal size to the array passed as the `mimeTypes` argument. Each entry in the array must be either null or a string value. If a string value is provided, then the value may be used by the `getExternalData` DIBO implementation as a request message for the resource whose allowable MIME types are specified by the corresponding index in the `mimeTypes` array.

Return Value

The `getExternalData` DIBO returns null or an array. If no resources were chosen by the User, then null is returned. Otherwise an array whose size is equal to the size of the array passed as the `mimeTypes` argument is returned. Each entry in the array will be null or a string value representing a Uniform Resource Locator (URL) locating a resource whose MIME type will be one of those as specified by the corresponding entry in the `mimeTypes` array.

Exceptions

If the `mimeTypes` argument is not an array of string values, a `DIPError` will be thrown with its DIP error code set to `INVALID_PARAMETER`.

If the `requestMessages` argument is not null and is not an array of the same size as `mimeTypes` with each array member being a string value or null, a `DIPError` will be thrown with its DIP error code set to `INVALID_PARAMETER`.

runDIM

Syntax

```
runDIM(itemIdType, itemId, componentIdType, componentId, arguments)
```

Semantics
A DIM author calls this DIBO to run an identified DIM. This allows a DIM to be invoked from within another DIM. The identified DIM is the DIM declared in the DIDL `Component` element that is identified by a combination of the `itemId` and `componentId`. Control is not returned to the invoking DIM until the invoked DIM has completed (normally or otherwise).

Arguments
The `itemIdType` argument is a string value that identifies the type of the `itemId` identifier. Possible values for this string are 'dii', which means the `itemId` parameter will be a DII Identifier or 'uri', which indicates a URI-based `itemId`. If the `itemId` is null, then only the `componentId` is used and the identified `Component` is run.

The `componentIdType` argument is a string value that identifies the type of the `componentId` identifier. Possible values for this string are 'dii', which means the `componentId` parameter will be a DII Identifier or 'uri', which indicates a URI-based `componentId`. If the `componentId` is null, then the DIBO implementer may choose any `Component` found in the identified `Item` containing a DIM and run that DIM.

The `arguments` argument is an array of zero or more DOM `Elements` representing the DID Objects that are the required arguments for this DIM.

Exceptions
If the DIBO implementation fails to locate a DIM based on the two identification arguments, a `DIPError` will be thrown with its DIP error code set to `NOT_FOUND`.

If either the `itemIdType` or `componentIdType` parameters do not specify a valid value, a `DIPError` will be thrown with its DIP error code set to `INVALID_PARAMETER`.

If both the `itemId` and `componentId` parameters are null, a `DIPError` will be thrown with its DIP error code set to `INVALID_PARAMETER`.

If both the `itemId` or `componentId` parameters are not string values or null values, a `DIPError` will be thrown with its DIP error code set to `INVALID_PARAMETER`.

If either the `itemId` parameter, if not null identifies an element that is not an `Item` or the `componentId` parameter, if not null, identifies an element that is not a `Component`, a `DIPError` will be thrown with its DIP error code set to `INVALID_PARAMETER`.

If both an `Item` is identified and a `Component` is identified, but the `Component` is not a child of the `Item`, a `DIPError` will be thrown with its DIP error code set to `INVALID_PARAMETER`.

If the `arguments` array does not contain objects of the required Argument Types for this DIM, a `DIPError` will be thrown with its DIP error code set to `INVALID_PARAMETER`.

If the DIM cannot be run for any other reason, a `DIPError` will be thrown with its DIP error code set to `GENERAL_EXCEPTION`.

runJDIXO
Syntax

```
runJDIXO(itemIdType, itemId, componentIdType, componentId, classname,
    arguments)
```

Semantics
A DIM author calls this DIBO to run an identified J-DIXO (for more information on DIXOs and J-DIXOs, see 9.6.4). This allows a J-DIXO to be invoked from within a

DIM. The identified J-DIXO is the J-DIXO with the given classname declared in the DIDL `Component` element that is identified by a combination of the `itemId` and `componentId`. Control is not returned to the invoking DIM until the invoked J-DIXO has completed (normally or otherwise).

Arguments

The `itemIdType` argument is a string value that identifies the type of the `itemId` identifier. Possible values for this string are 'dii', which means the `itemId` parameter will be a DII Identifier or 'uri', which indicates a URI-based `itemId`. If the `itemId` is null, then only the `componentId` is used and the identified `Component` is run.

The `componentIdType` argument is a string value that identifies the type of the `componentId` identifier. Possible values for this string are 'dii', which means the `componentId` parameter will be a DII Identifier or 'uri' indicates in a URI-based `componentId`. If the `componentId` is null, then the DIBO implementer may choose any `Component` found in the identified `Item` containing a J-DIXO with the given classname and run that J-DIXO.

The `classname` argument is a string containing the fully qualified Java class name of the J-DIXO.

The `arguments` argument is an array of zero or more DOM `Elements` representing the DID Objects that are the required arguments for this J-DIXO.

Exceptions

If the DIBO implementation fails to locate a J-DIXO based on the two identification arguments, a `DIPError` will be thrown with its DIP error code set to `NOT_FOUND`.

If the type of the objects passed in the `arguments` argument do not match those returned by the `getArgumentTypes()` method of the J-DIXO interface, a `DIPError` will be thrown with its DIP error code set to `INVALID_PARAMETER`.

If either the `itemIdType` or `componentIdType` parameters do not specify a valid value, a `DIPError` will be thrown with its DIP error code set to `INVALID_PARAMETER`.

If both the `itemId` and `componentId` parameters are null, a `DIPError` will be thrown with its DIP error code set to `INVALID_PARAMETER`.

If both the `itemId` or `componentId` parameters are not string values or null values, a `DIPError` will be thrown with its DIP error code set to `INVALID_PARAMETER`.

If either the `itemId` parameter, if not null identifies an element that is not an `Item` or the `componentId` parameter, if not null, identifies an element that is not a `Component`, a `DIPError` will be thrown with its DIP error code set to `INVALID_PARAMETER`.

If both an `Item` is identified and a `Component` is identified, but the `Component` is not a child of the `Item`, a `DIPError` will be thrown with its DIP error code set to `INVALID_PARAMETER`.

If the J-DIXO cannot be run for any other reason, a `DIPError` will be thrown with its DIP error code set to `GENERAL_EXCEPTION`.

Alert

Syntax

```
alert(message, messageType)
```

Semantics

A DIM author calls this DIBO to alert the User of some circumstance via a specified message. The manner by which the User is alerted of the circumstance via the message is a DIBO implementation choice. For example, a GUI might be presented to a human user displaying the `message` in a format based on the `messageType` argument. In another implementation that might be part of an automated test environment, the `message` might be appended to a log file with time-stamp and prefix based on the `messageType` argument.

Arguments

The `message` argument is a string value that is the textual message to be used to alert the User.

The `messageType` argument is an integer value that indicates the nature of the message and must be one of the message type codes defined as value properties of the DIML global property (as described in Section 9.6.2.2). Specifically, the value must be one of `MSG_INFO`, `MSG_WARNING`, `MSG_ERROR`, or `MSG_PLAIN`.

Exceptions

If `message` is not a string value or `messageType` does not specify a valid value, a `DIPError` will be thrown with its DIP error code set to `INVALID_PARAM`.

Wait

Syntax

```
wait(timeInterval)
```

Semantics

A DIM author calls this DIBO to pause execution of the invoking DIM. When the `wait` DIBO is invoked, control is not returned to the DIM until the specified time interval has elapsed.

The accuracy to which the time interval for the pause in execution is able to be adhered to should be specified by DIBO implementers. The DIBO implementer should also indicate any circumstances in which the pause might be interrupted.

Arguments

The `timeInterval` argument is an integer value specifying the number of milliseconds defining the time interval for which the DIM execution should be paused.

Exceptions

If `timeInterval` is not a positive number value, a `DIPError` will be thrown with its DIP error code set to `INVALID_PARAM`.

Example

Example 9.17 demonstrates the use of most DIP DIBOs. The `alert` DIBO is used to print the message 'This is nice!'. Afterwards, a combination of the `getValues`, `play`, `wait` and `release` DIBOs is used to allow playback of movie01.mpg for a predefined length of time. The `print` DIBO is used to print the 'Print me!' *descriptor*. After printing, the Java program defined in coolprogram.class is run by calling the `execute` DIBO. The `runDIM` DIBO is used to call test_DIP_DIBOs, resulting in an infinite loop of the test_DIP_DIBOs DIM.

9.6.3.6 REL-Related DIBOs

These DIBOs provide access to functionality that is related to part 5 of MPEG-21, REL. REL is described in Chapter 5 of this book. These REL-related DIBOs are function properties of the global `REL` object.

```xml
<?xml version="1.0" encoding="UTF-8"?>
<DIDL xmlns="urn:mpeg:mpeg21:2002:02-DIDL-NS"
      xmlns:dip="urn:mpeg:mpeg21:2005:01-DIP-NS" >
  <Item>
    <Descriptor id="descriptor_01">
      <Statement mimeType="text/plain">Print me!</Statement>
    </Descriptor>
    <Component id="component_01">
      <Resource mimeType="video/mpeg" ref="movie01.mpg"/>
    </Component>
     <Component id="component_02">
      <Resource mimeType="application/java" ref="coolprogram.class"/>
    </Component>
    <Component id="component_03">
      <Descriptor>
        <Statement mimeType="text/xml">
          <dip:Label>urn:mpeg:mpeg21:2005:01-DIP-NS:DIM</dip:Label>
        </Statement>
      </Descriptor>
      <Resource mimeType="application/mp21-method">
          function test_DIP_DIBOs()
          {
          DIP.alert("This is nice!", MSG_INFO);
          var component = didDocument.getElementById( "component_01" );
          var dataTypes=["Number"];
          var messages=["How long (in milliseconds) do you want to
                                              listen?"];
          var time=DIP.getValues(dataTypes,messages);
          var playstatus=DIP.play(component,true);
          DIP.wait(time [0]);
          DIP.release(playstatus);
          var descriptor = didDocument.getElementById("descriptor_01");
          DIP.print(descriptor);
          var program = didDocument.getElementById( "component_02" );
          DIP.execute(program);
          DIP.runDIM("uri",null,"uri","#component_03",new Array());
          }
      </Resource>
    </Component>
  </Item>
</DIDL>
```

Example 9.17   Calling DIP related DIBOs

getLicense
Syntax

```
getLicense(resource)
```
Semantics

A DIM author calls this DIBO to request an attempt to retrieve any licenses associated with a specified *resource* be done.

The licenses retrieved by the `getLicense` DIBO will be those associated with the specified *resource* and that are expressed as specified in REL. Such a license can be

- located within the DI,

- referenced from the DI,
- accessed via a service referenced from the DI, or
- none of the above.

More than one license may be retrieved since the *resource* may have more than one associated license for different rights/actions.

It is important to understand that the getLicense DIBO should be considered as providing the DIM author the capability to retrieve license information, that is, it enables license information to be available to the DIM author. The getLicense DIBO is not intended to be used to actually protect a *resource*. Rights associated with a *resource*, or any part of a DI, must be checked and enforced regardless of whether a User interacts with the DI using a DIM or not. Recall that it is not mandatory for a User to use the DIMs to interact with the DI. The DIMs provide a DI author an interoperable means of providing their suggested interactions with the DI. Therefore, a User might interact with the DI without using any DIMs. In all cases in which a *resource* is acted upon, any associated rights must be checked and enforced. If the User does interact with the DI using the available DIMs, then all of the underlying DIBO implementations that act upon parts of the DI must also do rights checks and enforcement.

Implementer's note: Since an MPEG-21 Peer is required to do rights checks and enforcement regardless of how the interaction with the DI takes place, a possible implementation option for a Peer is to utilize a library of rights checking and enforcement functions. This library of rights-related functions could then be used in a modular way. They could be called by the DIBO implementer from the implementation of any DIBOs that require implementing rights related functionality (possibly including the implementation of getLicense DIBO), as well as from any other components of the Peer that act upon parts of the DI. In referring back to the DIP-enabled Peer architecture in Section 9.3, this might be represented by the REL engine.

Arguments
The resource argument is a DOM Element object representing the DIDL Resource element for which associated licenses will be retrieved.

Return Value
The getLicense DIBO returns null or an array of DOM Element objects. If no licenses associated with the specified resource could be retrieved, then null is returned. Otherwise an array of DOM Element objects is returned with each Element representing the licenses associated with the specified resource.

Exceptions
If the resource argument is not a DOM Element representing a DIDL Resource element, a DIPError will be thrown with its DIP error code set to INVALID_PARAMETER.

queryLicenseAuthorization
Syntax

```
queryLicenseAuthorization(license, resource, rightNs, rightLocal,
    additionalInfo)
```

Semantics
A DIM author calls this DIBO to request an attempt to check for the existence, at the time of calling the DIBO, of an authorization proof for an authorization request based on the DIBO arguments as described below.

REL defines an authorization model that includes the concepts of an authorization request and an authorization proof. To answer yes to an authorization request, an authorization proof must exist to prove the answer is yes. Otherwise, if no authorization proof exists then the answer to the authorization request is no. For further information on the REL authorization model, see Chapter 5 of this book.

The `queryLicenseAuthorization` DIBO provides the DIM author with the capability to check for the existence of an authorization proof for an authorization request containing the following members:

1. A principal representing the User running this DIBO

2. A right as specified by the `rightNs` and `rightLocal` arguments to this DIBO

3. The resource specified by the `resource` argument to this DIBO

4. A time interval of zero length that occurs during the execution of this DIBO

5. An authorization context determined nonnormatively by the DIBO implementation, possibly with the guidance of the `additionalInfo` argument to this DIBO

6. A set of licenses with one member, that member being the license specified in the `license` argument to this DIBO

7. A set of grants determined nonnormatively by the DIBO implementation to serve as root grants, possibly with the guidance of the `additionalInfo` argument to this DIBO

The DIBO only supports authorization requests containing a single license. It is not intended to support authorization requests with more than one license, for example, the case where one license allows a group to play a *resource*, and a second license that assigns a User to that group.

The DIBO will return a boolean value of true if an authorization proof exists for a yes answer to the authorization request formulated as described above. Otherwise, if an authorization proof does not exist or cannot be found, a boolean value of false is returned.

It is important to note that the return value of the `queryLicenseAuthorization` DIBO depends on the authorization context and trust root determined by the DIBO implementation. Therefore, a DIM author should not rely on the return value of this DIBO unless the mechanism used by the DIBO implementation to determine the authorization context and trust root are known to the DIM author.

Authorization context information can include such information as a User's location, payment history and usage history. This information could be obtained, for example, automatically by examining information stored on the Peer, or querying a server. Another example of how the information could be obtained is by asking assistance from the User via a user input window. In other cases, the DIM author might provide some of this information using the `additionalInfo` argument.

It is also possible that a `queryLicenseAuthorization` DIBO implementation might choose to knowingly construct a false authorization context in some circumstances. A DIBO implementation might do this, for example, to provide preliminary screening of a User's intentions. As an example, if the license is conditional on a per-use payment, the DIBO implementation might ask a human if they are willing to make the payment to exercise the associated right on the associated resource. Without actually charging for the payment, the DIBO implementation might still use an authorization context that says the User did make the payment so that the result reflects whether the User would be authorized if the payment was made.

Implementer's note: To assist DIM authors, it might be helpful for implementers of the `queryLicenseAuthorization` DIBO to make known to DIM authors such information as how the authorization context and trust root is determined (possibly including circumstances in which a false authorization context will be used, if that is part of the DIBO implementation).

Note also that the return value of the `queryLicenseAuthorization` DIBO reflects the status of the existence of the authorization proof for the authorization request at the time and in the authorization context determined when the check is evaluated during the running of the DIBO. Thus, it is possible that the authorization proof exists at the time the check is evaluated during the running of the DIBO, but not at a time subsequent to the running of the DIBO, due to conditions included in the license information. Hence, the return value of the `queryLicenseAuthorization` DIBO should be used by the DIM author with this in mind. For example, consider the case where the DIM author calls the `queryLicenseAuthorization` DIBO and receives a return value of true, then calls the `play` DIBO to play the resource. While playing the resource, the Peer might be required to update the authorization context and trust root. This might be, for example, to enable checking that the play is still allowable over a time interval. If the play extends beyond the allowable time interval, an authorization proof might no longer exist, and the Peer can stop the play in accordance with the conditions of the license. However, if this occurs, then the previous true value returned by the initial call to the `queryLicenseAuthorization` DIBO is now out of date and no longer reflects the current status.

It is important to understand that the `queryLicenseAuthorization` DIBO should be considered as providing the DIM author the capability to check from within a DIM and at the time that the DIBO is called, whether an authorization proof exists for the authorization request on the basis of the DIBO parameters. The `queryLicenseAuthorization` DIBO is not intended to be used to actually protect a resource. If rights are associated with a resource, or any part of a DI, then existence of such an authorization proof for a suitable authorization request must be checked regardless of whether a User interacts with the DI using a DIM or not. Recall that it is not mandatory for a User to use the DIMs to interact with the DI. The DIMs provide a DI author an interoperable means of providing their suggested interactions with the DI. Therefore, a User might interact with the DI without using any DIMs. In all cases in which a resource is acted upon, any associated rights must be checked and enforced. If the User does interact with the DI using the available DIMs, then all of the underlying DIBO implementations that act upon parts of the DI must also do rights checks and enforcement.

Arguments

The `license` argument is a DOM `Element` object representing the license information.

The `resource` argument is a DOM `Element` object representing the DIDL `Resource` element.

The `rightNs` argument is null or a string value representing the namespace of the right to be checked.

The `rightLocal` argument is a string value. If the value of the `rightNs` argument is not null, then the value of the `rightLocal` argument represents the localname of the right to be checked. If the value of the `rightNs` argument is null, then the value of the `rightLocal` argument is the value of the definition attribute of `sx:rightUri`.

The `additionalInfo` argument is null or an array of DOM `Element` objects representing additional information that can be considered when validating the license.

Return Value

The `queryLicenseAuthorization` DIBO returns a boolean value. If a corresponding authorization proof is found, true is returned. Otherwise if a corresponding authorization proof does not exist or cannot be found, false is returned.

Exceptions

If the `license` argument is not a DOM `Element` or does not contain any license information, a `DIPError` will be thrown with its DIP error code set to `INVALID_PARAMETER`.

If the `resource` argument is not a DOM `Element` representing a DIDL `Resource` element, a `DIPError` will be thrown with its DIP error code set to `INVALID_PARAMETER`.

If the `rightNS` argument is not a string value or null value, a `DIPError` will be thrown with its DIP error code set to `INVALID_PARAMETER`.

If the `rightLocal` argument is not a string value or null value, a `DIPError` will be thrown with its DIP error code set to `INVALID_PARAMETER`.

If the `additionalInfo` argument is not an array of DOM `Element` objects or a null value, a `DIPError` will be thrown with its DIP error code set to `INVALID_PARAMETER`.

Example

Example 9.18 demonstrates the use of the `getLicense` and the `queryLicense-Authorization` DIBOs. During the first steps, a license for the movie01.mpg *resource* is retrieved. On the basis of this license, it is checked if it is allowed to exercise the play right on that *resource*. If it is allowed, a message 'Playing operation is authorized' is shown. Note that the result of the `queryLicenseAuthorization` is an informative result, it does not give any guarantee that the playing when called upon will actually succeed.

9.6.4 DIGITAL ITEM EXTENSION OPERATIONS

A DIXO is an action allowing nonstandardized operations to be invoked by a DIM. The DIXO mechanism realizes the extension of the set of standardized DIBOs in an interoperable way.

Apart from being a mechanism to access native functionality, DIBOs also serve to abstract out complicated operations and provide a high-level interface to DIMs. On a number of occasions, one might need access to operations that are not normatively defined

```xml
<?xml version="1.0" encoding="UTF-8"?>
<DIDL xmlns="urn:mpeg:mpeg21:2002:02-DIDL-NS"
      xmlns:dip="urn:mpeg:mpeg21:2005:01-DIP-NS"
      xmlns:r="urn:mpeg:mpeg21:2003:01-REL-R-NS"
      xmlns:mx="urn:mpeg:mpeg21:2003:01-REL-MX-NS">
  <Item>
    <Component id="component_01">
      <Descriptor>
        <Statement mimeType="text/xml">
          <r:license>
            <r:grant>
              <mx:play/>
              <r:digitalResource>
                <r:nonSecureIndirect URI="movie01.mpg"/>
              </r:digitalResource>
            </r:grant>
            <r:issuer/>
          </r:license>
        </Statement>
      </Descriptor>
      <Resource mimeType="video/mpeg" ref="movie01.mpg"/>
    </Component>
    <Component>
      <Descriptor>
        <Statement mimeType="text/xml">
          <dip:Label>urn:mpeg:mpeg21:2005:01-DIP-NS:DIM</dip:Label>
        </Statement>
      </Descriptor>
      <Resource mimeType="application/mp21-method">
        function test_REL_DIBOs()
         {
          var resource=didDocument.getElementsByTagNameNS(
               "urn:mpeg:mpeg21:2002:02-DIDL-NS","Resource").item(1);
          var license = REL.getLicense(resource);
          var isOK = REL.queryLicenseAuthorization(license[0],
           resource,"urn:mpeg:mpeg21:2003:01-REL-MX-NS", "play", null);
          if (isOK) DIP.alert("Playing operation is
             authorized",MSG_INFO);
         }
      </Resource>
    </Component>
  </Item>
</DIDL>
```

Example 9.18   Calling REL-related DIBOs.

as DIBOs. Such operations may be highly specific to the application area of the DI and hence they need not to be normatively defined. The standard set of DIBOs was selected on the basis of a set of use cases and requirements [22] that were defined for DIP. A detailed procedure for selecting those DIBOs is described in [23]. In such situations, implementing the required functionality using DIML is a possibility. However, this may neither be easy nor optimal in terms of execution efficiency and size. Thus, there is a need for a mechanism enabling the User to easily extend the set of base operations without compromising efficiency or interoperability. DIXOs address this need.

The language in which DIXOs are implemented is not limited (in theory) by the MPEG-21 DIP specification. This means that a DIXO may be implemented in any programming language. DIP normatively defines how DIXOs are included in DIDs and the set of APIs that can be used when programming a DIXO. This can be done for every DIXO language proposed to MPEG. At the time of writing, only Java has been proposed as a DIXO language. Therefore, the DIP specification provides this information for DIXOs written in Java, called *J-DIXOs*. Information about the inclusion, calling and execution of DIXOs written in other languages can be added to the MPEG-21 DIP specification in the future.

9.6.4.1 J-DIXOs

Although theoretically speaking, it is possible to write DIXOs in any language, as stated earlier, at the time of writing Java is the only language for which all the necessary interfaces for usage in DIP have been standardized. Now let us have a look at what is actually standardized for the J-DIXO language.

The first part of J-DIXOs that is standardized is the calling mechanism. Calling a DIXO is done just as calling any other functionality of the multimedia platform, that is, using a DIBO. The DIBO for calling J-DIXOs is the `runJDIXO` DIBO. This DIBO has been discussed in Section 9.6.3.5.

The second part that is standardized is the type of arguments that can be passed to a J-DIXO. A J-DIXO designated method may take arguments of the following types only:

- org.w3c.dom.Document
- org.w3c.dom.Element
- org.iso.mpeg.mpeg21.mpegj.dibo.ObjectMap
- org.iso.mpeg.mpeg21.mpegj.dibo.PlayStatus
- java.lang.String
- java.lang.Object
- the primitive Java data types byte, int, long, char, float, double, boolean.
- array [] instances of any of the data types above

Besides an invocation mechanism, it is also necessary for the DIXOs to be able to call the DIBOs. To realize this, MPEG has defined interfaces for DIBO sets and utility classes for calling them. Example 9.19 shows the interface for the DID DIBOs.

Implementer's note: The DIP specification contains an informative part helping the implementer to create a secure environment in which the J-DIXOs can be run. The execution environment consists of:

- a Java Virtual Machine,
- platform (java.*) packages: java.io, lang and util.
- Required ISO/IEC JTC 1/SC 29/WG 11 defined APIs.
- mappings for normative DIBO APIs. (org.iso.mpeg.mpeg21.dibo.*)

```
package org.iso.mpeg.mpeg21.mpegj.dibo;

import org.w3c.dom.*;

/**
 * Java Bindings for the DID related DIBOs
 */
public interface DID {

    public boolean areConditionsSatisfied(Element element)
    throws DIPError;

    public boolean configureChoice(Element choice)
    throws DIPError;

    public void setSelection(Element selection, String state)
    throws DIPError;

}
```

Example 9.19   Java bindings for DID DIBOs.

Example 9.20 demonstrates how a J-DIXO calling the `configureChoice` DIBO can be created. All J-DIXOs must implement the normatively defined J-DIXO interface. This interface requires an implementation of the `setGlobalEnv` method, which sets the `GlobalEnv` object. This object is one of the utility classes that helps getting access to the DIBOs and the DIDL document with which the J-DIXO is interacting. Besides that method, the J-DIXO interface requires an implementation of the `callJDIXO` method. This makes it possible for the DIP engine to know which method to call when the `run-JDIXO` DIBO is invoked. Finally, the J-DIXO interface requires a `getArgumentTypes` method and a `getReturnType` method. These methods are used by the DIP engine to find out the data types of the arguments and return value. The `JDIBOFactory`, used in the example, is a utility class that gives access to the Java DIBO bindings.

```
import org.iso.mpeg.mpeg21.mpegj.*;
import org.iso.mpeg.mpeg21.mpegj.dibo.*;
import org.w3c.dom.*;

public class TestJDIXO implements JDIXO {
    GlobalEnv env;

    /**
     * Constructor
     */
    public TestJDIXO() {
    }

    // JDIXO methods
    /**
     * Sets globalenv
     */
```

```java
  public void setGlobalEnv(GlobalEnv env) {
    this.env = env;
  }

/**
  * callJDIXO method
  *
  * @param args
  *              Object array
  * @return Boolean object
  */
  public Object callJDIXO(Object[] args) throws DIPError {

    Element element = (Element) args[0];
    DID didOps = (DID) (env.getJDIBOFactory().getJDIBOObject("DID"));
    return didOps.configureChoice(element);
  }

  /**
  * callJDIXO takes one argument - an object of type
  * org.w3c.dom.Element.
  * Therefore, return a Class array with a single element of type
  * org.w3c.dom.Element
  */
  public Class[] getArgumentTypes() {
    Class[] argTypes = null;
    try {
      argTypes = new Class[] { Class.forName("org.w3c.dom.Element") };
    } catch (Exception e) {
      e.printStackTrace();
    }
    return argTypes;
  }

  /**
  * returns a Class object of type java.lang.Boolean
  */
  public Class getReturnType() {
    Class retType = null;
    try {
      retType = Class.forName("java.lang.Boolean");
    } catch (Exception e) {
      e.printStackTrace();
    }
    return retType;
  }
}
```

Example 9.20   a J-DIXO example.

9.7 USAGE OF DIGITAL ITEM PROCESSING INFORMATION

9.7.1 A REAL-WORLD EXAMPLE – PART TWO

This section extends the example from Section 9.5.3 and discusses how MPEG-21 DIP can be used in a real-world example. It illustrates how the different DIP constructs can be integrated in a DI and how the DIP information can be used to create an interoperable framework in which the processing of DIs can be realized.

Example 9.21 is constructed using a *container*. The *container* contains three *items*. The first two *items* represent a declaration of the multimedia data of the movie collections, that is, two declarations of movies and their accompanying posters. The third *item* contains the DIP information.

```xml
<?xml version="1.0" encoding="UTF-8"?>
<DIDL xmlns="urn:mpeg:mpeg21:2002:02-DIDL-NS"
      xmlns:dip="urn:mpeg:mpeg21:2005:01-DIP-NS">
  <Container>
    <Item>
      <Descriptor>
        <Descriptor>
          <Statement mimeType="text/xml">
            <dip:ObjectType>urn:foo:MoviePoster</dip:ObjectType>
          </Statement>
        </Descriptor>
        <Component>
          <Resource mimeType="image/jpeg" ref="poster01.jpg"/>
        </Component>
      </Descriptor>
      <Descriptor>
        <Statement mimeType="text/xml">
          <dip:ObjectType>urn:foo:Movie</dip:ObjectType>
        </Statement>
      </Descriptor>
      <Component>
        <Resource mimeType="video/mpeg" ref="movie01.mpg"/>
      </Component>
    </Item>
    <Item>
      <Descriptor>
        <Descriptor>
          <Statement mimeType="text/xml">
            <dip:ObjectType>urn:foo:MoviePoster</dip:ObjectType>
          </Statement>
        </Descriptor>
        <Component>
          <Resource mimeType="video/mpeg" ref="movie02.mpg"/>
        </Component>
      </Descriptor>
    </Item>
    <Item>
      <Component>
        <Descriptor>
          <Statement mimeType="text/xml">
            <dip:MethodInfo>
              <dip:Argument>urn:foo:Movie</dip:Argument>
            </dip:MethodInfo>
          </Statement>
        </Descriptor>
        <Descriptor>
          <Statement mimeType="text/xml">
```

```
            <dip:Label>urn:mpeg:mpeg21:2005:01-DIP-NS:DIM</dip:Label>
          </Statement>
        </Descriptor>
        <Resource mimeType="application/mp21-method">
function playPreview( arg1 )
 {
    var component=arg1.getElementsByTagNameNS("urn:mpeg:mpeg21:2002:02-DIDL-
       NS","Component").item(1);

    var playstatus=DIP.play(component, true);
    DIP.wait(5000);
    DIP.release(playstatus);
 }
        </Resource>
      </Component>
      <Component>
        <Descriptor>
          <Statement mimeType="text/xml">
            <dip:MethodInfo>
              <dip:Argument>urn:foo:Movie</dip:Argument>
            </dip:MethodInfo>
          </Statement>
        </Descriptor>
        <Descriptor>
          <Statement mimeType="text/xml">
            <dip:Label>urn:mpeg:mpeg21:2005:01-DIP-NS:DIM</dip:Label>
          </Statement>
        </Descriptor>
        <Resource mimeType="application/mp21-method">

 function playMovie( arg1 )
 {
   DIP.runJDIXO( "uri", null, "uri", "#dixo_01",
      "com.company.authentication.PerformCustomAuthentification" , null );
   var component=arg1.getElementsByTagNameNS("urn:mpeg:mpeg21:2002:02-DIDL-
      NS","Component").item(1);

   var playstatus=DIP.play(component, false);
 }
        </Resource>
      </Component>
      <Component>
        <Descriptor>
          <Statement mimeType="text/xml">
            <dip:Label>urn:mpeg:mpeg21:2005:01-DIP-NS:DIM</dip:Label>
          </Statement>
        </Descriptor>
        <Resource mimeType="application/mp21-method">
function copyMovie()
 {
    var itemlist=didDocument.getElementsByTagName("Item");
    var item=itemlist.item(0);
    var newItem= item.cloneNode(true);
```

```
        item.parentNode.insertBefore( newItem, item);
        var registry=DOMImplementationRegistry.getDOMImplementation("LS");
        var domWriter = registry.createLSSerializer();
        domWriter.writeToURI(didDocument,"http://server/modifiedDID.xml");
}
        </Resource>
      </Component>
      <Component>
        <Descriptor>
          <Statement mimeType="text/xml">
            <dip:MethodInfo>
              <dip:Argument>urn:foo:MoviePoster</dip:Argument>
            </dip:MethodInfo>
          </Statement>
        </Descriptor>
        <Descriptor>
          <Statement mimeType="text/xml">
            <dip:Label>urn:mpeg:mpeg21:2005:01-DIP-NS:DIM</dip:Label>
          </Statement>
        </Descriptor>
        <Resource mimeType="application/mp21-method">

function printPoster( arg1 )
{
   DIP.runJDIXO( "uri", null, "uri", "#dixo_01",
      "com.company.authentication.PerformCustomAuthentification" , null );
   var component=arg1.getElementsByTagNameNS("urn:mpeg:mpeg21:2002:02-DIDL-
      NS","Component").item(0);

   DIP.print(component);
}
        </Resource>
      </Component>
      <Component>
        <Descriptor>
          <Statement mimeType="text/xml">
            <dip:MethodInfo>
              <dip:Argument>urn:foo:MoviePoster</dip:Argument>
            </dip:MethodInfo>
          </Statement>
        </Descriptor>
        <Descriptor>
          <Statement mimeType="text/xml">
            <dip:Label>urn:mpeg:mpeg21:2005:01-DIP-NS:DIM</dip:Label>
          </Statement>
        </Descriptor>
        <Resource mimeType="application/mp21-method">

function removePoster( arg1 )
{
   var parent=arg1.parentNode;
   parent.removeChild(arg1);
   var registry=DOMImplementationRegistry.getDOMImplementation("LS");
   var domWriter = registry.createLSSerializer();
```

```
domWriter.writeToURI(didDocument,"http://server/modifiedDID.xml");
}
          </Resource>
        </Component>
        <Component id="dixo_01">
          <Descriptor>
            <Statement mimeType="text/xml">
              <dip:JDIXOClasses>
                <dip:Class>
                  com.company.authentication.PerformCustomAuthentification
                </dip:Class>
              </dip:JDIXOClasses>
            </Statement>
          </Descriptor>
          <Descriptor>
            <Statement mimeType="text/xml">
              <dip:Label>
                urn:mpeg:mpeg21:2005:01-DIP-NS:DIXO:Java
              </dip:Label>
            </Statement>
          </Descriptor>
          <Resource mimeType="application/java"
            ref="authentication.jar"/>
        </Component>
      </Item>
    </Container>
</DIDL>
```

Example 9.21   A movie collection DID including DIP information – Part Two.

Figure 9.6 shows the structure of an *item* containing a movie. The movie is declared using an *item* with two *descriptors* (i.e., the poster *descriptor* and the Object Type *descriptor*) and a *component* (i.e., the movie *component*). The first *descriptor* contains the poster. The second *descriptor* associates the Object Type urn:foo:Movie with the movie *item*. The poster *descriptor* of the *item* contains a *descriptor* associating the Object Type urn:foo:MoviePoster with the poster *descriptor*, and a *component* with a reference to the

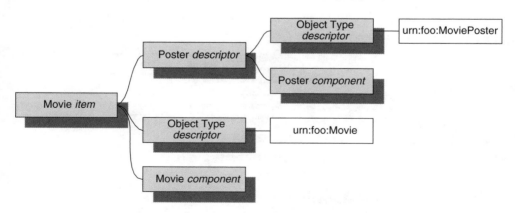

Figure 9.6   The structure of an *item* containing a movie

actual poster data. Finally, the movie *component* of the movie *item* contains a reference to the actual movie.

It should be noted that the movie *item* from the example is only a very simple example of how a movie *item* could look like. In a more complex scenario, this *item* would most likely contain DRM-related expressions, MPEG-7 expressions describing the content of the movie, and so on. For the purpose of illustrating DIP, this example has been reduced to the basics that are needed for creating a functional movie collection.

Figure 9.7 gives an overview of the DIM structure in the *item* containing the DIP information. For the sake of simplicity of the example, the *components* containing the DIMs and DIXOs are grouped together in one *item*. It should be noted that this is not prescribed by the DIP specification. The DIM *item* contains six different *components*, of which the first five declare DIMs. The first *component* contains the playPreview DIM. It contains a `MethodInfo` *descriptor* and a `Label` *descriptor*. The `MethodInfo` *descriptor* indicates that the Argument Type of the argument for the playPreview DIM is of the type urn:foo:Movie. The other DIM *components* share the same structure.

At the end of the example DID, the declaration of an included J-DIXO is done. Figure 9.8 contains an overview of the structure of the *component* containing the J-DIXO. It should be noted once again that the declaration of the J-DIXO was done in the same *item* as the DIMs for the sake of simplicity of the example and that this is not required by the DIP specification.

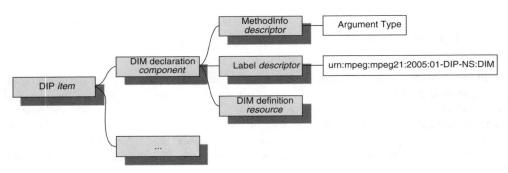

Figure 9.7   The structure of the item containing the J-DIXOs

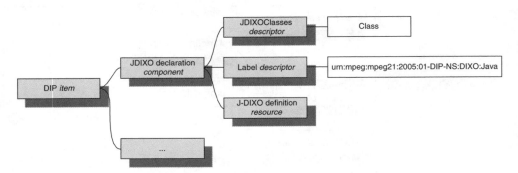

Figure 9.8   The structure of the item containing the digital item methods

The *component* containing the J-DIXO contains a `JDIXOClasses` *descriptor* stating the fully qualified Java name of the J-DIXO com.company.PerformCustomAuthentication, a `Label` *descriptor* with the value urn:mpeg:mpeg21:2005:01-DIP-NS:DIXO:Java and the J-DIXO definition *resource*. This J-DIXO is used in the definition of the playMovie DIM and the printPoster DIM.

9.8 DIP REFERENCE SOFTWARE AND OTHER IMPLEMENTATIONS

Just as with any other MPEG standard, the DIP specification also has reference software. The DIP reference software was built on the basis of the architecture discussed in Section 9.3 and the walk-through discussed in Section 9.5.2. It contains an example implementation of a DIP engine, an implementation of the full DIBO set and a mechanism for loading and executing J-DIXOs. This software implementation has not been discussed in this chapter since it was not part of the MPEG-21 Part 8: Reference Software specification at the time of writing. In the future, this software will be added to MPEG-21 Part 8 and therefore it will be available as a part of this specification.

Although, at the time of writing, the DIP specification is still a rather new specification, it is already used in the(Dynamic and distributed Adaptation of scalable multimedia coNtent in a context-Aware Environment) (DANAE) project [24]. This project, which is a 30-month IST European cofunded project, proposes to address the dynamic and distributed adaptation of scalable multimedia content in a context-aware environment. Its objectives are to specify, develop, integrate and validate in a test bed a complete framework able to provide end-to-end quality of (multimedia) service at a minimal cost to the end-user. DIP plays a central role in DANAE both at client and at server side. The main task of the DANAE DIP engines is to provide the communication between the different modules used throughout the content delivery chain.

9.9 SUMMARY

This chapter discusses the processing of so-called 'DI' within the MPEG-21 Multimedia framework. It illustrates that the DIP specification provides the author of DIs a means by which they can express their suggested processing of DIs. The terms and definitions used in DIP, how DIP information can be included in DIs, and how DIBOs and DIML can be used to write DIMs, are described.

This chapter provides a set of examples that are used in real-world use cases and ends with an overview of existing software implementations, which were available at the time of writing.

REFERENCES

[1] ISO/IEC, "ISO/IEC 21000-2 Information technology – Multimedia framework (MPEG-21) – Part 2: Digital Item Declaration," 2nd Edition, 2005.
[2] World Wide Web Consortium, "Extensible Markup Language (XML) 1.0," 2nd Edition, W3C Recommendation, 6 October 2000.
[3] F. De Keukelaere, S. De Zutter, R. Van de Walle, IEEE Computer Society, "MPEG-21 Digital Item Processing," *IEEE Transactions on Multimedia*, vol. **7** no. 3, June 2005, pp. 427–434.

[4] ISO/IEC, "ISO/IEC 21000-10 Information technology – Multimedia framework (MPEG-21) – Part 10: Digital Item Processing," 2005.

[5] I. Burnett, R. Van de Walle, K. Hill, J. Bormans, F. Pereira, IEEE Computer Society, "MPEG-21: Goals and Achievements," *IEEE Multimedia*, vol. **10** no. 4, Oct–Dec 2003, pp. 60–70, 2003.

[6] ISO/IEC, "ISO/IEC 11172-3:1993 Information technology – Coding of moving pictures and associated audio for digital storage media at up to about 1,5 Mbit/s – Part 3: Audio," 1993.

[7] Microsoft, "Windows Media 9 Series," (http://www.microsoft.com/windows/windowsmedia/default.aspx), 2005.

[8] ISO/IEC, "ISO/IEC TR 21000-1:2004 Information technology – Multimedia framework (MPEG-21) – Part 1: Vision, Technologies and Strategy," November 2004.

[9] World Wide Web Consortium, "HyperText Markup Language (HTML) 4.01 Specification," W3C Recommendation, 24 December 1999.

[10] World Wide Web Consortium, "Cascading Style Sheets, level 2 revision 1," W3C Candidate Recommendation 25 February 2004.

[11] World Wide Web Consortium, "Synchronized Multimedia Integration Language (SMIL 2.0)," W3C Recommendation 07 August 2001.

[12] F. Pereira, T. Ebrahimi, "The MPEG-4 Book," Prentice Hall, NJ, 2002, pp. 103–148.

[13] ISO/IEC, "ISO/IEC 16262"2002 Information technology – ECMAScript language specification," June 2002.

[14] Netscape Devedge, "Core JavaScript Guide," (http://devedge.netscape.com/library/manuals/2000/javascript/1.5/guide/), 2004.

[15] Sun Microsystems, "Java Programming Language," (http://java.sun.com/), 2004.

[16] Internet Engineering Task Force, "Uniform Resource Identifier (URI): Generic Syntax," RFC 3986, January 2005.

[17] World Wide Web Consortium, "Document Object Model (DOM) Level 3 Core Specification Version 1.0," W3C Recommendation, 07 April 2004.

[18] World Wide Web Consortium, "Document Object Model (DOM) Level 3 Load and Save Specification Version 1.0," W3C Recommendation 07 April 2004.

[19] Internet Engineering Task Force, "Multipurpose Internet Mail Extensions (MIME) Part One: Format of Internet Message Bodies," RFC 2045, November 1996.

[20] Internet Engineering Taks Force, "Multipurpose Internet Mail Extensions (MIME) Part Two: Media Types," RFC 2046, November 1996.

[21] Moving Picture Experts Group, "Requirements for Digital Item Processing: Digital Item Methods, Digital Item Method Engine, Digital Item Base Operations, and Digital Item Method Language," ISO/IEC JTC1/SC29/WG11 N5330, Awaji, December 2002.

[22] Moving Picture Experts Group, "Procedure for Accepting DIBOs in MPEG-21 DIP,"/IEC JTC1/SC29/WG11/N5854, Trondheim, July 2003.

[23] DANAE, "Dynamic and distributed Adaptation of scalable multimedia coNtent in a context-Aware Environment", http://danae.rd.francetelecom.com/, 2005.

10

Event Reporting

FX Nuttall, Andrew Tokmakoff and Kyunghee Ji

10.1 INTRODUCTION[1]

MPEG-21 is a global toolkit that can be used to trade content in a secure and predictable manner. In the early days of MPEG-21, representatives from different industries ranging from those in consumer electronics to telecom operators and rights holders gathered to define the scope of the tools needed. It quickly became apparent that a monitoring tool was needed to understand what Users or Peers were doing or in what state they were. This is especially true for the entertainment industry (record companies, collective rights management's societies, film studios), where copyrighted content needs to be monitored in order to process the royalties derived from its usage.

In the specific case of the entertainment industry, proper reporting of content usage is the basis for its economy. It is needed, for instance, for the calculation of royalty splits to all the parties involved in the life cycle of a song or a movie. Because of huge disparities in the reporting formats, the reporting process has always been a tedious task for all those managing the rights. It was typical to find as many reporting formats as they were parties involved in a transaction. Imagine a record label licensing its content to 50 on-line music stores; the record label would receive 50 reports in 50 different formats. The formats to report content usage range from spreadsheets with different layouts and CSV[2] (comma-separated values file format) files to more specialized formats such as the CLL[3] format (Central Licensing Label Information), not to mention specific layouts on paper listings. It is easy to imagine the overhead costs induced by processing such a wide range of formats

---

[1] At the time of writing this chapter, the ISO standard on Event Reporting is still at FCD (Final Committee Draft) status, and some parts are subject to change, although major changes are unlikely. The reader should refer to the official ISO standard (ISO/IEC 21000:15), when available, for full and complete specification.

[2] CSV: The comma-separated values (CSV) file format is a tabular data format that has fields separated by the comma character and quoted by the double-quote character.

[3] CLL: Central Licensing Label Information – CT90-2121 – a data format designed to transmit label data, in computer-processible form, between Record Companies and Licensing Societies.

The MPEG-21 Book   Ian S Burnett, Fernando Pereira, Rik Van de Walle, Rob Koenen
© 2006 John Wiley & Sons, Ltd

in order to consolidate data. As it is anticipated that a large-scale growth in the usage of content in digital form will happen, it is now crucial to define a standardized format that allows for the cost of processing data to be significantly reduced.

Further exploring the future of content distribution, scenarios like super-distribution would become a nightmare without a standardized usage-reporting format. Super-distribution is an e-commerce model where each and every consumer becomes a potential retailer. In order to comply with the legal and financial environment of copyrighted content, every transaction involves not only delivering the content but also a billing transaction with a financial service, a revenue split amongst the parties in the value chain, including the various rights holders. This may involve as many as six or seven message exchanges between different parties for each transaction. This is obviously unrealistic unless a standard shared by all parties truly exists, and this is where Event Reporting comes in.

The telecommunication industry has also expressed some specific needs. It is critical for the telecommunications operators to manage their bandwidth properly. If content can be adapted to the receiving device before it is transmitted, it will result in optimal usage of the bandwidth. Operators would want to ask a device to report on its current state, such as how much bandwidth is currently available, its display or storage capacity. The dynamic nature of Event Reporting makes it an ideal tool for these kinds of operations.

On the transactional side, telecom operators are interlinked via roaming agreements. They pass on telecommunications, be it voice or data, from one network to another. To track and reconsolidate traffic and billing, they use a variety of data-exchange standards like CIBER[4] (Cellular Intercarrier Billing Exchange Roamer) or TAP[5] (Transferred Account Procedure). By wrapping existing data-exchange formats, Event Reporting may be advantageously used in the telecom environment in fully automated processes.

10.2 SCOPE OF THE EVENT REPORTING STANDARD

The requirements for reporting events from MPEG-21 Peers span across many industry sectors, therefore the specification for Event Reporting has to be quite broad. Retailers of e-books will have different requirements from pay-per-view services, and telecom operators delivering ring tones will have different needs. Trying to define the exact structure and content of reports for every industry would have been a tedious task and would have rendered the standard obsolete on the day of its publication. Furthermore, it is impossible to predict all of the specific fields of applications of MPEG-21.

Event Reporting does not specify the format of the reports but rather the mechanism by which a User will request and receive reports. As an example, a User wants to receive all the titles of the songs that have been played by a Peer in the last 14 days. But this User does not know the name of the XML element in the Digital Item Declaration [1] (DID; Chapter 3) that contains the title of the song. This element may be simply called <Title>

---

[4] CIBER: the Cellular Intercarrier Billing Exchange Roamer record is a proprietary protocol and specification developed by Cibernet® corporation for the exchange of roaming billing information, for voice and data, among wireless telecommunication companies, their billing vendors and data clearing houses. See http://www.cibernet.com.

[5] TAP: The Transferred Account Procedure developed by the GSM Association is the world standard for billing of roaming services. See http://www.gsmworld.com/using/billing/whatis.shtml

and also <Titre> for the Digital Items (DIs) that were created in France or <название> for the Russian ones.

It is beyond the scope of Event Reporting to standardize a specific syntax for the metadata elements describing DIs. Event Reporting will allow Users to retrieve *any* metadata and send it in *any* format.

Event Reporting may also be used as a wrapper to exchange other standardized data formats such as Music Industry Integrated Identifier Project (MI3P)[6] [2] messages, for example.

10.3 THE GENERAL EVENT REPORTING MECHANISM

Event Reporting [4] is based on three basic concepts: Event Report Requests (ER-Rs), Events and Event Reports (ERs). Each of these will be discussed in detail in the following sections, but in general terms, ER-Rs are specialized DIs where a User can send a request to a Peer to receive specific information when a given Event occurs. An Event happens when something 'special' happens inside a Peer. It may be that somebody has opened a DI to render its content, like watch a movie or play a game. It may also be that the Peer state has changed, for example, its internal clock has turned midnight or its storage is full. When the Event, as defined by the ER-R, has been detected, the Peer will then generate and send an ER. The ER may describe the Event itself, like sending a message indicating that it is midnight on the Peer clock, or might return other types of information, like the number of DIs contained in the Peer and their respective descriptive metadata.

Event Reporting is therefore a dynamic system (Figure 10.1) where a Peer first receives a request (an ER-R) asking him/her to send a report (an ER) when certain conditions are fulfilled. The Peer then enters into a 'watch' mode, where it tests whether the condition specified by the ER-R is fulfilled. When the condition(s) occur(s), the Peer creates and sends the ER as specified by the ER-R.

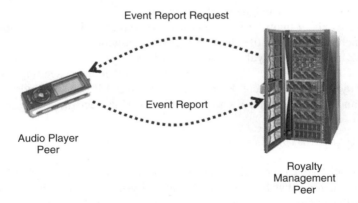

Event Report Request

Event Report

Audio Player
Peer

Royalty
Management
Peer

Figure 10.1 Event Reports are defined and requested by Event Report Requests

---

[6] MI3P: Music Industry Integrated Identifier Project. A technical Standard to support the management of on-line music e-commerce, jointly developed by the IFPI (International Federation of the Phonographic industry), the RIAA (Recording Industry Association of America), CISAC (Confédération Internationale des Sociétés d'Auteurs et Compositeurs) and BIEM (Bureau International des Sociétés Gérant les Droits d'Enregistrement et de Reproduction Mécanique).

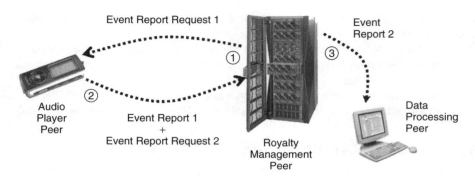

Figure 10.2   Example of an Event Report embedding an Event Report Request

ER-Rs may be received at any time. They can be embedded within the DI containing the resource to be monitored but may also be received as a stand-alone DI that is sent asynchronously. A Peer containing 50 DIs with music can on one day receive one ER-R saying 'Between 15 September 2006 and 31 December 2006, send me the titles of all the movies that are played'. As of 15 September, at midnight, the Peer would then enter into a watch mode and gather all the titles of the movies rendered by the Peer.

To further allow for the automation of Event Reporting, other dynamic mechanisms have been specified.

The first one is that ERs can be sent to multiple recipients at once. This allows for 'mass mailing' when multiple parties need to be informed of a specific Event. But a more powerful mechanism is that Event Report Requests can be embedded within ERs. As a result, when a Peer receives an ER from another Peer, it can contain another ER-R asking the receiving Peer to take further action like forwarding the ER to somebody else or sending back an acknowledgment of receipt.

These features allow many different parties involved in a given transaction to be automatically synchronized at all the steps of a given transaction.

Figure 10.2 illustrates a Royalty Management Peer sending a first ER-R (1) to an Audio Player Peer. The Audio Player Peer embeds another ER-R when sending his/her ER (2). When the Royalty Management Peer receives his/her ER, he/she finds a new ER-R, processes it and sends a second ER to a Data Processing Peer (3).

10.4 EVENTS

By definition, an Event is the 'the occurrence of a reportable activity'. In other words, an Event is something that happens inside an MPEG-21 Peer and that *can be predicted and unambiguously described*. The fact that a User is acting upon a given DI to play its content, for instance, is an Event. The fact that the Peer internal clock indicates midnight is an Event.

Events are described as a set of conditions that must be met. There can be one or more conditions. It is also possible to apply Boolean logic to the conditions and combine them. For example, an Event will have occurred when either one of the specified condition has been fulfilled or it may be required that all conditions are fulfilled. The description of

Events is specified using the XML element called `EventConditionDescriptor` (see Section 10.5 for more detail).

It is up to the User of Event Reporting to properly describe the Event, so that the Peer can unambiguously detect the Event.

Event Reporting differentiates two classes of 'reportable events': *DI-related* Events and *Peer-related* Events.

The DI-related Events encompass all events that occur when a DI is acted upon. For example, if a User opens a DI to play the song that it contains, it triggers a DI-related Event. The action of playing (rendering) the song is the Event. An ER may be sent to another Peer to inform it that the song has been played. This class also contains all the Events identified when DIs are moved, copied or transformed.

A specific XML element called `DIOperationCondition` is used to specify the type of operation and the description or identifier of the DI(s) to be monitored (see Section 10.5 for more detail). It is also possible to limit the condition to a given User or Peer.

In the field of copyrighted music for example, a DI-related Event would be used by a rights holder to monitor the usage of his repertoire. Translated into plain English, the `DIOperationCondition` element would look like: 'Send me a Report when anybody plays a song AND the value of the element <author> equals "Fred Smith"'.

The second class of Events, Peer-related Events, is used to monitor the status or the capabilities of a Peer regardless of whether they contain or process DIs. The ability of a Peer to process a DI is dependent on its technical capabilities or on its configuration. It may be important to know the bandwidth capacity of a Peer, its rendering capability – whether it has the proper decoder, for instance – or its geographical location (if the Peer can know where it is of course, using Global Positioning System (GPS) technology, for instance). The list of Peer-related conditions is obviously not exhaustive, and it is not predefined either by the standard. It would be either impossible or too restrictive to list all the Peer characteristics. Not to mention that different device manufacturers may use different naming conventions to describe similar characteristics. One manufacturer may use country codes and another country names to identify territories for instance.

The description of Peer Events is specified in the XML element called `PeerCondition`. This element is very flexible, as it allows for manufacturer-specific characteristics to be defined using the namespace convention. It may be possible, for instance, to specify `<nokia:7211:ScreenResolution>` to request the display capacity as stored by the 7211 model of Nokia® phones.

Some foreseen applications of DI-related Events would be, for example, to assess the territory in which the Peer is located. This could be used for selecting the proper language for rendering of content. It could also be used to adapt the resolution of a movie to the actual capacity of the Peer's screen in order to optimize bandwidth or storage usage.

It is important to note that it is anticipated that Peer conditions will be Peer specific and that proper knowledge of the internal Peer architecture and configuration be known before the Event can be properly described. It would be quite useless to ask an audio-only player to return its screen resolution if it doesn't have a display.

10.5 EVENT REPORT REQUESTS

As discussed in Section 10.3, ER-R is the mechanism used within Event Reporting for ERs to be specified by a User that is interested in obtaining information on either DI

```
<xsd:element name="ERR">
 <xsd:complexType>
   <xsd:sequence>
     <xsd:element ref="erl:ERRDescriptor"/>
     <xsd:element ref="erl:EventConditionDescriptor"/>
     <xsd:element ref="erl:ERSpecification"/>
   </xsd:sequence>
 </xsd:complexType>
</xsd:element>
```

Figure 10.3   Core structure of Event Report Requests

usage or Peer state. This section provides a more detailed overview of how Users can request ERs to be generated.

There are two ways by which an ER-R can be delivered to a Peer. The first is within a DI, as discussed in Section 10.3. This mechanism may itself have two different permutations, that is, the bundling of an ER-R with the DI that it is to report on or, alternatively, issuing a stand-alone ER-R, independently of the DI that it may report on.

The second mechanism for issuing ER-Rs to a Peer is to simply embed them inside the Peer, which is in some way 'hardcoded'. This mechanism assumes that a Peer will always wish to report certain classes of Events and, thus, may simply have the ER-R inside it. In fact, using this mechanism, the Peer may not actually have an explicit ER-R embedded inside it but may instead simply generate predetermined ERs. In this case, the ER-R becomes 'virtual'. However, this is purely an implementation decision and, thus, is beyond the scope of the Event Reporting [4] specification.

The core structure of an ER-R is composed of three elements, as shown by the XML fragment in Figure 10.3: ERRDescriptor, EventConditionDescriptor and ERSpecification.

10.5.1 DESCRIBING THE EVENT REPORT REQUEST

The first element is the ERRDescriptor containing all the information describing the ER-R. The User can define the following parameters:

LifeTime: Used to control the life cycle of an ER-R. This is important since an ER-R should not be 'active' for an infinite amount of time since this would have a strong impact on the resources. Peers need to process Event Reporting. The Peer knows when an ER-R can be safely discarded. It is yet possible to assign an infinite value to LifeTime, therefore providing a permanent validity period to the ER-R.

Modification: Used to maintain a record of the creation and modification history of an ER-R. This is largely an administrative convenience for ER-R creators since it allows ER-Rs to store a (self-contained) history of any changes that take place. Modification is typically used at least once to reference the time and date of creation of the ER-R.

Priority: Used to allow an ER-R to be allocated a priority that influences how it will be processed by a Peer. Lower-priority ER-Rs will be serviced by a Peer after the higher-priority ones. In this context, the term 'serviced' refers to handling of Events and generation of related ERs for a given ER-R. This may be used when more than one ER-R needs to be processed at the same time, therefore helping the Peer to decide which ones are considered more important.

10.5.2 CONDITIONS OF THE EVENT REPORT REQUEST

The second element is the EventConditionDescriptor. This element is very important as it specifies the set of conditions that must be fulfilled before the Event is deemed to have occurred and therefore when the ER should be generated and then sent. The Event-ConditionDescriptor may contain multiple conditions combined using Boolean operators, but creators of ER-Rs must ensure that the combination of multiple conditions remains consistent, therefore allowing for the Peer to properly interpret the ER-R.

There are three types of conditions:

DIOperationCondition: Defines conditions that are related to the direct manipulation or usage of a DI. This is one of the most fundamental condition types since it allows an ER-R to be used to monitor DI usage. These conditions only refer to terms describing actions as defined within the Rights Data Dictionary (RDD) [5]. Translated in plain English, a DIOperationCondition may be 'If you PLAY the song identified by the ISRC [6] code #:US-IR2-78-00041, ...'. Here, the term [PLAY] is the action of 'Transforming a Fixation into a Perceivable and Transient representation of its contents' as defined within the RDD [5].

PeerCondition: Defines conditions that are related to 'internal' Peer operations or states. This may, for example, consider the network status of the Peer or the location/region of the Peer. The syntax of a PeerCondition element depends on the design of the hardware and/or the firmware of the Peer and therefore needs to be defined within a specific namespace.

TimeCondition: Defines conditions that are based upon time or date specifications. These can be defined to be Periodic (a time specification that is based upon a periodic recurrence such as 'Every Monday'), Specific (a time specification that defines a single point in time) or Elapsed (a time specification that indicates that a certain amount of time needs to have elapsed from a given time point).

10.5.3 DESCRIBING THE EVENT REPORT

The last element of the ER-R is the ERSpecification. It describes the ER that must be generated and sent when all the specified conditions are met. The ERSpecification element indicates to the Peer what the ER should 'look like', what it should contain and to whom it must be sent, how and when. The elements contained in the ERSpecification are:

ERIdentifier: Each ER that is created as the result of processing this ER-R should use this specified identifier either as a base id, which is extended on a per ER basis, or as an absolute value. This allows ER-Rs to specify the identity of the resulting ERs, thereby easing traceability.

ERDescription: A free form string that can be used to annotate the ERs that are created through processing of this ER-R.

AccessControl: Describes an access control restriction that should be applied to all ERs that result from processing this ER-R. The access control condition is defined as a

Right Expression Language (REL) [7] in the form of a license granted to a principal to perform an action on a resource. The principal is the User to whom the right is granted, the action will again be an RDD [5] term such as READ or WRITE and the resource is all or part of the ER. It is an important feature since it may be necessary to ensure that ERs remain confidential (i.e., are not 'readable' by Users or Peers that are not the intended ER recipient) or trusted (i.e., are not modified from their original state).

ReportData: Describes what the ER should contain. It is the 'payload' of the ER. There are a number of normalized reportable items that may be specified here and that are relatively self-explanatory:

PeerID: The identifier of the Peer.

UserID: The identifier of the User.

Note that the UserID or the PeerID are optional since Users may not have identifiers or the privacy policies may prevent the identifier from being reported.

Time: Time of the Event (when all conditions have been fulfilled). This is not the time at which the ER is to be delivered as specified in another element further below.

Location: The geographic location of the Peer. The syntax of the location element is taken from the MPEG-7 [8] specification.

DIOperation: The ER should specify the action that has been performed on a DI (if any). It may be important, for example, to differentiate when a song has been played or copied.

DomainData: A flexible and extensible mechanism that allows domain-specific data items to be included in the ER that may be, for example, manufacturer-specific metadata items such as model descriptions and firmware details.

In addition, the ER-R must specify the DI that must be accessed and the information that must be retrieved.

SelectionOfMetaDataElement contains the reference to the DI from which information is to be retrieved. Given the wide variety of formats and methods for storing metadata within a DI, Event Reporting provides a variety of methods to identify DIs.

SelectionViaDII: This is the simplest method as it contains the unique Digital Item Identification (DII) [9] identifier of the DI.

SelectionViaMetaDataElement: Used when the metadata structure of the DI is known. For example, a User knows that all the DIs he/she wants to monitor contain a field called <Label> whose value is 'Blue Label'. He/she can therefore identify these specific DIs using the element SelectionViaMetaDataElement.

SelectionViaXPath: Used to specify an XPATH flag that is contained within a DI.

The selection of the metadata elements that are to be reported from the above selected DI is specified by the element MetadataElementsToBeReported. This element lists the names of the fields whose values will be included in the ER.

ReportFormat: Describes the format that should be used to 'package' the reportable items into the ER. The format may be specified by an external schema definition, may conform to a well-defined mime-type data definition or may use the 'built-in' default ER data format. This default format is a quite simple XML format, which is discussed further in Section 10.6.

EmbeddedERR: An embedded ER-R that should be included in all ERs that are generated as a result of processing this ER-R. As discussed in Section 10.3 and shown in Figure 10.2, this mechanism allows an ER to be propagated from one Peer to another in a controlled and deterministic manner. The ER-R that is to be included can either be embedded inline or just be a reference to an externally obtainable ER-R.

DeliveryParameters: Having created an ER, the Peer needs to know how to deliver the ER to the recipient. There are, of course, many different ways of getting data from one place to another, and MPEG-21 does not specify how the delivery of an ER must be done. Instead, it provides a 'hook' onto which the ER-R creator may specify the mechanism to be used to deliver the ER to the destination Peer. In fact, the ER specification makes use of the REL's [7] `ServiceReference` construct, which is technology-neutral. Other delivery aspects, such as the time at which delivery should occur (`DeliveryTime`) and the intended `Recipient` (a specified Peer and User), are included here.

10.6 EVENT REPORTS

The result of processing an ER-R is an ER or a set of ERs, depending on the conditions that have been specified in the ER-R. This section provides a more detailed description of ERs and their composition. The XML fragment in Figure 10.4 illustrates the three main elements of an ER, which are described in further detail below: `ERDescriptor`, `ERData`, and `EmbeddedERR`.

10.6.1 DESCRIBING THE EVENT REPORT

The first element of an ER is the `ERDescriptor`. As with an ER-R, each ER needs to be able to provide some descriptive metadata. In the case of an ER, the `ERDescriptor` element is used to bundle this metadata together and includes the following fields:

Description: A free form descriptive text of the ER, as defined by the originating ER-R within its `ERSpecification` element.

```
<xsd:element name="ER">
 <xsd:complexType>
    <xsd:sequence>
     <xsd:element ref="erl:ERDescriptor"/>
     <xsd:element ref="erl:ERData"/>
     <xsd:element ref="erl:EmbeddedERR" minOccurs="0"
        maxOccurs="unbounded"/>
    </xsd:sequence>
 </xsd:complexType>
</xsd:element>
```

Figure 10.4 Core structure of Event Reports

Recipient: This is the intended recipient of the ER (an MPEG-21 Peer and User). It was also specified in the originating ER-R within its `ERSpecification` element. It is up to the receiving Peer to ensure that the ER is delivered to the specified user on that Peer. In addition, there may be other IPMP [10] techniques and tools used to ensure that only the intended recipient on the specified Peer can access the contents of this ER (e.g., use of cryptographic techniques to ensure confidentiality).

Status: Indicates the 'processing' status of this ER. In this context, the status refers to the ability of the generating Peer to fully evaluate all the conditions that were specified in the parent ER-R and its ability to extract the required metadata elements. In some cases, optional conditions may not be fulfilled but all mandatory conditions are fulfilled, or, similarly, not all of the reportable fields could be obtained when creating the ER. This would result in a 'FALSE status for the ER since not all of the conditions were fulfilled or some of the reportable data items are missing from the ER. As an example, an ER-R requested that both the director's and the writer's names of a specific movie be reported. The Peer that processed the ER-R only found the director's name. It will still send the ER but indicate in its status field that something went wrong.

Modification: Includes the creation and, potentially, the modification history of the ER that may be altered as it passes through different Peers (perhaps for the purpose of statistical aggregation). This field is used at least once to indicate the creation time of the ER.

ERSource: Provides a reference to the original ER-R that created the ER. It may be via an ER-R (either embedded in the ER or obtainable via an external reference) or simply from an external Program/Application, which has an embedded 'virtual' ER-R. In this case, a generic Uniform Resource Identifier (URI) is used to refer to the external application that was the source of the ER.

10.6.2 THE EVENT REPORT ITSELF

The second element, `ERData`, contains the report itself. Depending on the specified `ReportFormat`, the contents of this element may be vastly different. For example, the ER-R may specify that an externally defined schema (which could be application-domain specific) is to be used to represent the reportable data items.

Alternatively, a predefined mime-type such as comma-separated values (CSV) (`text/csv`) or Microsoft® Excel™ (`application/x-msexcel`) may be used to represent the reportable data items. In this latter example, the `ERData` would contain an Excel™ file. This file may simply be stripped out of the XML and directly imported into the Excel™ application for viewing and editing. It is important to note that Event Reporting does not define these mime types and, therefore, that the Peer generating the ER must know how to generate the requested format. However, as a default, a simple XML format is used that uses the reportable data field name as the tag and includes the values of the data item to be reported as the text of that element.

10.6.3 EMBEDDING AN EVENT REPORT REQUEST

The last element, `EmbeddedERR`, may contain an embedded ER-R, which in turn must be evaluated and processed by the recipient Peer. The embedded ER-R may be a reference to an externally obtainable ER-R or an inline ER-R itself. As explained in Section 10.3, the embedded ER-R will be processed by the Peer receiving the ER. This mechanism may be used to send an acknowledgment of receipt or to forward the ER to another Peer.

10.7 RELATIONSHIP WITH OTHER PARTS OF MPEG-21

Event Reporting has the fairly unique characteristic of interacting or using many other parts of MPEG-21. This not only is a consequence of the requirements for Event Reporting but also a goal in order to ensure maximum integration within MPEG-21.

The first parts to be referenced by Event Reporting were REL [7] and RDD [5], respectively, Parts 5 and 6 of MPEG-21. This was a necessity, as Event Reporting needs to identify actions taken upon DIs, such as PLAY or COPY. These actions need to be terms that are unambiguously understood by a Peer and therefore referenced by the Data Dictionary. These actions must further be allowed or granted by the REL. Therefore, all actions that are defined as a condition of an Event are specified as an RDD [5] term using the REL [7] syntax. If an action is not already defined within the RDD [5], it must be added to it, as permitted by the RDD [5] standard.

The REL [7] has been further used to define a Users' right to create, read or modify an ER or an ER-R. ERs may contain very critical data such as royalty statements or even financial statement. In order to use the data received by means of an ER, it is essential that this data can be trusted and that the recipient be assured that the Report has not been tampered with. Event Reporting therefore supports a set of expressions that defines specific access rights to specific Users or Peers. You may, for instance, grant a READ access only to a specific User for confidentiality purposes. This access right is defined using the REL [7] standard.

It is important to understand that REL [7] defines the rights but does not enforce them. This principle is valid not only for Event Reporting but also throughout MPEG-21. The enforcement of an access right is done by an IPMP [9] (Intellectual Property and Management) tool.

Two other important parts that are extensively used by Event Reporting are the DID [1] and Digital Item Identification DII [9], respectively, Parts 2 and 3 of MPEG-21. Since ERs and ER-R were represented as DIs, it was necessary to use the standardized syntax to describe, define and identify these DIs.

Event Reporting is also able to interface with Digital Item Adaptation [10] (DIA) tools for 'trapping' DIA-related Events (such as, the adaptation of a DI) and for obtaining DIA-related metrics that need to be included in an ER. For example, an ER-R may require that the current bandwidth of the Peer is reported within an ER. In this case, Event Reporting makes use of the DIA [11] Usage Environment Description tool, which is able to provide detailed network metrics. Other Peer characteristics can also be obtained from such a tool for inclusion in an ER.

In order to 'trap' the execution of RDD [5] ActTypes, Event Reporting can interact with a Digital Item Processing [11] (DIP) engine. Such an engine is used to execute DIMs (Digital Item Methods), which in turn, are composed of Digital Item Base Operations (DIBOs) and Digital Item Extended Operations (DIXOs). When a DIP [2] Engine executes DIBOs or DIXOs that map onto REL [7] ActTypes, it signals this to the Event Reporting infrastructure. This infrastructure is then able to act on the Event according to the ER-Rs that it is currently servicing.

```
  <ERR>

<ERRDescriptor>
    <LifeTime>
      <StartTime>2005-09-01T00:00:00</StartTime>
      <EndTime>2005-09-30T23:59:59</EndTime>
    </LifeTime>
    <Priority>3</Priority>
  </ERRDescriptor>

<ERSpecification>
    <ReportData>
      <PeerId/><UserId/><Time/><Location/><DII/><DIOperation/>
    </ReportData>
    <ReportFormat>
      <Ref> xmlns:xsi="http://www.w3.org/2001/XMLSchema-instance"</Ref>
    </ReportFormat>
    <DeliveryParams>
      <Recipient>
        <PeerId>MAC:00-08-E3-AE-1E-62</PeerId>
        <UserId>John</UserId>
      </Recipient>
      <DeliveryTime>
        <SpecificTime><beforeOn>2005-09-30T00:00:00</beforeOn>
        </SpecificTime>
      </DeliveryTime>
    </DeliveryParams>
  </ERSpecification>

<EventConditionDescriptor>
    <TimeCondition>
      <TimeEvent>
        <SpecificTime>
          <afterOn>2005-09-10T00:00:00</afterOn>
          <beforeOn>2005-09-20T00:00:00</beforeOn>
        </SpecificTime>
      </TimeEvent>
    </TimeCondition>
    <DIOperationCondition>
      <UserId>Ann</UserId>
      <PeerId>MAC:04-36-A5-C4-28-7D</PeerId>
      <Operation>urn:mpeg:mpeg21ra:RDD:156:735</Operation>
      <DII>urn:mpegra:mpeg21:dii:di:0824108241</DII>
      <Operator kind="AND" location="prefix"/>
    </DIOperationCondition>
  </EventConditionDescriptor>
  </ERR>
```

Figure 10.5   Example of an ER-R

```
<ER>
 <ERDescriptor>
   <Status value="true"/>
   <Modification>
     <UserId>Administrator</UserId>
     <PeerId>MAC:04-36-A5-C4-28-7D</PeerId>
     <Time>2005-09-23T12:31:00</Time>
     <Description>Date and time of the creation of the ER</Description>
   </Modification>
   <ERSource>
     <ERRReference>mpeg:mpeg21:dii:ERRID:4385736754</ERRReference>
   </ERSource>
 </ERDescriptor>

<ERData>
   <UserId>Ann</UserId>
   <PeerId>MAC:04-36-A5-C4-28-7D</PeerId>
   <Time>2005-09-23T12:30:25</Time>
   <Location>UK</Location>
   <DII>urn:mpeg:mpeg21:DII:DI:0824108241</DII>
   <DIOperation>urn:mpeg:mpeg21:ra:RDD:156:735</DIOperation>
 </ERData>

</ER>
```

Figure 10.6   Example of an Event Report

10.8 EXAMPLES

Figure 10.5 is an example of an ER-R with its three distinct sections:

- The ERRDescriptor specifies that the ER-R is valid throughout the month of September 2005.
- The ERSpecification indicates that the report should contain the Identifiers of the Peer and the User, the time and location at which a DI has been acted upon. The ER shall be in XML and delivered to John on the Peer whose Media Access Control (MAC)[7] address is 00-08-E3-AE-1E-62 before 30 September 2005.
- The EventConditionDescriptor specifies that the ER must be sent if Ann Plays (the RDD [5] term referenced as 156:735[8]) the song contained in the DI numbered 0824108241 during the month of September 2005 on the Peer whose MAC address is 04-36-A5-C4-28-7D.

Upon processing the above ER-R, the resulting ER would look as follows (Figure 10.6):

- The status element of the ERDescriptor indicates that the ER has been generated compliantly with the ER-R number 4385736754 at 12:31:00 on 23 September 2005 by the Administrator of the Peer.

---

[7] MAC address, short for Media Access Control Address, is one method for uniquely identifying Peers. MPEG-21 does not specify how Peers are uniquely identified. Other methods may be used such as Internet IP (Internet Protocol) address or CPU-IDs.

[8] At the time of writing this chapter, the numbering of the RDD [5] terms is not finalized. The final value for the specific word PLAY is subject to change.

- The ERData element reports that Ann has played the song contained in the DI number 0824108241 at 12:30:25 on 23 September 2005 from the United Kingdom.

10.9 CONCLUSION

As the consumption of digital content increases, accurate and automated processes to monitor and manage the usage of content become a strict necessity. While the number of retailers or broadcasters of digital content increases from a few hundred to tens of thousands, it is no longer possible to support the current disparity of reporting formats. Event Reporting does not pretend to standardize an ultimate data format that could fulfill the needs of various industries for the years to come, but rather a common platform that can support all of the various initiatives. It is hoped that Event Reporting, through its flexible approach, will allow for greater interoperability between existing and forthcoming infrastructures of many industries.

REFERENCES

[1] DID (Digital Item Declaration) – ISO/IEC 21000:2,2005
[2] DIP Digital Item Processing) – ISO/IEC 21000:10.
[3] MI3P (Music Industry Integrated Identifier Project) – see http://www.mi3p-standard.org.
[4] ER (Event Reporting) – ISO/IEC 21000:15.
[5] RDD (Rights Data Dictionary) – ISO/IEC 21000:6,2004
[6] ISRC (International Standard Recording Code) – ISO 3901:2003 – see http://www.ifpi.org/isrc/index.html, 2003.
[7] REL (Rights Expression Language) – ISO/IEC 21000:5,2004
[8] MPEG-7 Part 5 (Multimedia content description interface/Multimedia description schemes) – ISO/IEC 15938-5, 2003.
[9] DII (Digital Item Identification) – ISO/IEC 21000:3,2003
[10] IPMP (Intellectual Property and Management) – ISO/IEC 21000:4.
[11] DIA Digital Item Adaptation – ISO/IEC 21000:7,2004

11

Future MPEG Developments

Fernando Pereira and Ian S Burnett

Since 1988, MPEG has been a major player in the multimedia technological landscape. While some MPEG standards have achieved great success, others have been less successful. Overall, it is probably most important that the right standards are available even if, in the process, less popular standards get produced; these latter standards are easily forgotten while the former rise quickly to success and transform people's lives. Fortunately, most MPEG standards have proven to be very useful and thus are largely deployed, even if this deployment often takes more time than expected.

Since digital media is now a mature area in which the most obvious needs are covered, less obvious needs require sometimes a long investment to clarify what if anything is needed and useful. As a body that tries to evolve and react to the technological evolution, MPEG is always in search of new multimedia areas in need of standardization. For this reason, MPEG starts the so-called 'exploration activities', which typically study the potential of new technologies to address specific application demands, for example, in terms of the new or improved functionalities that they offer. For example, a rather long exploration activity on 3D video coding led, on July 2005, to a Call for Proposals on Multi-view Video Coding tools [1]. This was the result of MPEG concluding that there were technologies potentially justifying standardization and determining that there was a clear need following requests from the relevant industry.

This chapter will review current MPEG activities relevant to MPEG-21; we have chosen to organize these into two main clusters: MPEG-21 specifications and other relevant MPEG specifications related to MPEG-21 but not framed under the MPEG-21 umbrella.

11.1 MPEG-21 UNDER DEVELOPMENT

While the previous chapters in this book have addressed MPEG-21 technologies for which the specifications have been mostly finalized (at least FCD[1] stage was reached),

---

[1] See Chapter 1 for definition.

The MPEG-21 Book   Ian S Burnett, Fernando Pereira, Rik Van de Walle, Rob Koenen
© 2006 John Wiley & Sons, Ltd

this section will briefly review and summarize the most relevant MPEG-21 standardization efforts ongoing at the time of writing.

11.1.1 DIGITAL ITEM STREAMING

As stated in Chapter 2, the major objective of MPEG-21 is the definition of a multimedia framework to enable transparent and augmented use of multimedia resources across a wide range of networks and devices used by different communities. Within MPEG-21, a Digital Item is defined as a structured digital object with a standard representation, identification and description. This entity is also the fundamental unit of distribution and transaction within this framework.

Digital Items (DIs) – and Digital Item Declarations (DIDs) – have been considered to be static objects in terms of delivery, where the complete DID is first downloaded, and the associated resources optionally accessed by some unspecified means [2]. However, it has been recognized that this scenario is inadequate in several use cases. One such use case is the broadcasting use case described in Chapter 2 in which a broadcaster is looking to repurpose content. Often the repurposing will be to transports and platforms that require a stream of content rather than one large container.

According to [2], DI Streaming targets 'the incremental delivery of a DI (DID, metadata, resources) in a piecewise fashion with temporal constraints such that a receiving Peer may incrementally consume the DI'. The delivered Digital Item can potentially be a derivation of the original DI, that is, in which some parts may have been removed or transformed. In other words, DI Streaming brings temporal delivery of the complete Digital Item into the MPEG-21 standard.

During the DI Streaming exploration process, a few relevant use cases were identified [2], notably: (i) conditional DI Streaming, (ii) multi-channel streaming of a DI, (iii) dynamic and distributed, gBSD-based adaptation (in unicast scenarios, and in a multicast or broadcast scenario), and (iv) streaming of a DI with fragment-level protection.

For example, the multi-channel streaming use case in Figure 11.1 considers the broadcasting of content to users across multiple channels (MPEG-2 TS, DVB-H, T-DMB[2], IP-based network) and is quite similar to that considered in Chapter 2. It requires a DI (containing a DID, a rights expression, a video resource, and timed metadata) to be prepared for delivery via each channel [2]; users receive the content via just one of the channels. The final DI that users will receive may be adapted from the master DI so as to be suitable for the channel in use, for example, a video resource may be reduced in resolution for DVB-H or the metadata constrained owing to memory limitations on a mobile. The broadcaster wishes to avoid duplicating any part of the original content (for instance, by creating a sub-item version of the content for each delivery channel) within the DI. Thus, it is necessary to provide mechanisms to 'translate' the DI and its constituent resources and metadata to the temporal stream. The DID may also contain references to other resources, potentially from other sources or from the local machine that may be delivered by some out-of-band means. These other resources may be, for example, referenced by a Hypertext Transfer Protocol (HTTP) address and are not necessarily streamed.

---

[2] T-DMB (Terrestrial Digital Multimedia Broadcasting) is a new broadcasting system that has been proposed in WorldDAB forum, ETSI and ITU-R.

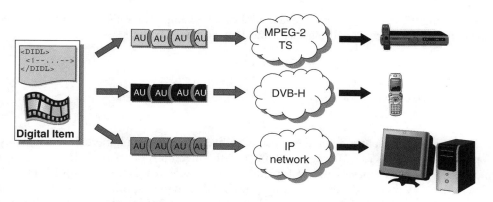

Figure 11.1 Multi-channel streaming of a DI [2][3]. Copyright ISO. Reproduced by permission

This latter facility is intended to allow a broadcaster to add, for example, Web page references or extra 'on-demand' content to the Digital Item.

As another example, the conditional DI Streaming use case in Figure 11.2 considers a DI that contains two versions of the same video (e.g. at different resolutions) [2]. This use case relies on the existence of a back channel, as the user (or the terminal as an MPEG-21 User) must make a selection between the two versions on the basis of a DID Choice. The information required to resolve the Choice, for example, a combination of MPEG-7 descriptions of resource resolution and DIA descriptions of terminal capabilities, will be streamed to the user's terminal. Additionally, the DID contains inline proprietary XML format subtitles with timing information and an MPEG-7 MDS program description [3]. As a first step, the server will deliver a fragment of the DID containing the information required by the client to resolve the Choice. The DID may contain some information requiring the client to resolve the Choice and send its Selection to the originating node. The server then delivers the selected resource and relevant temporal metadata (subtitles + MPEG-7 MDS program description).

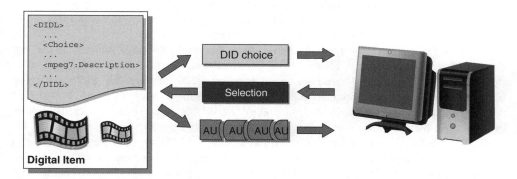

Figure 11.2 Conditional DI Streaming [2]. Copyright ISO. Reproduced by permission

---

[3] An Access Unit (AU) is a segment of data for transport purposes that is atomic in time, that is, to which a unique time can be attached; an AU is specific to the encoding method.

Following the analysis of the relevant application examples, the DI Streaming requirements have been derived and clustered into four groups [2]:

- DI fragmentation for progressive consumption
- Progressive delivery of DIs
- Progressive consumption of DIs
- General requirements.

Following the conclusion that the streamed delivery of DIs is an important functionality for adding to the MPEG-21 framework [4] and the identification of the relevant requirements, a Call for Proposals on DI Streaming technology was issued in April 2005 [5]. The technology proposed was evaluated in July 2005, and the specification process started targeting finalization during 2006. The DI Streaming specification is expected to become part 18 of MPEG-21. However, it is already clear that DI Streaming offers useful new standardization opportunities that will allow MPEG-21 DIs to be delivered as a temporal user experience. This has clear applications, for example, the use case introduced in Chapter 2, and is very relevant to the current content-repurposing trend in the media industry.

11.1.2 PROFILING

Since the very beginning of MPEG, it has always been clear that, for a standard to attain longevity in the market, it should specify only the minimum necessary to provide the required interoperability. This approach increases the lifetime of the standard by allowing new technology to be introduced and encourages competition. Also, on the basis of the developers' increasing knowledge of how best to exploit the coding tools, the standard's performance may increase over time, and retain a competitive position against newer, competing technologies.

While simplicity has always been a target for MPEG standards, breadth has always been another major target; this means that MPEG standards should address as many applications and functionalities as possible under the targeted domain. This avoids the need for many specific standards in the same technical area. These two targets – simplicity and breadth – tend to be contradictory since simplicity indicates lower complexity while breadth suggests more tools, and thus higher complexity.

MPEG's requirement for breadth results in standards with many tools that are not essential for all applications. Thus, MPEG has always had to find ways of creating solutions that represent adequate trade-offs between simplicity and breadth while achieving both interoperability and complexity goals. On the one hand, adopting a single standard solution including all the tools in the standard would maximize interoperability, while, on the other, the resulting complexity would likely exclude many applications.

The trade-off is clear in the very first MPEG standard, MPEG-1, both for the Audio and the Video parts. In MPEG-1 Video, the so-called 'constrained parameter set' is defined to limit the complexity of streams and decoders, guaranteeing interoperability within that set. Meanwhile, MPEG-1 Audio specified three audio coding layers with different coding tools; each one of these provides a trade-off in terms of quality, bitrate and complexity.

When, with MPEG-2, the range of applications to be addressed increased (e.g., digital television at any spatial resolution, video storage, stereoscopic video, etc.), the trade-off problem became even more complex. In MPEG-2 Video, and later standards, the need to compromise interoperability with complexity as well as simplicity and breadth was addressed by defining the so-called 'profiles' and 'levels'. The combination of these two concepts allows the definition of a set of standard solutions, each solution being adequate in terms of functionality and complexity for the relevant application class. Profiles are concerned with 'functionality' and correspond to a set of tools that provide the required functionalities for a certain class of applications. Levels address 'complexity' and correspond to a set of constraints (e.g., on bitrate) on the usage of the tools within a certain profile; in this way, they control the associated complexity.

MPEG-2 Video extensively exploited the notions of profiles and levels, defining several, while MPEG-2 Audio (Part 2) retained the notion of layers that is part of MPEG-1 Audio. Of all the MPEG-2 Video profiles, the 'Main' profile is most popular while most of the remaining have not been much used. Later, both MPEG-4 and MPEG-7 adopted the notions of profile and level; for example, MPEG-4 defines several profiling dimensions such as Visual, Audio, Graphics, Scene Graph, Text, and 3D Compression.

Thus, returning to the topic of this book, in MPEG-21, MPEG is again using the profiling approach to trade off complexity and interoperability. The first profile defined in MPEG-21, in July 2005, was a REL profile, the so-called Base REL profile, specified through an amendment to the REL specification (Part 5) [6]. As stated in Chapter 5, the MPEG-21 Rights Expression Language 'support(s) new business models through extensions and optimization through profiles'. The first mechanism, extensions[4], adds language elements needed to address the requirements of a particular application domain. Profiles, on the other hand, identify a subset of the full set of elements in the language (including extensions) sufficient to address the needs of a particular application domain.

While the specification of the Base REL profile was originally targeted at the mobile domain, it has been improved during the standardization process so as to address the needs of other domains, such as the optical media domain, without burdening the other domains that have been targeted; this is valuable for increasing interoperability. Because the lack of interoperability among different DRM (Digital Rights Management) systems is a key issue, the Base REL profile has also been developed to allow easy interoperability with the Open Mobile Alliance (OMA) DRM v2.0 specification, which is an important solution in mobile environments.

Summarizing, the Base REL profile adds the following extensions to the REL specifications [6]: `IdentityHolder` as an extension to Principal, `GovernedMove`, `Store`, `Inherit` and `GovernedCopy` as an extension to Right, `ProtectedResource` as an extension to Resource, and `DrmSystem`, `EnhancementConstraint`, `SeekPermission`, `StartCondition` and `OutputRegulation` as an extension to Condition.

It is expected that the specification of profiles in MPEG-21 will continue for other parts of MPEG-21 when profiling will meet the industry demands. The procedure for defining profiles consists of two main stages. First, profile proposals are collected; these must include details of the application domain(s) addressed, a list of tools in the proposed profile, a list of functionalities compared to the closest existing profile(s), conformance and verification test plans, supporting companies and references. When a profile proposal

---

[4] An Extension is a set of new XML Schema elements, XML Schema types, QNames and/or URIs usable within Rights Expressions. Some of these new XML Schema elements or XML Schema types might be derived from existing elements or types already defined in MPEG-21 REL or in other extensions.

is mature, a decision will be made about its inclusion in the standard; such a choice will be made on the basis of two important criteria: the necessary functionality is not supported (or at least well supported) by existing profiles; a declared and strong interest exists in the industry to deploy the profile in services and/or products. Whenever relevant, a similar approach is followed for the definition of any levels that might be created for MPEG-21 Parts.

11.2 BEYOND MPEG-21

Since it started in 1988, MPEG's activities have been organized around large projects, typically driven by major application domains, or functionalities. This approach led to the following set of MPEG standards:

- ISO/IEC 11172 (MPEG-1), 'Coding of Moving Pictures and Associated Audio at up to about 1.5 Mbit/s', mostly addressing CD-ROM digital storage;
- ISO/IEC 13818 (MPEG-2), 'Generic Coding of Moving Pictures and Associated Audio', mostly addressing digital television and DVD digital storage;
- ISO/IEC 14496 (MPEG-4), 'Coding of Audio-Visual Objects', providing new object-based functionalities, synthetic and natural integration, new forms of interaction, and so on;
- ISO/IEC 15938 (MPEG-7), 'Multimedia Content Description Interface', providing multimedia content description capabilities for a large range of applications;
- ISO/IEC 21000 (MPEG-21), 'Multimedia Framework', providing an integration framework for the previous MPEG standards and missing technologies such as IPMP, rights expression and content adaptation.

With the exception of MPEG-21, which does not include any media representation technologies, all the previous standards were structured and addressed their objectives on the basis of a 'trilogy' of Systems, Video/Visual and Audio specifications.

However, in July 2005, MPEG acknowledged that new, large projects such as those mentioned above were not being created, but instead several smaller, individual standards were being developed. Many of the latter are not directly related to any of the large MPEG projects already in existence. This fact raised questions regarding the best place to fit these new standards; following some discussion, a new type of MPEG standards organization was created to complement the existing set of large projects.

The new organization of MPEG's future standards acknowledges the following facts:

- Many of the new standards under development are concerned with tools that can be used across all of the large MPEG standards or are, at least, not directly linked to them.
- Most of the new standards target the provision of individual systems, audio and visual functionalities.

Thus, MPEG decided to start a new set of standards organized as follows:

- *ISO/IEC 23000 (MPEG-A), 'Multimedia Application Formats'*, defining application-driven file formats across MPEG standards (Section 11.2.1).

- *ISO/IEC 23001, 'MPEG Systems Technologies'*, defining systems tools that can be used across all MPEG large standards or at least are not directly linked to them; Part 1 specifies the so-called Binary XML standard, which provides a generic binary format to encode XML documents in a compact bitstream that can be efficiently parsed, and allows the streaming of XML files so that a large document can be progressively sent or carouselled.

- *ISO/IEC 23002, 'MPEG Visual Technologies'*, defining visual coding tools that can be used across all large MPEG standards or are, at least, not directly linked to them; Part 1 includes a specification for the IDCT accuracy that can be referenced in lieu of the IEEE 1180 standard, which has been withdrawn by IEEE [7].

- *ISO/IEC 23003, 'MPEG Audio Technologies'*, defining audio coding tools that can be used across all large MPEG standards or are, at least, not directly linked to them. Part 1 specifies the MPEG Surround standard, a coding standard for multi-channel, spatial (typically, 5.1 channel) sound, which requires the transmission of a compressed stereo (or even mono) audio program and an additional low-rate side-information channel [8].

- *ISO/IEC 23004 (M3W), 'MPEG Multimedia Middleware'*, improving application portability and interoperability through the specification of a set of APIs (syntax and expected execution behaviour) dedicated to multimedia as well as providing a standard way of delivering the implementation(s) of these APIs in the form of software components and a standard for a component infrastructure that allows dependable dynamic integration of these components in a system [9] (Section 11.2.2).

The fact that this set of standards has been started does not mean that in parallel the existing large standards do not grow whenever this is an adequate approach. For example, the most active MPEG effort in July 2005 targets the specification (in collaboration with ITU-T) of the so-called Scalable Video Coding (SVC) standard, which will be defined as an amendment to MPEG-4 Part 10 since it is, technologically speaking, an extension of the solutions that are already included in that part.

The next two sections will detail the objectives and status (August 2005) of MPEG-A and M3W owing to their conceptual novelty and advanced status.

11.2.1 MULTIMEDIA APPLICATIONS FORMATS

It is well recognized that MPEG standards have had a huge impact in determining the current multimedia landscape, not only in terms of technologies, services and products but also in terms of new user habits and behaviours. All of this has been underpinned by the notion of interoperability, which refers to the ability of a system, or a product, to work with other systems or products without special effort on the part of the user. Providing the user with benefits and advantages in the area of interoperable multimedia representation has always been the major goal of MPEG standards. Indirectly, industries benefit since happy users should result in happy (and wealthy) industries.

MPEG-1, MPEG-2, MPEG-4, MPEG-7 and MPEG-21 are a wide ranging set of multimedia technology standards with different objectives and defined at different points over a near twenty-year period. However, during 2004, MPEG embarked on a new standard formally known as ISO/IEC 23000, Multimedia Application Formats and also known as MPEG-A. The new MPEG-A standard targets the definition of Multimedia Application Formats (MAFs), which are basically 'super-formats' combining tools defined across existing MPEG standards or parts of standards. These combination 'super-formats', for example, a combination of audio and video coding formats with some metadata, take the notion of MPEG interoperability to a new dimension. Instead of interoperability associated with single media, MPEG-A associates interoperability with complete applications.

11.2.1.1 Context and Motivation

While MPEG has defined profiles and levels within parts of the various standards with the objectives stated in Section 11.1.2, for example, MPEG-2 Video and MPEG-4 Audio, it has never defined any standard combinations of tools or profiles across different standards or parts of standards. For example, MPEG never specified a combination of a MPEG-2 Video profile@level with an MPEG-2 Audio layer as the combined solution for digital television. In this sense, the approach has always been to leave to the industry and industry consortia, for example, DVB [10], ATSC [11] and ISMA [12], the decisions on how to make the best combinations of MPEG tools – in addition to other non-MPEG technologies – ideally in terms of standard profiles and levels so as not to jeopardize interoperability. However owing to the growing number of tools in the standards, and also the number of profiles and levels, it has become increasingly difficult for industries to select combinations of tools or profiles. This situation has also led to different industry consortia in related application domains picking different solutions, resulting in decreased interoperability.

Moreover, MPEG now provides a range of technical solutions that are not only related to coding but also to metadata, DRM, content adaptation, and so on; this makes it even more difficult to select the right combination of tools. Also, many users of MPEG standards, not familiar with the details of all these standards and parts of standards, do not just need standard solutions targeting a specific media, for example, video coding, but rather more complete solutions that target a specific application. This is emphasized by the growing number of applications for which a simple combination of audio and video coded streams is no longer enough and where multimedia representation involves a more complex, integrated usage of tools and streams. In this context, the MPEG-21 DI concept provides the perfect tool to create a structured combination of resources, descriptions, and so on, coded according to any of the MPEG standards. Also, existing transport and delivery standards such as the MPEG File Format and MPEG-2 Transport Stream (TS) are important ingredients in this context.

It is in this context that the notion of MPEG super-formats, addressing full-fledged applications while still targeting maximum interoperability, was born. These super-formats represent a desire from MPEG to make sure that its component standards are actually used and to take responsibility for defining adequate combinations of tools across MPEG standards and parts of standards. MPEG no longer believes that it can only rely on others outside MPEG to make those choices.

11.2.1.2 Definition and Objectives

The standard ISO/IEC 23000, Multimedia Application Formats, also known as MPEG-A, targets the specification of formats combining tools/formats mostly from the various

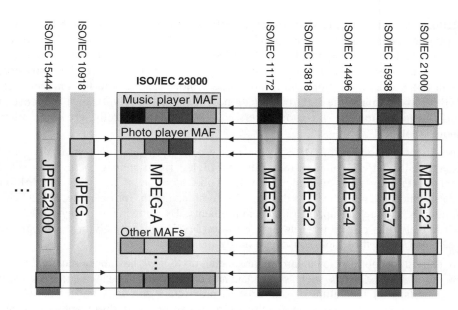

Figure 11.3   MAF conceptual overview [13]. Copyright ISO. Reproduced by permission

MPEG standards, but also including non-MPEG tools to address the needs of a certain application domain. This means that MPEG-A does not target the definition of new MPEG tools but mostly targets combinations of previously defined tools, ideally in terms of profiles and levels (Figure 11.3). However, if needed, new tools, profiles and levels may be defined and added to the right MPEG standard to be used in the relevant MAF.

The MPEG-A standard will consist of various parts, each defining one or more related application formats. However, Part 1 is a Technical Report explaining the purpose of these novel formats [13]. The specification of a MAF includes not only the drafting of a textual definition but also complete MAF reference software to ease the adoption of the MAF in question. Further, marketing material must be produced with the aim of explaining the MAF's capabilities and advantages to the relevant industry sector and other adopters.

The major objectives of specifying MAFs as defined above are as follows:

- Provide complete solutions to the identified market needs by facilitating the development of standards-based multimedia applications and services. This should contribute to the quick deployment of rich multimedia applications and services with interoperability at the application level.

- Support a fast track to standardization by selecting readily tested and verified tools taken from the MPEG body of standards and combining them to form a MAF. This solution is one layer above the typical MPEG way of building solutions using a toolbox approach.

- Combine MPEG with non-MPEG tools for the areas not addressed by MPEG, to provide users with complete solutions. Where MPEG cannot provide a certain tool, then additional tools provided by other organizations can be included by reference to make the target MAF more complete and powerful.

- Reduce the effort involved in selecting MPEG technologies and combining them to meet the specific needs of an application class. Industries and adopters interested in an application area for which a MAF exists can benefit from a ready-made format specification that has been put together by a team of experts. This can be easily used as a starting point for development of products, without need to wait for someone outside MPEG to determine the right combination of tools (especially when it is increasingly difficult to put together the right expertise to do that owing to the growing variety of tools involved).

- Increase the adoption of MPEG standards especially in technical areas that are less traditionally MPEG, such as metadata and digital rights management. It is expected that MAFs may have an important role in the adoption of MPEG-7 and MPEG-21 tools, which often require that the tools be combined with, for example, MPEG coding standards to meet an application's requirements. By showing the way MPEG-7 and MPEG-21 tools can be used at an application level, and not only at the stream level, users can adopt these solutions more easily.

- Increase the adoption of standards in areas that, up to now, have been mostly dominated by proprietary solutions, such as metadata and digital rights management. This adoption is important for the future of standards as proprietary solutions may, otherwise, take the lead in the multimedia landscape with a host of resulting disadvantages for users.

- Provide a reference software implementation, which can be used for either experimenting with the standard or for the speedy development of compliant products and services.

- Offer more guidance and also normative specifications to potential adopters on how to use MPEG standards; MPEG is thus taking a more proactive role. This guidance includes textual specifications, reference software and explanatory material.

MAFs have a role to play in providing integrated solutions for novel applications and existing applications. Even if proprietary solutions exist for certain applications, MAFs may bring something that proprietary solutions cannot provide: one example is interoperability across implementations in different companies through an open technical specification.

11.2.1.3 Creating a MAF

It is clear that the jump associated with MAFs whereby MPEG goes from specifying tools and profiles to specifying integrated solutions targeting the application layer requires more knowledge of the market than is required for single tools. Since MPEG intends to only define MAFs to address well identified market needs, the process of defining a MAF must be started with a request from industry since it is the companies who know about the market needs, trends, and so on.

To define a MAF, proposers of a new MAF should bring to MPEG the following elements [14]:

- Evidence of market need, notably a description of the targeted application scenario and value of a MAF in that context.

- A list of requirements associated with the relevant application scenario.
- A list of the MPEG tools/profiles (and also possible non-MPEG tools) to be included in the proposed MAF to address the identified requirements.
- Assessment of the relationship between the proposed MAF and any related solutions that already exist, be they standards or proprietary based.
- Demonstration of sufficient industry support for the proposed MAF including commitment to successfully complete the specification work, develop the reference software, crosscheck conformance streams, and produce marketing material explaining the benefits of the new MAF. To avoid developing too many MAFs, the industry also needs to demonstrate a commitment to deploy the format as a measure of the 'enthusiasm' surrounding a candidate MAF.

On the basis of the above elements, MPEG decides by consensus on the commencement of the specification of a new MAF. At any moment during that process, there will be a document containing the information associated with the MAFs under consideration [14]; for MAFs already under specification, there will be the usual standardization process involving WD, CD, FCD and FDIS (Chapter 1 for the definitions).

11.2.1.4 MAFs Defined and Under Work

At any moment in time, there are three types of MAFs: those already specified (a final technical specification is already available), those under development (the decision to create that MAF has been taken but technical work is still ongoing), and those under consideration (a decision on the specification of that MAF has not yet been made).

At the time of writing (August 2005), only one MAF has achieved complete specification – the Music Player MAF [15] – another MAF is under development – the so-called Photo Player MAF [16] – and a few other MAFs are under consideration [14].

Since MPEG will only specify those MAFs for which there is clear industry support, it may happen that some of the MAFs that appear in the document with the MAFs under Consideration [14] may never be specified.

In the following, only the Music Player MAF will be addressed since this is the single MAF that has already been finalized and published as ISO/IEC 23000-2 [15]. Its specification was completed over a short period of time since all the tools involved were already available in previous MPEG standards.

The application scenario targeted by the Music Player MAF was (annotated) digital music libraries, where each music asset is defined as a combination of audio, metadata and images, for example, the cover image associated with the relevant record. MPEG recognized the interest in providing to users the capability to interoperate not only at the audio, metadata and image individual levels but also at the level associated with the combination of the elementary levels. Not having a MAF for this scenario prevents interoperability since different companies/users may define (even if slightly) different combinations of the relevant streams.

Regarding MPEG and non-MPEG tools, this MAF adopted the following solutions:

- MP3, which means MPEG-1/2 Audio Layer III, to code the audio data since it is nowadays the most used music format.

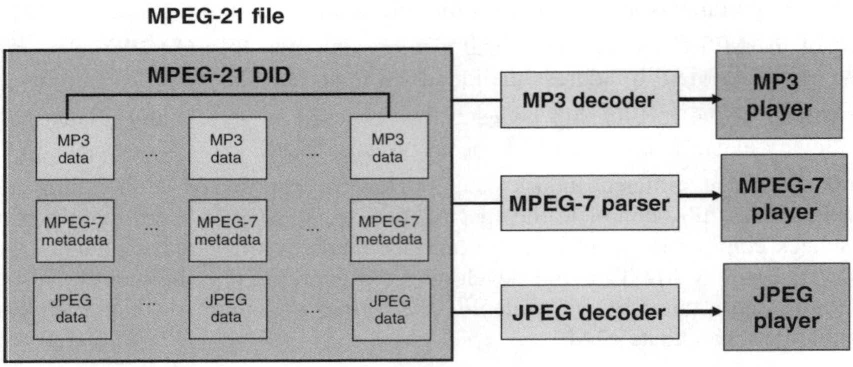

MPEG-21 file

Figure 11.4 Simplified architecture of a player for the Music Player MAF with multiple tracks

- MPEG-7 for the metadata (including the binary coding), notably important descriptors such as Artist, Album and Song Title; since ID3 is the most used metadata format for MP3 music, MPEG-7 has been updated to include the necessary tools in order that ID3 compatibility is available.
- JPEG for the images.
- MPEG-21 Digital Item Declaration to represent the music library structure.
- MPEG-4 or MPEG-21 file formats as the file formats; MPEG-4 is utilized solely for single tracks, while the MPEG-21 file format version can also store multiple tracks, JPEGs and metadata with the logical structure detailed in a DID.

The Music Player MAF combines the tools above in three different ways providing three MAF configurations within a single MAF[5]:

- A single music track with MPEG-7 metadata and a JPEG image in an MPEG-4 File Format.
- A single music track with MPEG-7 metadata and a JPEG image structured in an MPEG-21 DID with everything wrapped in an MPEG-21 File Format.
- Multiple music tracks with MPEG-7 metadata and JPEG images structured using an MPEG-21 DID with everything wrapped in an MPEG-21 File Format.

Figure 11.4 shows the simplified architecture of the Music Player MAF player when multiple tracks are considered. For the playback of the content, it is necessary to decode the set of streams: an MP3 player to listen to the music, an MPEG-7 player to consume the metadata, and a JPEG player to look at the pictures. This normative super-format will allow users to exchange music libraries without caring about the details of the combined format since the rules for the combination are also now set by MPEG. This ease of use can only contribute to users having an improved and richer multimedia experience.

---

[5] These are not three MAFs but a single MAF with three different data structures which means that a player complaint to this MAF has to be able to 'decode' all the three 'flavours' within this MAF.

11.2.2 MPEG MULTIMEDIA MIDDLEWARE

The MPEG Multimedia Middleware (M3W) activity was motivated by the acknowledgment that more and more applications, such as (secure) multimedia players/browsers, enterprise applications (e.g., e-learning, corporate information), and consumer applications (games, interactive DVD, digital television, etc.), make intensive usage of multimedia functionalities [17]. For this, the market offers many choices in terms of Operating Systems (OSs) and versions of these OSs, run-time environments (PocketPC, Smartphones, Brew, Java VM), application development and software development formats (C++ compiled for specific OSs, multiple Java versions, C#, etc.), and APIs' definition (MHP, JSR, MPEG, proprietary APIs, Linux CE, Java libraries). This situation makes it hard for application developers to create and maintain applications, makes it costly for application and service providers to deliver applications for their installed base of devices, and finally makes it difficult for device manufacturers to offer compelling platforms capable of being quickly provisioned with multimedia content and applications.

The ongoing trend in multimedia products is one of increased software dependency and a corresponding reduction in specific hardware. Also, product parts that do not create product differentiation are increasingly obtained from external manufacturers (and are not produced by product developers). This situation has led semiconductor vendors to focus on providing a general applicable platform that can serve a big market and end-user product developers to focus on offering features and services that differentiate their products from those of the competitors. This context created space and opportunity in the middle for the so-called 'middleware', which is bridging the gap between platforms and applications.

MPEG decided to acknowledge the importance of middleware in the multimedia landscape by starting the MPEG Multimedia Middleware (M3W) project with the goal of improving application portability and interoperability through the specification of a set of APIs (with normative syntax and expected execution behaviour) dedicated to multimedia (Figure 11.5). These APIs will enable the following:

1. application developers to quickly develop and easily maintain multimedia applications;
2. application and service providers to limit the cost of application provisioning;
3. middleware vendors to develop their middleware for multiple multimedia platforms using a single API;

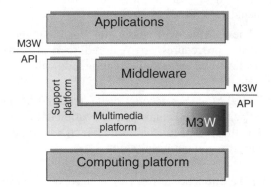

Figure 11.5 M3W positioning in a multimedia-enabled device [9]. Copyright ISO. Reproduced by permission

4. multimedia platform providers to offer a platform implementation that is attractive because it can be used by various existing middleware and application implementations;

5. (software) manufacturers to provide compelling and interoperable application development environments.

In technical terms, the objective of the M3W standard is the definition of a set of APIs that will allow [9] (i) multimedia applications to execute functions provided by the middleware in a standard way without the need for an in-depth knowledge of the middleware; and (ii) multimedia applications to trigger the update/upgrade or extension of the API, for example, because the required functions are not present or are outdated.

To satisfy these objectives, the following M3W capabilities have been identified:

- 'functional' APIs to allow the applications to access the multimedia functions;
- APIs to manage the execution behaviour, for example, resource (memory, power, processor, network usage) of the multimedia functions;
- APIs to manage the isolation (e.g., memory protection) of the multimedia functions;
- APIs to manage the 'life cycle' of the components of the M3W (e.g., identification, download, installation, activation, de-activation, un-installation).

While MPEG has traditionally focused on the first of the above-described capabilities (APIs for functionality used in multimedia applications), the M3W project intends to go beyond these capabilities. M3W is to be positioned at the platform layer interface and provides an interface definition that abstracts, at a relatively high level, from the platform specifics (See Figure 11.5). As such, M3W provides a common development platform for middleware providers that can serve a variety of platforms with one design.

Following the objectives stated above, the major M3W requirements address [9]:

- *Multimedia APIs*: These APIs are intended to give access to the functionalities provided by the multimedia platform, for example, media processing services (including coding, decoding and transcoding), media delivery services (through files, streams, messages), DRM services, access to data (which could be media content), access to metadata, metadata edit, and metadata search.
- *Management APIs*: These APIs will allow the handling and management of the middleware components; procedures are required to deal with essential processes such as service localization, decision/validation, downloading, instantiation and de-installation/removal.
- *In-operation APIs*: These APIs take account of the required 'extra' functionalities that are essential for reliable, robust and continuing execution of an application that depends on services provided by the multimedia platform, which, in turn, depends on the services provided by the underlying computing platform.

Although other middleware functionalities will be needed, for example, execution environment APIs, composition APIs, rendering APIs, user interface APIs, communication APIs and protocols, these will be defined outside M3W.

The current plan foresees that the M3W standard will include seven parts named *Architecture, Multimedia API, Component Model, Resource and Quality Management, Component Download, Fault Management and System Integrity Management.*

11.3 CONCLUSIONS

For almost two decades, MPEG has been providing the industry with standards that efficiently solve many problems in the area of multimedia representation. Until recently, MPEG decided not to mandate how some of these tools should (normatively) operate together.

This approach has now changed – first, in MPEG-21, where a framework that combines existing and newly developed tools is addressed, and more recently in MPEG-A, which standardizes explicit and useful combinations of available tools from all MPEG standards, and in M3W, which acknowledges the importance of middleware in the multimedia landscape with the goal of improving application portability and interoperability. This broader approach to the multimedia landscape, which commenced with MPEG-21, represents a challenge to MPEG because it is now vital for MPEG (and hence its participants) to possess deep knowledge of not only single technologies, for example, video and audio codecs, but also of the complete applications, the market and overall user needs.

Especially, the deployment of MPEG-21 and MPEG-A will measure the success of MPEG in making this jump from providing the 'pieces' to providing the 'complete engine'.

REFERENCES

[1] MPEG Video Subgroup, "Call for Proposals on Multi-View Video Coding", Doc. ISO/MPEG N7327, Poznan, Poland, July 2005.
[2] MPEG Requirements Subgroup, "Requirements, Terminology and Use Cases for Digital Item Streaming", Doc. ISO/MPEG N7278, Poznan, Poland, July 2005.
[3] B.S. Manjunath, P. Salembier, T. Sikora (eds), "Introduction to MPEG-7: Multimedia Content Description Language", John Wiley & Sons, 2002.
[4] ISO/IEC JTC 1/SC 29/WG 11. ISO/IEC 21000-1, Vision, Technologies and Strategy, 2004.
[5] MPEG Requirements Subgroup, "DI Streaming Call for Proposals", Doc. ISO/MPEG N7066, Busan, Korea, April 2005.
[6] MPEG MDS Subgroup, "MPEG-21 REL Profiles", PDAM-1, Doc. ISO/MPEG N7429, Poznan, Poland, July 2005.
[7] MPEG Video Subgroup, "Introduction to Accuracy Requirements for 8×8 IDCT", Doc. ISO/MPEG N7326, Poznan, Poland, July 2005.
[8] MPEG Audio Subgroup, "Tutorial on MPEG Surround", Doc. ISO/MPEG N7390, Poznan, Poland, July 2005.
[9] MPEG Systems Subgroup, "White Paper on Multimedia Middleware", Doc. ISO/MPEG N7510, Poznan, Poland, July 2005.
[10] Digital Video Broadcasting (DVB), Home Page, http://www.dvb.org/, accessed 2005.
[11] Advanced Television Systems Committee (ATSC) Home Page, http://www.atsc.org/, accessed 2005.
[12] Internet Streaming Media Alliance (ISMA), Home Page, http://www.isma.tv/, accessed 2005.
[13] MPEG Requirements Subgroup, "Purpose of Multimedia Application Formats", PDTR 23000-1, Doc. ISO/MPEG N7281, Poznan, Poland, July 2005.
[14] MPEG Requirements Subgroup, "MAFs under Consideration", Doc. ISO/IEC JTC1/SC29/WG11/N7280, Poznan, Poland, July 2005.

[15] ISO/IEC JTC, 1/SC 29/WG 11. ISO/IEC 23000-2, MPEG Music Player Application Format, 2005.

[16] MPEG Video Subgroup, "Photo Player Multimedia Application Format", Working Draft, Doc. ISO/MPEG N7324, Poznan, Poland, July 2005.

[17] MPEG Systems Subgroup, "MPEG Multimedia Middleware: Context and Objective", Doc. ISO/MPEG N6335, Munich, Germany, March 2004.

Index